AF361733

Ecological Status Assessment of Transitional Waters

Ecological Status Assessment of Transitional Waters

Editor

Chiara Facca

MDPI • Basel • Beijing • Wuhan • Barcelona • Belgrade • Manchester • Tokyo • Cluj • Tianjin

Editor
Chiara Facca
University of Ca' Foscari Venice
Italy

Editorial Office
MDPI
St. Alban-Anlage 66
4052 Basel, Switzerland

This is a reprint of articles from the Special Issue published online in the open access journal *Water* (ISSN 2073-4441) (available at: https://www.mdpi.com/journal/water/special_issues/Ecological_Assessment_Transitional_Waters).

For citation purposes, cite each article independently as indicated on the article page online and as indicated below:

LastName, A.A.; LastName, B.B.; LastName, C.C. Article Title. *Journal Name* **Year**, *Volume Number*, Page Range.

ISBN 978-3-03943-973-7 (Hbk)
ISBN 978-3-03943-974-4 (PDF)

Cover image courtesy of Chiara Facca.

Contents

About the Editor

Chiara Facca obtained her Ph.D. in Environmental Science and Ecology at Ca' Foscari University of Venice (Italy) and at the University of Montpellier II (France) in 2003. She is a technical assistant at the Department of Environmental Sciences, Informatics and Statistics (DAIS), Ca' Foscari University of Venice (Italy). At DAIS, beyond academic activities with courses in Environmental Sciences, Informatics and Chemical Sciences for the Conservation of Cultural Heritage, the following main research topics are investigated: ecology, analytical and environmental chemistry, earth sciences, environmental risk assessment, chemistry for the restoration of cultural heritage, informatics, and statistics. Dr. Facca is mainly involved in the study related to the ecology of transitional and coastal waters. She specializes in the taxonomic identification of microalgae and in their use as ecological quality indicators. In the framework of the European Framework Water Directive (2000/60/EC), together with the Italian National Council of Research and the Italian Institute for Environmental Protection and Research, she implemented the Multimetric Phytoplankton Index, used to assess transitional water status in Italy and Greece. Her most recent research activities have been dedicated to trophic chain investigation, namely the study of the role of microalgae in the diet of some target fish species. She is also carrying out studies on nekton to observe reproductive behavior in different environmental conditions, simulating climate change scenarios. Moreover, she has been involved in the technical and administrative management of European projects (i.e., the LIFE Programme).

Editorial

Ecological Status Assessment of Transitional Waters

Chiara Facca

Department of Environmental Sciences, Informatics and Statistics, Ca' Foscari University of Venice, via Torino 155, 30172 Venice, Italy; facca@unive.it; Tel.: +39-(0)-41-234-7733

Received: 26 October 2020; Accepted: 6 November 2020; Published: 12 November 2020

Abstract: Transitional Waters are worldwide high valuable ecosystems that have undergone significant anthropogenic impacts. The ecological assessment is therefore of fundamental importance to protect, manage and restore these ecosystems. Numerous approaches can be used to understand the effects of human pressures, and, in case, the effectiveness of recovery plans. Eutrophication, climate change and morphological loss impacts can be assessed by means of aquatic vegetation, benthic fauna, and nekton. Moreover, before planning new infrastructures or interventions, predictive approaches and statistical analyses can provide indispensable tools for management policies.

Keywords: CO_2 flux; salinity; desalinization; trophic status; eutrophication; aquatic angiosperms; benthic fauna; nekton; uncertainty analysis

---

1. Introduction

The ecosystems that can be found between the land and the sea are characterized worldwide by a wide range of different natural conditions [1]. Therefore, several definitions are used to describe the habitats along coasts [2]. The United Nations glossary of environment statistics defines coastal lagoons as "Sea-water bodies situated at the coast, but separated from the sea by land spits or similar land features. Coastal lagoons are open to the sea in restricted spaces" [3]. The European Water Framework Directive legally defines Transitional waters (TWs) the "bodies of surface water in the vicinity of river mouths which are partly saline in character as a result of their proximity to coastal waters but which are substantially influenced by freshwater flows" [4]. The term "transitional waters" is now consolidated as a scientific term [2]. In the transition between the land and the sea, and hence, from freshwater to marine environment, a number of ecosystems can be found: rias, fjords, fjards, estuaries, lagoons [5,6]. TWs are often characterized by shallow waters [1] that are rapidly influenced by external changes, and, so, they are considered to be naturally stressed [7]. Moreover, the species' taxonomic richness tends to be limited compared to the adjacent sea or freshwater environments [5,8]. However, the morphological structure and the isolation from the sea provide a sheltered habitat for numerous species of flora and fauna, and a high productivity, making TWs valuable ecosystems [9]. Beyond the natural variability, TWs have been long exploited by human activities to settle cities and harbors, for land-reclamation, aquaculture and more recently, tourism, making them more and more vulnerable [1,9] and deteriorated. Among the 27 European Member states, only six have Coastal lagoon habitats with Favorable conservation status [10]; in China major urban and economic developments are causing loss of coastal wetlands and serious environmental problems [11]; in North America, where there is one of the most extensive surfaces of coastal lagoons (17.6% of coastline), eutrophication represents one of the greatest long-term threats to the ecological integrity [1].

The degradation of such ecosystems of exceptional ecological, recreational, and commercial value [1] has brought up the need to adopt restoration and/or protection measures, that, to be adequately planned, require an accurate ecological status assessment. Since the second half of the 20th century, when Chlorophyll *a* was first proposed as an index of productivity and trophic conditions in transitional and coastal waters [12,13], uncountable methods, based on both biotic and abiotic parameters, have been validated to provide a reliable assessment of environmental conditions. In the framework of National and International regulations aiming at preventing further degradation, the number of indices is continuously increasing. The necessity to have reliable tools, assessing TW status, responds not only to the implementation of management policies but also to improve the knowledge on biological, ecological, and anthropogenic interactions in these very complex ecosystems. Therefore, a constant update of the research on the effects that human pressures have on TWs is of fundamental importance.

The present Special Issue collects eight original research papers and one review, that describe completely different methods of assessing the ecological status of TWs. The number of approaches is different, certainly in proportion to the ecosystem complexity.

2. Overview of this Special Issue

Zhang et al. [14] describe the role of biological processes in carbon dioxide fluxes. A balance mass model has been used to calculate the net ecosystem production (NEP) and the CO_2 flux caused by biological processes and its contribution to the air–sea CO_2 exchange flux. Results show that seawater in the near-shore region of the Changjiang estuary (China) is a source of atmospheric CO_2, and the front and offshore regions generally serve as atmospheric CO_2 sinks. This procedure can provide interesting information on the assessment of trophic status and of the potential carbon stock of coastal waters, above all in the framework of ongoing climate change.

Considering again primary producers, Sfriso et al. [15] describe the role of aquatic angiosperms in TWs, where they are demonstrated to favor the maintenance of good ecological conditions. Seagrass transplantations are an important tool for restoring TWs, and they can be successful in areas where the effects of eutrophication are under control. After a year from the first transplantations, some indices used to assess the ecological status highlight an improvement of water quality due to the increase in seagrass beds.

Benthic fauna has been long used to assess environmental conditions [16], and in the present issue, three papers present data on the relationship between the zoobenthos of TW bottoms and different anthropogenic pressures.

Among zoobenthic organisms, meiofauna has been poorly studied due to the small size and the necessity of time and appropriate analysis techniques. However, it has a well-recognized role in the food webs connecting microbial components to higher trophic levels that contributes to the overall carbon fluxes and organic matter mineralization. Moreover, Semprucci et al. [17] support the hypothesis that meiofaunal organisms are good indicators of the spatial heterogeneity in TWs. Meiofauna displays, in fact, significant spatial variations in relation to environmental conditions, mainly salinity, dissolved oxygen and trophic components, and no changes on the temporal scale.

Benthic epifauna community composition, together with in-laboratory bioassay on brittle stars, were studied by Petersen et al. [18] to verify the impact of desalination brine discharges. In fact, seawater desalination by reverse osmosis is increasing due to the shortage of freshwater supply. Human population growth, agricultural expansion and environmental changes are causing the decline of natural freshwater, and, hence, Desalination Plants are becoming the main response. However, continuous discharge of high-salinity brine into coastal environments may have a severe impact on the ecosystem. Significant salinity anomalies can be detected even some hundred meters from the shore, but a careful assessment of site characteristics, when the desalination plants are constructed, can avoid changes in biological communities. Therefore, to ensure adequate mixing of the discharge brine, desalination plants should be located at high-energy sites with sandy substrates, and discharges should occur through diffusor systems.

Macrozoobenthic fauna was also studied to assess the ecological conditions of TWs on long term scale by Magni et al. [19]. The proposed studied case was a Chinese TW, severely damaged by domestic and industrial pollution and land reclamation already in the 1970s. In late 1980s, restoration interventions started to recover a healthy status and the related ecosystem services. Decade by decade, the continuous improvement of the environmental conditions was reflected in the major recovery and revitalization of the soft-bottom benthic assemblage. Beyond highlighting the importance and success of a good restoration plan, the data further confirm the effectiveness of macrozoobenthic fauna as biological tool to follow the recovery progress and the future evolution of TWs. However, the monitoring activities revealed another important threat for TWs, that is the role of invasive species.

The effectiveness of restoration was assessed also by Scapin et al. [20] who investigated the distribution of nekton communities along a salinity gradient. The long-lasting exploitation of TWs often has determined morphological alterations and hydrodynamism changes. The general pattern of TWs in recent and future decades is toward the homogenization of the physical characteristics with a tendency to marinization [21]. Therefore, the creation of new freshwater inputs is expected to restore the lost transitional attributes mitigating the negative effects of climate change and rising the nekton biomass thanks to the increase in oligo-mesohaline species. The use of a predictive approach with a functional perspective was demonstrated to provide tools to forecast the effects of salinity alteration regimes for both ecologists and ecosystem management [20].

Likewise, Castillo [22] studied the role of seasonal freshwater outflow on abundance of delta smelt and on the entire aquatic community. Qualitative community models were used to describe the effects of salinity variations on community interactions and stability patterns. It turned out that the overlap among pelagic and benthic species and trophic levels was significantly related to their salinity-dependent and geography-dependent distributions. In particular, the relative position of the near-bottom 2 salinity-isohaline along the estuary had different effects on the delta smelt subadults. Therefore, the management of hydrological conditions could be of help to mitigate the declining trends of delta smelt and of other native fish populations.

Focusing on the morphological structures that characterized TWs (i.e., salt marshes, small channels, isolated pools, etc.), Facca et al. [23], by literature review, describe how the presence and abundance of lagoon fish species may provide important indications on TW conservation status. In fact, the occurrence, distribution and biology of resident fish fauna tightly depend on salt marsh complexity, habitat connectivity within the lagoon or among adjacent lagoon systems. This complex system of small creeks and pools has a relevant role, not only for lagoon residents, but also for migrant species, that usually use TWs as a nursery. The review also highlighted the risk connected to the potential impact of alien species.

Being aware that an ecosystem assessment requires the continuous collection of data, as also demonstrated in all the above described studies, Cacciatore et al. [24] propose a tool to optimize the sampling effort in TWs monitoring activities, by applying a multi-approach. The combination of inferential statistics, spatial analyses and expert judgment allows us to optimize the monitoring effort ensuring, at the same time, a high reliability of the achieved information. This approach can be particularly useful for the routine monitoring activities aiming at classifying the water bodies status.

3. Conclusions

The effects of eutrophication have become particularly evident in the TWs worldwide from the 1970s [15,19]. The progressive deterioration of TWs and the concurrent loss of ecosystem services induced National and International authorities to act, protecting and restoring these habitats (US Clean Water Act, European Water Framework Directive, and the National Water Act in South Africa). At present, the positive results of the interventions counteracting the eutrophication phenomenon can be observed and confirmed by the good responses of aquatic vegetation, benthic fauna and nekton [15,17,19,20]. However, other significant threats to TWs are recently requiring attention, such as (i) the impact of climate changes determining alterations of hydrodynamism and of freshwater supply [14,18,22], (ii) the constant loss of morphological structures [23], (iii) the impact of alien and invasive species [19,23]. Climate changes are altering salinity regimes causing unbalances in the biological communities [20,21] and, hence, requiring interventions to manage the freshwater inputs [20,22]. In the framework of climate changes, the understanding of air–sea CO_2 exchange flux is of particular importance [14], because TWs can either be a source or a sink of atmospheric CO_2. The loss of habitat morphological features is related to both sea level rise and anthropogenic exploitation and requires urgent actions to save the vocational habitat of numerous aquatic species [23].

TWs require concrete management policies to prevent further deterioration and to plan the needed recovery interventions. Aiming at supporting this decisional process, the usefulness of biological communities to assess ecological conditions under anthropogenic impacts are widely described [17–19,23,24].

Funding: This research received no external funding.

Acknowledgments: I am thankful to all authors that contribute with their research articles to enrich this special issue. A special thanks goes to the MDPI *Water*'s Editorial Office, to Vanessa Sun and to all the reviewers and Academic Editors that dedicated their time and expertise in the manuscript revisions.

Conflicts of Interest: The author declares no conflict of interest.

References

1. Kennish, M.J.; Paerl, H.W. Coastal Lagoons: Critical Habitats of Environmental Change. In *Coastal Lagoons: Critical Habitats of Environmental Change*; Kennish, M.J., Paerl, H.W., Eds.; CRC Press: Boca Raton, FL, USA, 2010; pp. 1–15.
2. Tagliapietra, D.; Sigovini, M.; Volpi Ghirardini, A. A review of terms and definitions to categorise estuaries, lagoons and associated environments. *Mar. Freshwater Res.* **2009**, *60*, 497–509. [CrossRef]
3. United Nations Statistics Division. Environment Glossary (Last Update in UNdata: 2016). Available online: http://unstats.un.org/unsd/environmentgl/ (accessed on 30 October 2020).
4. European Community 2000. *Directive 2000/60/EC of the European Parliament and of the Council of 23 October 2000 Establishing a Framework for Community Action in the Field of Water Policy*; Official Journal of the European Communities 43 (L327): Bruxelles, Belgium, 2000.
5. Basset, A.; Barbone, E.; Elliott, M.; Li, B.-L.; Jorgensen, S.E.; Lucena-Moya, P.; Pardo, I.; Mouillot, D. A unifying approach to understanding transitional waters: Fundamental properties emerging from ecotone ecosystems. *Estuar. Coast. Shelf Sci.* **2013**, *132*, 5–16. [CrossRef]
6. Wiser: Water Bodies in Europe. Available online: http://wiser.eu/ (accessed on 12 October 2020).
7. Elliott, M.; Quintino, V. The Estuarine Quality Paradox, Environmental Homeostasis and the difficulty of detecting anthropogenic stress in naturally stressed areas. *Mar. Pollut. Bull.* **2007**, *54*, 640–645. [CrossRef] [PubMed]
8. Levin, L.A.; Boesch, D.F.; Covich, A.; Dahm, C.; Erséus, C.; Ewel, K.C.; Kneib, R.T.; Moldenke, A.; Palmer, M.A.; Snelgrove, P.; et al. The function of marine critical transition zones and the importance of sediment biodiversity. *Ecosystems* **2001**, *4*, 430–451. [CrossRef]
9. Newton, A.; Icely, J.; Cristina, S.; Brito, A.; Cardoso, A.C.; Colijn, F.; Riva, S.D.; Gertz, F.; Hansen, J.W.; Holmer, M.; et al. An overview of ecological status, vulnerability and future perspectives of European large shallow, semi-enclosed coastal systems, lagoons and transitional waters. *Estuar. Coast. Shelf Sci.* **2014**, *140*, 95–122. [CrossRef]
10. European Environment Agency. European Topic Centre on Biological Diversity—Report under the Article 17 of the Habitats Directive Period 2013–2018. Available online: https://nature-art17.eionet.europa.eu/article17/ (accessed on 19 October 2020).
11. Chen, L.; Ren, C.; Zhang, B.; Li, L.; Wang, Z.; Song, K. Spatiotemporal dynamics of coastal wetlands and reclamation in the Yangtze estuary during past 50 years (1960s–2015). *Chin. Geogr. Sci.* **2018**, *28*, 386–399. [CrossRef]
12. Steele, J.H. Environmental control of photosynthesis in the sea. *Limnol. Oceanogr.* **1962**, *7*, 137–150. [CrossRef]
13. Boyer, J.N.; Kelble, C.R.; Ortner, P.B.; Rudnik, D.T. Phytoplankton bloom status: Chlorophyll a biomass as an indicator of water quality condition in the southern estuaries of Florida, USA. *Ecol. Indic.* **2009**, *9*, 56–67. [CrossRef]
14. Zhang, Y.; Li, D.; Wang, K.; Xue, B. Contribution of biological effects to the carbon sources/sinks and the trophic status of the ecosystem in the Changjiang (Yangtze) river estuary plume in summer as indicated by net ecosystem production variations. *Water* **2019**, *11*, 1264. [CrossRef]

15. Sfriso, A.; Buosi, A.; Tomio, T.; Juhmani, A.S.; Facca, C.; Sfriso, A.A.; Franzoi, P.; Scapin, L.; Bonometto, A.; Ponis, E.; et al. Aquatic angiosperm transplantation: A tool for environmental management and restoring in transitional water systems. *Water* **2019**, *11*, 2135. [CrossRef]

16. Muxikaa, I.; Borja, A.; Bonnea, W. The suitability of the marine biotic index (AMBI) to new impact sources along European coasts. *Ecol. Indic.* **2005**, *5*, 19–31. [CrossRef]

17. Semprucci, F.; Gravina, M.F.; Magni, P. Meiofaunal dynamics and heterogeneity along salinity and trophic gradients in a Mediterranean transitional system. *Water* **2019**, *11*, 1488. [CrossRef]

18. Petersen, K.L.; Heck, N.; Reguero, B.G.; Potts, D.; Hovagimian, A.; Paytan, A. Biological and physical effects of brine discharge from the Carlsbad desalination plant and implications for future desalination plant constructions. *Water* **2019**, *11*, 208. [CrossRef]

19. Magni, P.; Como, S.; Gravina, M.F.; Guo, D.; Li, C.; Huang, L. Trophic features, benthic recovery, and dominance of the invasive *Mytilopsis sallei* in the Yundang Lagoon (Xiamen, China) following long-term restoration. *Water* **2019**, *11*, 1692. [CrossRef]

20. Scapin, L.; Zucchetta, M.; Bonometto, A.; Feola, A.; Boscolo Brusà, R.; Sfriso, A.; Franzoi, P. Expected shifts in nekton community following salinity reduction: Insights into restoration and management of transitional water habitats. *Water* **2019**, *11*, 1354. [CrossRef]

21. Ferrarin, C.; Bajo, M.; Bellafiore, D.; Cucco, A.; De Pascalis, F.; Ghezzo, M.; Umgiesser, G. Toward homogenization of Mediterranean lagoons and their loss of hydrodiversity. *Geophys. Res. Lett.* **2014**, *41*, 5935–5941. [CrossRef]

22. Castillo, G.C. Modeling the influence of outflow and community structure on an endangered fish population in the upper San Francisco Estuary. *Water* **2019**, *11*, 1162. [CrossRef]

23. Facca, C.; Cavraro, F.; Franzoi, P.; Malavasi, S. Lagoon resident fish species of conservation interest according to the Habitat Directive (92/43/CEE): A review on their potential use as ecological indicator species. *Water* **2020**, *12*, 2059. [CrossRef]

24. Cacciatore, F.; Bonometto, A.; Paganini, E.; Sfriso, A.; Novello, M.; Parati, P.; Gabellini, M.; Boscolo Brusà, R. Balance between the reliability of classification and sampling effort: A Multi-Approach for the Water Framework Directive (WFD) ecological status applied to the Venice Lagoon (Italy). *Water* **2019**, *11*, 1572. [CrossRef]

Publisher's Note: MDPI stays neutral with regard to jurisdictional claims in published maps and institutional affiliations.

Article

Aquatic Angiosperm Transplantation: A Tool for Environmental Management and Restoring in Transitional Water Systems

Adriano Sfriso [1,*], Alessandro Buosi [1], Yari Tomio [1], Abdul-Salam Juhmani [1], Chiara Facca [1], Andrea Augusto Sfriso [1], Piero Franzoi [1], Luca Scapin [1], Andrea Bonometto [2], Emanuele Ponis [2], Federico Rampazzo [2], Daniela Berto [2], Claudia Gion [2], Federica Oselladore [2], Federica Cacciatore [2] and Rossella Boscolo Brusà [2]

[1] Dipartimento di Scienze Ambientali, Informatica e Statistica (DAIS), Università Ca' Foscari Venezia, Via Torino 155, 30170 Mestre (Ve), Italy; alessandro.buosi@unive.it (A.B.); yari.tomio@unive.it (Y.T.); abdulsalam.juhmani@unive.it (A.-S.J.); facca@unive.it (C.F.); asfriso@unive.it (A.A.S.); pfranzoi@unive.it (P.F.); luca.scapin@unive.it (L.S.)

[2] Istituto Superiore per la Protezione e la Ricerca Ambientale (ISPRA), Loc. Brondolo, 30015 Chioggia (Ve), Italy; andrea.bonometto@isprambiente.it (A.B.); emanuele.ponis@isprambiente.it (E.P.); federico.rampazzo@isprambiente.it (F.R.); daniela.berto@isprambiente.it (D.B.); claudia.gion@isprambiente.it (C.G.); federica.oselladore@isprambiente.it (F.O.); federica.cacciatore@isprambiente.it (F.C.); rossella.boscolo@isprambiente.it (R.B.B.)

* Correspondence: sfrisoad@unive.it; Tel.: +39-041-234-8529

Received: 16 July 2019; Accepted: 9 October 2019; Published: 14 October 2019

Abstract: Since the 1960s, the Venice Lagoon has suffered a sharp aquatic plant constriction due to eutrophication, pollution, and clam fishing. Those anthropogenic impacts began to decline during the 2010s, and since then the ecological status of the lagoon has improved, but in many choked areas no plant recolonization has been recorded due to the lack of seeds. The project funded by the European Union (LIFE12 NAT/IT/000331-SeResto) allowed to recolonize one of these areas, which is situated in the northern lagoon, by widespread transplantation of small sods and individual rhizomes. In-field activities were supported by fishermen, hunters, and sport associations; the interested surface measured approximately 36.6 km^2. In the 35 stations of the chosen area, 24,261 rhizomes were transplanted during the first year, accounting for 693 rhizomes per station. About 37% of them took root in 31 stations forming several patches that joined together to form extensive meadows. Plant rooting was successful where the waters were clear and the trophic status low. But, near the outflows of freshwater rich in nutrients and suspended particulate matter, the action failed. Results demonstrate the effectiveness of small, widespread interventions and the importance of engaging the population in the recovery of the environment, which makes the action economically cheap and replicable in other similar environments.

Keywords: aquatic angiosperms; transitional waters; environmental restoration; ecological status; Venice Lagoon

---

1. Introduction

Seagrasses, and more generally the aquatic angiosperms [1], play a key role both in marine and transitional water ecosystems (TWS). These structuring plants are considered environmental engineers to which multiple functions are associated [2]. From the morphological point of view, they reduce the impact of winds and tides on sediment resuspension, favor the sedimentation of the suspended particulate, and contrast the erosion of the seabed and the morphological structures of shallow bottoms [2]. Aquatic angiosperms characterize the bottoms of habitat 1150* (coastal lagoons) and 1140 (muddy or sandy bottoms emerging during low tide) *sensu* Habitat Directive 92/43/EEC, contribute to CO_2 sequestration [3], and form the natural habitat for the biological communities [4] providing shelter and food for fish and macrofaunal organisms [5,6]. However, coastal areas and TWS are often very degraded and aquatic angiosperms have disappeared or are in rapid regression; that is mainly due to anthropogenic impacts such as eutrophication or pollution [7]. This was the case of the Venice Lagoon [8,9] and the lagoons of the Po Delta [10,11]. In both the TWS, aquatic angiosperms suffered from two main impacts: the overgrowth of nuisance macroalgae due to the eutrophication increase during the 1960s–1980s and the harvesting of the Manila clam *Ruditapes philippinarum* (Adams and Reeve) with hydraulic or mechanical rakes. Those activities destroyed the bottoms, uprooted the plants, and resuspended considerable quantities of sediments reducing water transparency and the growth of the plants which had survived [12].

Currently, the trophic conditions of the lagoons and ponds of the Po Delta are still bad/poor, because these environments are strongly affected by the waters of the Po River that drains the Po Valley [10,11]. Furthermore, due to the high trophic conditions, bivalves are abundant, clam fishing activities occur on a large scale, and water remains turbid. The aquatic angiosperms recorded in the past have disappeared and have been replaced by tionitrophilic macroalgae. The dominant taxa are Ulvaceae and the non-native Rhodophyceae, *Agarophyton vermiculophyllum* (Ohmi) Gurgel, J.N. Norris *et* Fredericq and *Solieria filiformis* (Kützing) P.W. Gabrielson.

In the Venice Lagoon, the effects of anthropogenic impacts have decreased since 2011 and the ecological status has started to improve [13]. Macroalgal biomass decreased significantly before the period of intense clam fishing [9] and in the last decade the dominant algal species have also changed. Ulvaceae have been largely replaced by several taxa of good-high ecological value, especially Rhodophyceae, and aquatic angiosperms are recolonizing the lagoon bottoms [7,13].

However, in all the basins of the Po Delta and in many choked areas of the Venice Lagoon plant recolonization was hampered by the lack of seeds. To favor the recolonization in the Venice Lagoon, the European Union funded the restoration project (LIFE12NAT/IT/000331—SeResto; www.lifeseresto.eu). The objective was the recolonization of the northern basin by small diffuse triggers of aquatic angiosperms which are typical of that environment: *Cymodocea nodosa* (Ucria) Ascherson, *Zostera marina* Linnaeus, *Zostera noltei* Hornemann, and *Ruppia cirrhosa* (Petagna) Grande.

In order to provide useful information for the project replication in similar TWS, this paper reports the results of the transplanting activities after the first year of plant rooting, the most common environmental parameters and nutrient concentrations in the various environmental matrices (water column, surface sediments, and suspended particulate matter (SPM)), the transplantation methods, and the most suitable environmental conditions to ensure the success of species rooting and spread.

2. Materials and Methods

2.1. Study Area

The transplants of aquatic angiosperms took place in an area measuring approximately 36.6 km^2 (Figure 1) situated in the northern basin of the Venice Lagoon (sexagesimal coordinates: 45°30′–34′ N, 12°27′–33′ E).

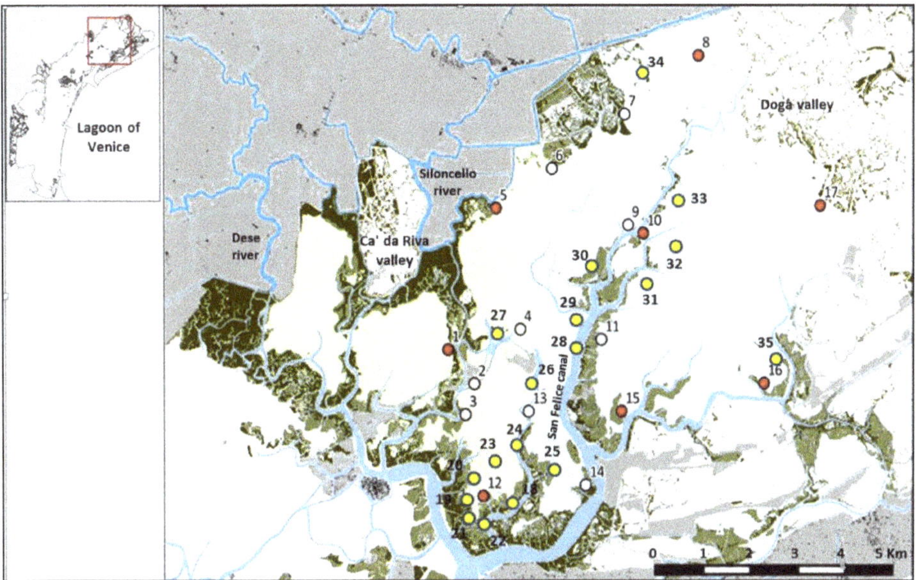

Figure 1. Map of the sampling area with the 35 transplanting stations. In red and white, the 17 stations transplanted in spring 2014. The stations in red were monitored monthly for one year for environmental parameters. In yellow, the 18 stations transplanted in spring 2015.

Thirty-five stations characterized by shallow waters were identified in the study area, along the edges of the salt marshes and lagoon canals. The transplanting area is naturally divided by a deep canal (San Felice) which flows in a south-northern direction. On the east side of the lagoon, bottoms are shallower, there is no source of freshwater, and the trophic status is low. On the contrary, the trophic status of the west side is quite high due to the waters of some branches of the Silone river (flow 5 m^3 s^{-1}) and the Sile river overflows in rainy periods (flow rates: from a few m^3 s^{-1} to 70 m^3 s^{-1}), which have an average frequency of 8–9 events per year [14]). The waters, rich in nutrients and suspended particulate, favor the growth of tionitrophilic algae and trigger phytoplankton blooms that hamper plant rooting.

2.2. Angiosperm Transplants

In spring 2014, transplanting occurred in 17 stations of 100 m^2 (10 × 10 m) and in spring 2015, in additional 18 stations (Figure 1). In the initial phase, the transplants took place using sods that were approximately 30 cm in diameter which were collected with a manual corer and arranged in groups of three for a total of nine sods per station, following the scheme in Figure 2. In order to avoid damaging the bottom, all operations were carried out by remaining on board of flat local boats or by divers. The depth of the intervention area was generally less than one meter on the average tide level and the boats were used during high tide to reach even the shallowest areas that emerge at low tide. Angiosperm sods and rhizomes were supplied by managers of closed fishing ponds (Dogà valley and Ca' da Riva valley) where ecological conditions are high and aquatic angiosperms are abundant. Hundreds of full-grown rhizomes were transplanted individually at each station using pliers with a handle of approximately 1 m length.

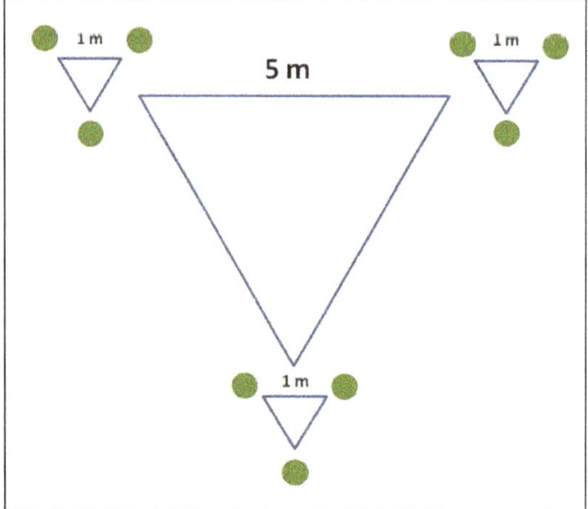

Figure 2. Scheme of sod transplants.

2.3. Angiosperm Monitoring

Sod rooting was verified on a monthly basis during the first months of spring 2014 and 2015 in order to replace those that had not taken root. The check was repeated, in summer and autumn. In addition, 100 rhizomes of *Zostera marina* that had been planted at each station were carefully monitored during the year to check the success rate of the single rhizome transplant.

Patch measurements reported in this paper refer to monitoring carried out one year after transplants. The growth of sods and rhizomes was mainly circular, therefore during each survey the average growth of patches was recorded by measuring their diameters and took into account only diameters >20 cm.

2.4. Physico-Chemical Parameter Determination

Before angiosperm transplants, the environmental conditions (physico-chemical parameters and nutrient concentrations) of the water column and surface sediments were monitored once in all the 35 stations. Out of them, eight stations (1, 5, 8, 10, 12, 15, 16, and 17), representative of the environmental conditions of the whole area, were chosen for the monthly detection of the environmental parameters of water, surface sediments, and the settled particulate material (SPM) collected by sedimentation traps placed on the bottom. The ecological status of the eight stations was also determined by sampling all the macrophytes (twice a year), fish fauna (twice a year), and the benthic macrofauna (once a year) according to the procedures used for the application of the macrophyte quality index (MaQI) [15], habitat fish biotic index (HFBI) [16], and multivariate-AZTI marine biotic index (M-AMBI) [17].

2.5. Statistical Analyses

The average values of (a) thirty eight physico-chemical parameters in the water column, surface sediments, and SPM collected monthly for one year in eight stations (1, 5, 8, 10, 12, 15, 16, and 17), (b) seven aquatic angiosperm variables: number, diameter, and growth of surviving sods and rhizomes, and (c) the results of the application of three indices of ecological status determined at the end of the sampling year were analyzed. The Shapiro–Wilk test differentiated non-normal data, and then Spearman's one-way ANOVA coefficients were calculated and summarized in Table 1. Significant values ($p < 0.05$) are marked in red.

Table 1. Spearman's non-parametric coefficients between sod/rhizome variables and physico-chemical parameters and nutrient concentrations in the water column and surface sediments. Significant values ($p < 0.05$) are in red. Legend:, Temp = water temperature, Transp = water transparency, Salin = salinity, %DO = percentage of dissolved oxygen, Eh = redox potential, w = water, s = sediment, Light-S = light at a depth of −5 cm, Light-B = light at bottom, FPM = filtered particulate matter, Si = silicates, RP = reactive phosphorus, NH_4^+ = ammonium, NO_2^- = nitrites, NO_3^- = nitrates, DIN = dissolved inorganic nitrogen, M-biom = macroalgal biomass, M-cover = macroalgal cover, Chl-*a* tot = total Chlorophyll-*a*, TP = total phosphorus, IP = inorganic phosphorus, OP = organic phosphorus, TN = total nitrogen, TC = total carbon, OC = organic carbon, SPM = settled particulate matter, part = particulate.

Spearman's Non Parametric Coefficients

	Temp	Depth	Transp	Salin	%DO	pHw	Ehw	pHs	Ehs	Light-S	Light-B	FPM
Survived sods	0.03	−0.21	−0.09	0.63	−0.33	0.43	0.18	0.30	0.39	−0.34	0.48	−0.31
Sod diameter	−0.29	−0.27	−0.17	0.73	−0.54	0.44	0.00	0.41	0.46	−0.56	0.54	−0.29
Sod growth	−0.29	−0.27	−0.17	0.73	−0.54	0.44	0.00	0.41	0.46	−0.56	0.54	−0.29
Survived rhizomes	0.07	−0.39	−0.29	0.66	−0.32	0.42	0.44	0.51	0.59	0.00	0.71	−0.27
Rhizome diameter	−0.14	−0.06	0.11	0.75	−0.71	0.69	0.02	0.34	0.20	−0.62	0.24	−0.67
Rhizome growth	−0.07	−0.12	0.05	0.67	−0.57	0.69	0.19	0.36	0.21	−0.50	0.40	−0.60
Patch surface	0.00	−0.32	−0.22	0.63	−0.34	0.39	0.29	0.37	0.51	−0.22	0.63	−0.20
MaQI	0.34	0.40	0.56	0.27	−0.38	0.87	0.39	0.46	−0.38	−0.08	−0.04	−0.95
M-AMBI	0.02	0.14	0.17	−0.71	0.26	0.10	0.24	−0.31	−0.67	0.21	−0.19	0.07
HFBI	0.33	0.10	0.19	0.26	0.00	0.19	0.02	0.00	0.00	−0.05	−0.17	−0.43

	Si	RP	NH_4^+	NO_2^-	NO_3^-	DIN	M-biom	M-cover	Chl-*a* tot
Survived sods	−0.65	−0.65	−0.74	−0.30	−0.50	−0.56	−0.33	−0.72	−0.33
Sod diameter	−0.61	−0.61	−0.61	−0.17	−0.27	−0.34	−0.27	−0.66	−0.41
Sod growth	−0.61	−0.61	−0.61	−0.17	−0.27	−0.34	−0.27	−0.66	−0.41
Survived rhizomes	−0.68	0.71	0.56	−0.41	−0.56	−0.56	−0.39	−0.68	−0.51
Rhizome diameter	−0.57	−0.60	−0.83	−0.42	−0.61	−0.68	−0.24	−0.66	−0.54
Rhizome growth	−0.71	−0.74	−0.79	−0.45	−0.60	−0.64	−0.41	−0.81	−0.40
Patch surface	−0.66	−0.66	−0.56	−0.22	−0.41	−0.44	−0.47	−0.71	−0.37
MaQI	−0.69	−0.69	−0.70	−0.95	−0.88	−0.84	−0.37	−0.61	−0.36
M-AMBI	0.12	0.05	0.48	−0.10	0.21	0.31	−0.18	−0.02	0.62
HFBI	−0.17	−0.19	−0.76	−0.48	−0.67	−0.76	0.26	−0.19	−0.24

	Fines	Density	Moisture	Porosity	TP sed	IP sed	OP sed	TN sed	TC sed	OC sed
Survived sods	−0.20	0.39	−0.39	−0.39	−0.73	−0.37	−0.86	−0.65	0.78	−0.05
Sod diameter	−0.15	0.49	−0.46	−0.37	−0.81	−0.49	−0.90	−0.65	0.75	−0.10
Sod growth	−0.15	0.49	−0.46	−0.37	−0.81	−0.49	−0.90	−0.65	0.75	−0.10
Survived rhizomes	−0.07	0.44	−0.51	−0.49	−0.56	−0.20	−0.83	−0.63	0.68	−0.20
Rhizome diameter	−0.18	0.74	−0.66	−0.57	−0.92	−0.61	−0.84	−0.61	0.66	−0.22
Rhizome growth	−0.26	0.57	−0.50	−0.43	−0.93	−0.64	−0.83	−0.54	0.66	−0.02
Patch surface	−0.24	0.39	−0.41	−0.32	−0.66	−0.29	−0.85	−0.65	0.75	−0.10
MaQI	−0.35	0.48	−0.40	−0.47	−0.64	−0.63	−0.25	−0.01	−0.07	0.12
M-AMBI	−0.07	−0.52	0.64	0.64	0.17	−0.31	0.74	0.93	−0.71	0.68
HFBI	0.21	0.26	−0.29	−0.60	−0.24	0.00	−0.33	−0.37	0.37	−0.20

	SPM	TP part	IP part	OP part	N part	TC part	OC part
Survived sods	−0.01	−0.51	−0.31	−0.74	−0.26	0.03	−0.16
Sod diameter	0.00	−0.56	−0.27	−0.81	−0.15	0.17	−0.17
Sod growth	0.00	−0.56	−0.27	−0.81	−0.15	0.17	−0.17
Survived rhizomes	0.05	−0.68	−0.37	−0.78	−0.32	−0.15	−0.34
Rhizome diameter	−0.30	−0.68	−0.63	−0.75	−0.30	0.30	−0.29
Rhizome growth	−0.31	−0.57	−0.57	−0.71	−0.24	0.29	−0.14
Patch surface	0.10	−0.61	−0.32	−0.76	−0.05	0.12	−0.12
MaQI	−0.64	−0.32	−0.91	−0.22	−0.56	0.21	−0.16
M-AMBI	−0.67	0.74	0.07	0.79	0.29	0.29	0.55
HFBI	−0.19	−0.17	−0.21	−0.31	−0.86	−0.55	−0.45

The principal component analysis (PCA) was applied to the same log-transformed data after removing redundancies to visualize the variances and associations between parameters, aquatic angiosperm variables, and index values of the eight stations. Results are reported in a bi-plot. Data were processed by Statistica software, release 10 (StatSoft Inc. Tulsa, OK, USA).

3. Results

Figure 3a shows per station, the average diameter of the patches formed by sods and rhizomes 12 months after transplants. The growth was very variable due to the different environmental conditions of the stations. The maximum average diameter of the patches was 143 cm at station 14. Patches over 100 cm in diameter were also found in six other stations (9, 11, 12, 15, 18, 33). In three stations, the sods disappeared completely despite frequent new transplants. In the other stations, the diameter increase was intermediate.

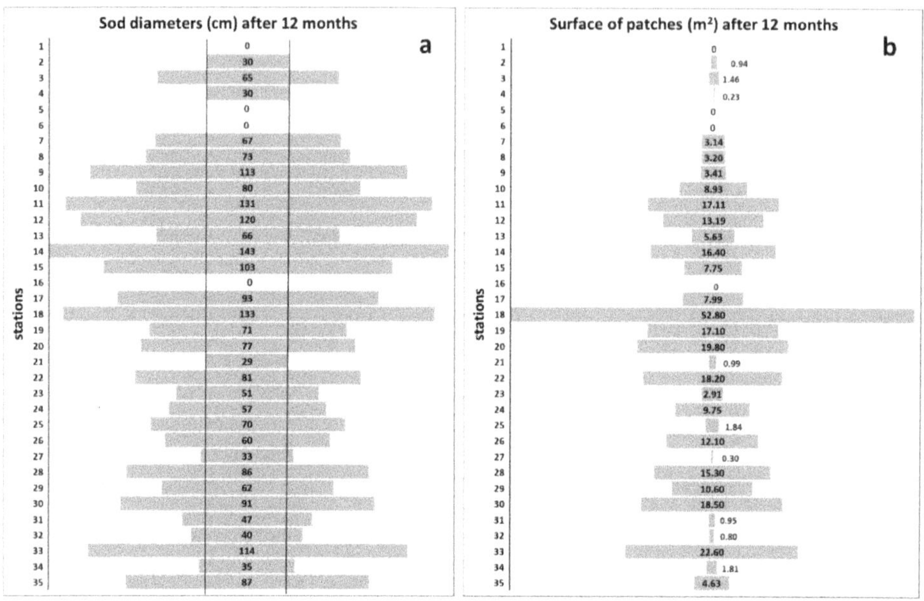

Figure 3. (**a**) Average diameters of the patches. Vertical bars represent the diameter of the transplanted sods; (**b**) Total surface of patches, formed by the growth of sods and rhizomes at the 35 sampling sites after 12 months. Only patches with a diameter >20 cm were considered.

The total surface of the patches is shown in Figure 3b. In most stations, there was no correspondence between the diameter and the surface of the patches. The patch surface, in fact, after one year did not distinguish between the sods and rhizomes that grew mixed. In some cases, the total patch surface was the result of a large quantity of patches, in other cases its size depended only on the number of those that had survived. Therefore, results were very different. The maximum growth occurred at station 18 with a colonization of 52% (52 m^2 out of 100 m^2) and a patch diameter that was among the biggest (133 cm). In contrast, at station 14, just less than 1 km away, the coverage was only 16.4%, even though the diameter of the patches was bigger (143 cm). No angiosperm rooting was observed at stations 1, 5, 6, and 16. Stations 1, 5, and 6 were affected by river outflows, whereas station 16 showed a higher trophic level than the other stations, which was probably due to the proximity of the Cavallino urban center.

The main metrics of sods and rhizomes one year after transplants are shown in Figure 4. The number of survived sods was, on average, 72 ± 38% with high differences between the stations. All the transplanted sods took root in 20 stations out of 35. No angiosperm rooting was found in four stations (1, 5, 6, and 16), whereas in the other stations, their rooting varied widely. On average, rhizome rooting was lower (37 ± 25%) than that of sods, but the number of single rhizome transplants per station was much higher (637) than the number of sods, and so was their potential for developing new rooting.

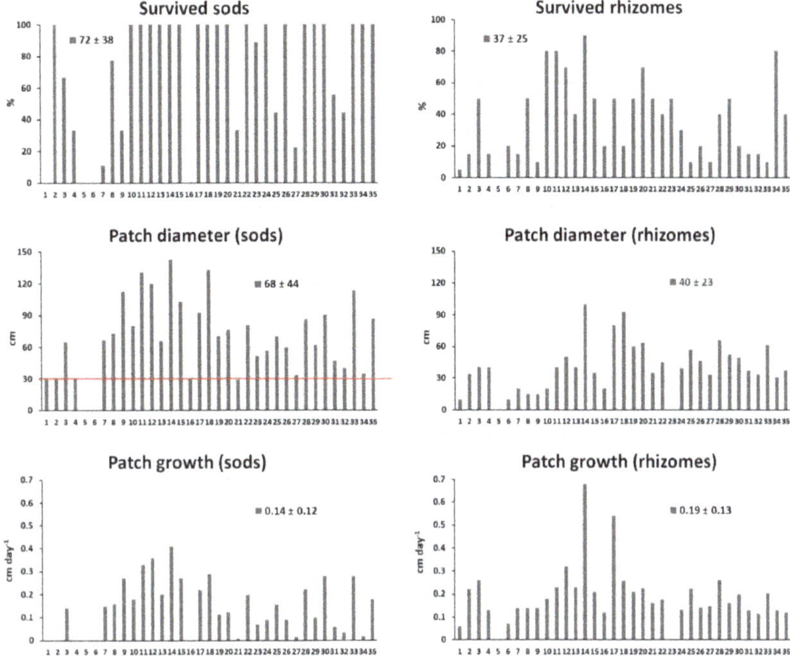

Figure 4. Percentage, patch diameter, and daily growth of survived sods and rhizomes.

The average diameter of the patches formed by the surviving sods was 68 ± 44 cm, whereas rhizomes exhibited, on average, a diameter of 40 ± 23 cm, a value very similar to that of sods if we take into account that the diameter of transplanted sods was 30 cm. Moreover, some rhizomes took root also at stations 6 and 16, whereas at station 23 some survived rhizomes formed patches with a diameter shorter than 20 cm. The stations 14 and 18 produced the largest patches.

By comparing growth, rhizomes showed values higher than sods: 0.19 ± 0.13 cm day^{-1} vs. 0.14 ± 0.12 cm day^{-1}. It is interesting to observe that the rhizomes transplanted at station 14 showed the highest growth with 0.68 cm day^{-1} and formed patches of approximately 1 m in diameter.

The analysis of nutrient concentrations in the samples of surface sediment collected in all 35 stations during the spring transplants shows that the area is divided into two parts: the one situated northwest of the San Felice canal is characterized by high nutrient concentrations due to river outflows; the other which is on the southeastern side of the canal is characterized by a higher seawater renewal and no river outflow (Figure 5).

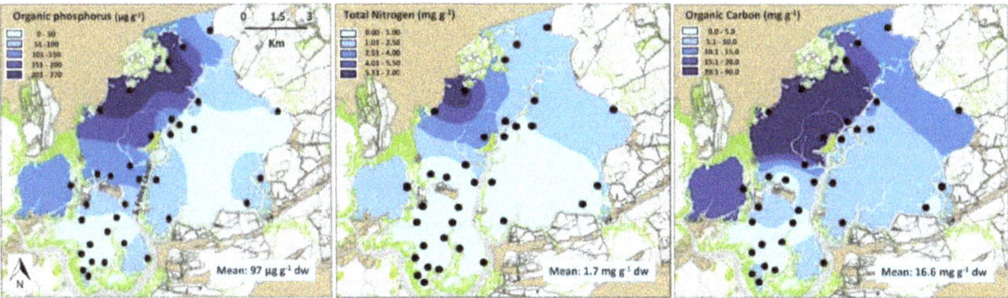

Figure 5. Concentration of organic phosphorus, total nitrogen, and organic carbon in surface sediments of the transplant area. Black dots represent transplant stations.

Stations 5 and 6 were the most affected by high nutrient concentrations that favored the growth of tionitrophilic macroalgae and phytoplankton blooms which prevent the rooting of aquatic angiosperms. In the northwestern area of the San Felice canal, the concentration of organic phosphorus (OP), total nitrogen (TN), and organic carbon (OC) were one order of magnitude higher than in the southeastern area.

Similarly, the analysis of the mean concentrations of nutrients in the water column of the eight stations selected for monthly monitoring shows that reactive phosphorus (RP), dissolved inorganic nitrogen (DIN), and silicates (Si) were significantly higher at stations 5 and 1 than in the others (Figure 6).

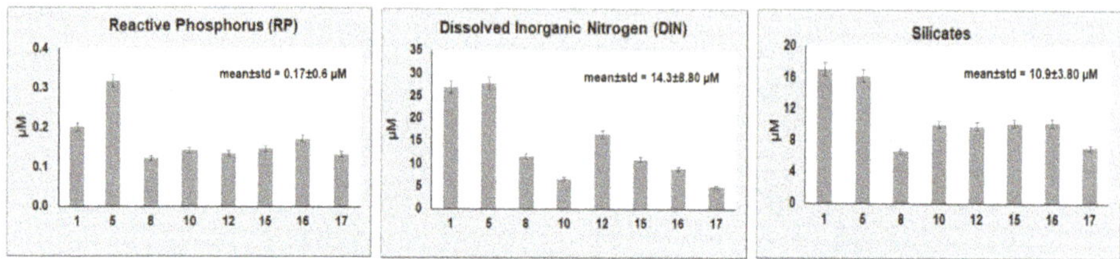

Figure 6. Concentration of reactive phosphorus, dissolved inorganic nitrogen, and silicates in the eight stations monitored monthly during one year.

On average, RP showed a concentration of 0.17 ± 0.6 µM with the highest value at station 5 (0.32 µM) near the mouth of the Siloncello river. The same results were found for DIN that averagely was 14.3 ± 8.8 µM, but at station 5, it was approximately twice as high (27.8 µM). Silicates were slightly higher at station 1 (17.0 µM) and the mean value was 10.9 ± 3.8 µM.

Salinity confirms the influence of the Siloncello freshwaters. Stations 5 and 1 exhibited values of approximately 21.0 and 22.4 psu, respectively, against 26.8 to 29.2 recorded in the others (Figure 7). Total chlorophyll-*a* (Chl-*a*) also showed the highest value (approximately 1.3 µg L^{-1}), although, on average, it was very low (0.94 ± 0.29 µg L^{-1}). The particulate collected by sedimentation traps (SPM) highlighted the environmental differences between the stations more clearly than Chl-*a*, because the sedimentation rates were collected and recorded constantly during the whole sampling year. Also, this parameter showed the highest value at station 5 (635 g dw m^{-2} d^{-1}) accounting for approximately 232 kg dw m^{-2} y^{-1}. On the contrary, extremely low values were found at stations 8 and 17, where aquatic angiosperms had widely taken root.

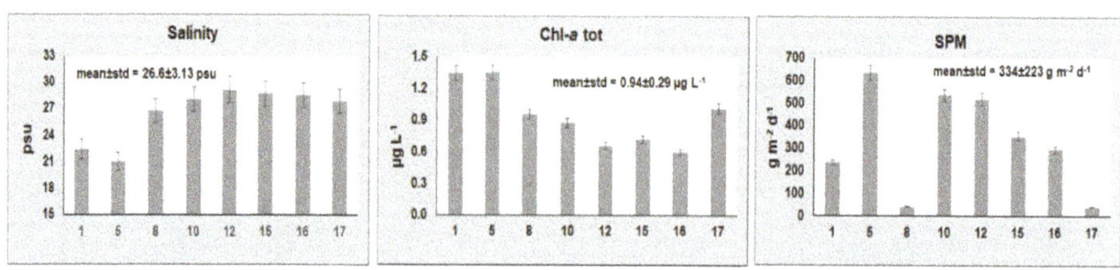

Figure 7. Values of salinity, total chlorophyll-*a* (Chl-*a* tot), and settled particulate matter (SPM) collected by sedimentation traps).

Two other important parameters that marked the differences between the stations were the percentage of sediment fraction <63 µm (fines) and the amount of dry sediment per volume unit (dry density) of surface sediments. Grain-size was >70% in all the stations (Figure 8), but with a different degree of compactness. Dry density values were overall very low (0.43 ± 0.14 g dw cm^{-3}), but the lowest were recorded at station 5 (0.16 g dw cm^{-3}) and station 1 (0.32 g dw cm^{-3}), where aquatic angiosperms had not taken root.

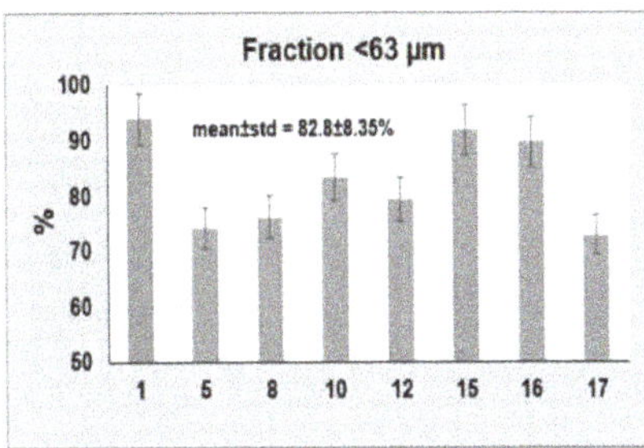
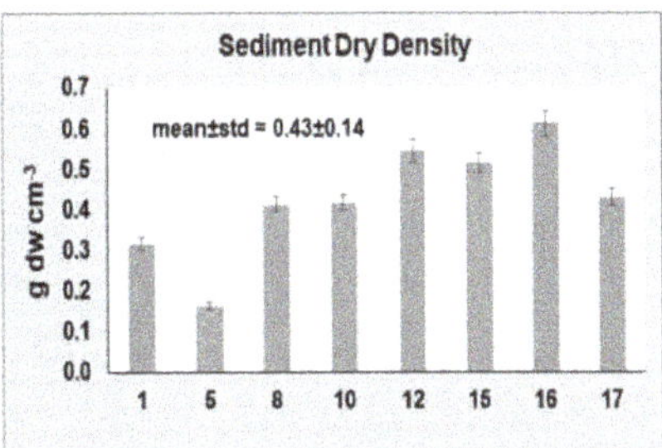

Figure 8. Percentage of fines (fraction < 63 μm) and dry density of surface sediments.

The non-parametric Spearman's coefficients calculated using 38 physico-chemical parameters, seven macrophyte variables, and three indices of ecological status in the eight stations showed that freshwaters coming from the river Siloncello affected the salinity of the water column seriously, deteriorating the ecological conditions and hindering the angiosperm rooting (Table 1). In fact, sod diameter (cm), sod surface (m^2), and sod growth (cm day^{-1}) were positively correlated to salinity, whereas rhizome survival (% yr^{-1}) was positively correlated to light availability at the bottom (Light-B), and inversely to SPM and the concentration of organic phosphorus (OP) in surface sediments.

Moreover, rhizome growth was counteracted by the concentrations of Si, RP, and ammonium (NH$_4^+$) in the water column by TP and OP in surface sediments and SPM. Macroalgal cover (M-cover) was also inversely and significantly correlated to angiosperm growth. The accumulation of tionitrophilic macroalgae, especially Ulvaceae and Gracilariaceae, on sod and rhizome transplants prevented their growth. These macroalgae have a much higher growth rate [18] than angiosperms [19]. The same inverse correlations were also found between sod growth and TP, OP in surface sediments, and OP in SPM. The indices of ecological status: MaQI, M-AMBI, and HFBI showed that the correlations with the physico-chemical parameters in the water column had a similar trend, whereas in surface sediments the correlations were not always clear, especially when M-AMBI was applied.

The principal component analysis (PCA) applied to the whole set of parameters and variables, after the exclusion of redundancies, confirmed the above relationships (Figure 9). About 66% of the total variance was explained by the first two components that clearly separate the variables of the aquatic angiosperms and most of the environmental parameters into two major groups. In the two minor groups, the variables of the first two components were intermediate. In group 1, both sod and rhizome rooting were associated with salinity, sediment density, sediment total carbon, sediment and water pH, and light availability at the bottom. The ecological quality ratio (EQR) of the indices of ecological status of the biological elements, macrophytes (MaQI) and fish fauna (HFBI), were also associated to aquatic angiosperm variables and the same environmental parameters.

In group 2, most of the nutrient concentrations in the water column, surface sediments, and SPM were plotted together with sediment moisture and porosity, water temperature, the %DO, Chl-*a*, FPM, and M-AMBI. Groups 3 and 4 highlighted some parameters which were intermediate between groups 1 and 2.

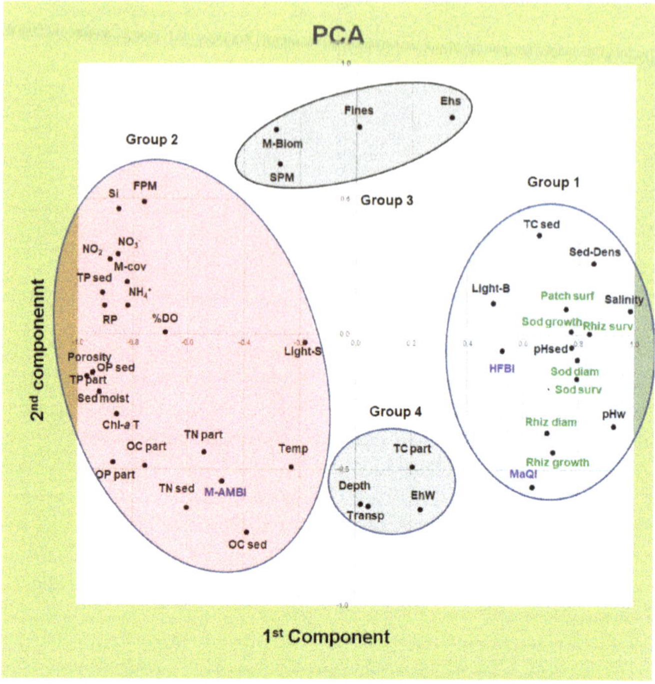

Figure 9. Principal component analysis (PCA) between physico-chemical parameters and macrophyte variables. Legend: Rhiz = rhizomes, Surv = survived, Diam = diameter, surf = surface, Temp = water temperature, Transp = water transparency, %DO = percentage of dissolved oxygen, w = water, sed = sediment, dens = density; moist = moisture, Light-S = light at a depth of −5 cm, Light-B = light at bottom, FPM = filtered particulate matter, Si = silicates, RP = reactive phosphorus, NH_4^+ = ammonium, NO_2^- = nitrites, NO_3^-, = nitrates, DIN = dissolved inorganic nitrogen, M-Biom = macroalgal biomass, M-cov = macroalgal cover, Chl-*a* T = total chlorophyll-*a*, TP = total phosphorus, OP = organic phosphorus, TN = total nitrogen, TC = total carbon, OC = organic carbon, SPM = settled particulate matter, part = particulate.

4. Discussion

The results of this study highlight the potential of small primers, especially diffused transplants of single rhizomes, for a rapid colonization of areas that have the suitable ecological conditions. The transplants of aquatic angiosperms by single rhizomes do not require large efforts and expertise and favors the diffusion of plants throughout the area. Therefore, the involvement of motivated groups of people such as fishermen, hunters, and sport associations who frequently spend part of their time in the lagoon was a great achievement for lagoon recovery. In the past, several transplants of sods measuring some square meters [20–22] were carried out at the edges of canals or in relatively deep areas. Such operations required great efforts, and also the use of large boats and mechanical means that cannot operate in shallow waters. The Project Life12 NAT/IT/000331—SERESTO (www.lifeseresto.eu) tested the possibility of recolonizing a large and shallow lagoon area (approximately 36.6 km^2), where the aquatic angiosperms had almost completely disappeared, through small, widespread transplants. The proposed strategy achieved a large-scale restoration through the transplants of a small number of plants (approximately 40 m^2 during the entire project), with advantages in terms of costs and impact on the donor sites.

Intense monitoring of plant growth and environmental parameters over the whole area during one year allowed to understand the reasons behind achievements and/or failures. Results showed that aquatic angiosperms took roots only in areas where nutrient concentrations were not too high and where the tionitrophilic macroalgae, especially Ulvaceae, Cladophoraceae, and Gracilariaceae prevailed, both sods and rhizomes chocked and died. Angiosperm transplants failed near river outflows because waters were turbid and rich in nutrients. The choice to transplant angiosperms as well in these areas and to monitor the environmental parameters allowed to identify a series of critical values that must be taken into account should these techniques be successfully applied to other areas. First of all, as reported in other studies [7], the presence and growth of aquatic angiosperms depended on the species considered, degree of trophic conditions, and clearness of the waters. The lagoon areas we studied are colonized by five species: three seagrasses (*Cymodocea nodosa* Ucria (Ascherson), *Zostera marina* Linnaeus, and *Zostera noltei* Hornemann) and two aquatic angiosperms (*Ruppia cirrhosa* Petagna (Grande) and *Ruppia maritima* Linnaeus) [23]. All of them have different ecological features. *Cymodocea nodosa* is a subtropical species that colonizes areas characterized by high salinity, coarse grain sediments, and low nutrient concentrations. It prefers the areas which are strongly influenced by marine waters near the lagoon mouths or high-flow canals, but it can also be present in choked areas [7,23] provided there are the conditions quoted above. However, in the shallow choked areas, where water renewal is low, it reaches much smaller sizes than in areas vivified by the sea. *Zostera marina* prefers areas characterized by fine sediments and high water exchange, because it is a species very sensitive to high temperatures. It reaches the highest growth in April–May, afterwards it decreases and disappears almost completely when the water temperature remains above 26–28 °C for long periods. *Zostera noltei* and *Ruppia cirrhosa* prefer shallow choked areas. The former grows mainly along saltmarsh edges and the latter in any shallow bottom, provided the waters are clear. Both species colonize areas with fine sediments where temperature and salinity are very variable, but *Ruppia cirrhosa* prefers the most choked areas and sediments with a higher percentage of fines. Finally, *Ruppia maritima* is the species which most easily adapts to extreme variations in salinity and temperature. It usually grows in the pools of salt marshes, which are in-flowed by new water only during particularly high tides [23].

Our transplants involved the first four species, particularly *Z. marina* and *Z. noltei*, whereas *R. cirrhosa* and *C. nodosa* were only transplanted in some stations with suitable ecological conditions. All the species took root, but *Z. noltei* was the most successful. It rapidly colonized all the saltmarsh edges of the whole transplanting area, where new plants were also produced through natural seed dispersion. *Zostera marina* took root in areas with a negligible tionitrophilic algal biomass, where good water renewal kept temperatures moderate, preventing the formation of sulfides that are very toxic and lead to the species local extinction [24]. Transplants of *Z. marina* carried out in the autumn were a great achievement, because they formed numerous patches, even though in the presence of the high summer temperatures (>26–28 °C), which were also recorded by [25] in the Denmark coasts, they started to regress and in some areas, disappeared altogether. Temperature was the most critical factor for the survival of this species. Some authors [26] who had carried out in-culture experiments fond that the best temperature for the growth of *Z. marina* was <25 °C. At 27 °C, the growth decreased significantly and at higher temperatures it died. This explains why it is widely distributed in the cold waters of the northern part of the Atlantic Ocean and the Pacific Ocean, up to the Arctic Circle. In the Mediterranean Sea, it is a relict species of the ancient Tetide Ocean that colonizes mainly areas characterized by moderate temperatures and salinities [10–25,27] such as the northern Adriatic Sea, the Aegean Sea, the Black Sea, and some areas of the Spanish and French coasts [28]. *Ruppia cirrhosa* was transplanted into 12 stations. In the three decades previous to transplant activities, there was no sign of the presence of this species in the lagoon which is open to tidal expansion, whereas it was abundant in the closed fishing ponds [28,29]. The year after transplanting its spread was explosive, as it rapidly colonized all the choked shallow areas, while no rooting was observed in areas vivified by large canals.

Cymodocea nodosa was transplanted only in a station and as expected, its spread was very limited (stations 7 and 17) because of the ecological conditions of the transplant areas.

The establishment and growth of these aquatic angiosperms are strongly affected by the excess of nutrients in the water column, surface sediment, and SPM, especially by the phosphorus compounds (Table 1) as reported by [30]. High sediment moisture and porosity also counteract plant rooting because sediments are less compact, anoxic, and rich of organic nutrients. In addition, sediments characterized by high moisture usually contain a high amount of ammonium [31] and sulfides [32,33] that act as phytotoxins and hamper plant rooting.

Despite the short period considered, the responses of some indices used to assess the ecological status were positive. The macrophyte quality index (MaQI) and habitat fish biotic index (HFBI) only one year after transplanting showed the same positive correlations with the presence of aquatic plants as by Spearman's non-parametric coefficients and PCA. On the contrary, M-AMBI did not respond positively to the presence of aquatic plants. In fact, the benthic macrofauna used for the index application lives inside the sediment, which requires longer time to change its physico-chemical characteristics than the water column. Therefore, we expect that benthic macrofauna will respond to the changes with a delay of a few years.

According to [34–36], pristine TWS should be colonized by aquatic angiosperms. These plants favor the maintenance of good ecological conditions. They control the change of water and sediments pH, favoring the development of calcareous algae, trap large amounts of CO_2 [37,38], and prevent sediment erosion and dispersion, favoring sedimentation processes. In addition, angiosperm meadows are natural nurseries where the fish macrofauna finds food and shelter [39–41]. Finally, the small aquatic angiosperms like *Z. noltei* and *Ruppia* spp. that colonize the emerging seabed at low tide are also the best environment for birds such as ducks, flamingos, and herons that feed on both plants and the organisms typical of those areas. Therefore, the recovery of TWS through recolonization of environments with aquatic angiosperms is of fundamental importance. But, the success of transplants largely lies also on the cooperation of the population interested in safeguarding the environments they attend for work or leisure.

Author Contributions: Conceptualization, A.S., P.F., A.B. (Andrea Bonometto) and R.B.B.; Formal analysis, A.B. (Alessandro Buosi), Y.T., A.-S.J., A.A.S., L.S., E.P., F.R., D.B., C.G., F.O. and F.C.; Funding acquisition, A.S., A.B. (Andrea Bonometto) and R.B.B.; Investigation, A.S. and P.F.; Methodology, A.S. and A.B. (Andrea Bonometto); Project administration, A.S. and C.F.; Resources, A.S. and A.B. (Andrea Bonometto); Supervision, A.S.; Visualization, A.S. and A.B. (Andrea Bonometto); Writing—original draft, A.S.; Writing—review and editing, A.S., A.B. (Alessandro Buosi), Y.T., A.-S.J., C.F., A.A.S., P.F., L.S., A.B. (Andrea Bonometto), E.P., F.R., D.B., C.G., F.O., F.C. and R.B.B.

Funding: This research was funded by the European Union's LIFE+ financial instrument (grant LIFE12 NAT/IT/000331—LIFE SERESTO, which contributes to the environmental recovery of a Natura 2000 site, SIC IT3250031—Northern Venice Lagoon).

Acknowledgments: The authors thank Orietta Zucchetta for reviewing the English language and are grateful to the anonymous reviewers that have revised the paper for useful suggestions to improve the manuscript.

References

1. Larkum, A.W.D.; McComb, A.J.; Shepherd, S.A. *Biology of Seagrasses: A Treatise of the Biology of Seagrasses with Special Reference to Australian Region*; Elsevier: Amsterdam, The Netherlands, 1989; p. 841.

2. Ondiviela, B.; Losada, I.J.; Lara, J.L.; Maza, M.; Galván, C.; Bouma, T.J.; van Belzen, J.V. The role of seagrasses in coastal protection in a changing climate. *Coast. Eng.* **2014**, *87*, 158–168. [CrossRef]

3. Duarte, C.M.; Losada, I.J.; Hendriks, I.E.; Mazarrasa, I.; Marba, N. The role of coastal plant communities for climate change mitigation and adaptation. *Nat. Clim. Chang.* **2013**, *3*, 961–968. [CrossRef]

4. Jones, C.G.; Lawron, J.H.; Shachak, M. Positive and negative effects of organisms as physical ecosystem engineers. *Ecology* **1997**, *78*, 1946–1957. [CrossRef]

5. Horinouchi, M. Horizontal gradient in fish assemblage structures in and around a seagrass habitat: Some implications for seagrass habitat conservation. *Ichthyol. Res.* **2009**, *56*, 109–125. [CrossRef]

6. Jackson, E.L.; Attrill, M.J.; Jones, M.B. Habitat characteristics and spatial arrangement affecting the diversity of fish and decapod assemblages of seagrass (Zostera marina) beds around the coast of Jersey (English Channel). *Estuar. Coast. Shelf Sci.* **2006**, *68*, 421–432. [CrossRef]

7. Sfriso, A.; Buosi, A.; Facca, C.; Sfriso, A.A. Role of environmental factors in affecting macrophyte dominance in transitional environments: The Italian Lagoons as a study case. *Mar. Ecol.* **2017**, *38*, e12414. [CrossRef]

8. Sfriso, A.; Pavoni, B.; Marcomini, A.; Orio, A.A. Macroalgae, nutrient cycles and pollutants in the lagoon of Venice. *Estuaries* **1992**, *15*, 517–528. [CrossRef]

9. Sfriso, A.; Facca, C. Distribution and production of macrophytes in the lagoon of Venice Comparison of actual and past abundance. *Hydrobiologia* **2007**, *577*, 71–85.

10. Sfriso, A.; Facca, C.; Bon, D.; Giovannone, F.; Buosi, A. Using phytoplankton and macrophytes to assess the trophic and ecological status of some Italian transitional systems. *Cont. Shelf Res.* **2014**, *81*, 88–98. [CrossRef]

11. Sfriso, A.; Facca, C.; Bon, D.; Buosi, A. Macrophytes and ecological status assessment in the Po delta transitional systems, Adriatic Sea (Italy). Application of Macrophyte Quality Index (MaQI). *Acta Adriat.* **2016**, *57*, 209–226.

12. Sfriso, A.; Facca, C.; Marcomini, A. Sedimentation rates and erosion processes in the lagoon of Venice. *Environ. Int.* **2005**, *31*, 983–992.

13. Sfriso, A.; Buosi, A.; Mistri, M.; Munari, C.; Franzoi, P.; Sfriso, A.A. Long-term changes of the trophic status in transitional ecosystems of the northern Adriatic Sea, key parameters and future expectations: The lagoon of Venice as a study case. In Italian Long-Term Ecological Research for Understanding Ecosystem Diversity and Functioning. Case Studies from Aquatic, Terrestrial and Transitional Domains. *Nat. Conserv.* **2019**, *34*, 247–272.

14. Available online: http://www.arpa.veneto.it/dati-ambientali/open-data/file-e-allegati/2017 (accessed on 10 July 2019).

15. Sfriso, A.; Facca, C.; Bonometto, A.; Boscolo Brusà, R. Compliance of the Macrophyte Quality Index (MaQI) with the WFD (2000/60/EC) and ecological status assessment in transitional areas: The Venice lagoon as study case. *Ecol. Indic.* **2014**, *46*, 536–547. [CrossRef]

16. ISPRA. *Manuale per la Classificazione Dell'elemento di Qualità Biologica "Fauna Ittica" Nelle Lagune Costiere Italiane Applicazione Dell'indice Nazionale HFBI (Habitat Fish Bio-Indicator) ai Sensi del D.Lgs 152/2006; Manuali e Linee Guida 168/2017*; ISPRA: Ispra, Italy, 2017.

17. Muxika, I.; Borja, A.; Bald, J. Using historical data, expert judgement and multivariate analysis in assessing reference conditions and benthic ecological status, according to the European Water Framework Directive. *Mar. Pollut. Bull.* **2007**, *55*, 16–29. [CrossRef]

18. Sfriso, A.A.; Sfriso, A. In situ biomass production of Gracilariaceae and Ulva rigida: The Venice Lagoon as study case. *Bot. Mar.* **2017**, *60*, 271–283. [CrossRef]

19. Sfriso, A.; Curiel, D.; Rismondo, A. The Venice Lagoon. In *Flora and Vegetation of the Italian Transitional Water Systems*; Cecere, E., Petrocelli, A., Izzo, G., Sfriso, A., Eds.; CORILA: Venice, Italy, 2009; pp. 17–80.

20. Consorzio Venezia Nuova, Cantieri Costruzioni Cementi, SELC. *Interventions to the Lagoon Inlets for the Regulation of Tidal Flows. Lido Mouth: S. Nicolò-Treporti. Lido-Arsenale Instrumentation and Control Systems. Morphological-Vegetation Restoration of the Area Occupied by the Artificial Island by Transplanting Seagrasses*; Final Report; Consorzio Venezia Nuova: Venice, Italy, 2016.

21. Venice Water Authority, THETIS, SELC. *Compensation, Conservation and Environmental Redevelopment Plan for SCI and SPA of the Venice Lagoon. Seagrass Transplant Operations*; Concessionaire Consorzio Venezia Nuova: Venice, Italy, 2010. (In Italian)

22. CORILA—Consortium for the Coordination of Research Related to the Venice Lagoon System. *Study B.6.72 B/13. Survey Activities for Monitoring the Effects Produced by the Construction of the Gates on the Lagoon inlets. Monitoring of Marine Phanerogam Transplants*; Final Report; CORILA: Venice, Italy, 2019.

23. Sfriso, A. *Chlorophyta Multicellulari e Fanerogame Acquatiche. Ambienti di Transizione Italiani e Litorali Adiacenti. I Quaderni di ARPA*; ARPA Emilia-Romagna: Bologna, Italy, 2010; p. 320.

24. Dooley, F.D.; Wyllie-Echeverria, S.; Rothc, M.B.; Warda, P.D. Tolerance and response of Zostera marina seedlings to hydrogen sulfide. *Aquat. Bot.* **2013**, *105*, 7–10. [CrossRef]

25. Höffle, H.; Thomsen, M.S.; Holmer, M. High mortality of Zostera marina under high temperature regimes but minor effects of the invasive macroalgae Gracilaria vermiculophylla. *Estuar. Coast. Shelf Sci.* **2011**, *92*, 35–46. [CrossRef]

26. Nejrup, L.B.; Pedersen, M.F. Effects of salinity and water temperature on the ecological performance of *Zostera marina*. *Aquat. Bot.* **2008**, *88*, 239–246. [CrossRef]

27. Mannino, A.M.; Menéndez, M.; Obrador, B.; Sfriso, A.; Triest, L. The genus Ruppia L. (Ruppiaceae) in the Mediterranean region: An overview. *Aquat. Bot.* **2015**, *124*, 1–9. [CrossRef]

28. Available online: https://it.wikipedia.org/wiki/Zostera_marina (accessed on 10 July 2019).

29. Sfriso, A. Ruppia maritima L. e Ruppia cirrhosa (Petagna) Grande (Helobiae, Spermatophyta) in laguna di Venezia. *Lavori Soc. Veneziana Sci. Nat.* **2008**, *33*, 41–46.

30. Brodersen, E.K.; Koren, K.; Moßhammer, M.; Ralph, P.J.; Michael Kühl, M.; Santner, J. Seagrass-Mediated Phosphorus and Iron Solubilization in Tropical Sediments. *Environ. Sci. Technol.* **2017**, *51*, 14155–14163. [CrossRef] [PubMed]

31. Sfriso, A.; Marcomini, A. Macrophyte production in a shallow coastal lagoon Part I. Coupling with physico-chemical parameters and nutrient concentrations in waters. *Mar. Environ. Res.* **1997**, *44*, 351–375. [CrossRef]

32. Koch, M.S.; Erskine, J.M. Sulfide as a phytotoxin to the tropical seagrass Thalassia testudinum: Interactions with light, salinity and temperature. *J. Exp. Mar. Biol. Ecol.* **2001**, *266*, 81–95. [CrossRef]

33. Pedersen, M.Ø.; Kristensen, E. Sensitivity of Ruppia maritima and Zostera marina to sulfide exposure around roots. *J. Exp. Mar. Biol. Ecol.* **2015**, *468*, 138–145. [CrossRef]

34. Orfanidis, S.; Panayotidis, P.; Stamatis, N. Ecological evaluation of transitional and coastal waters: A marine benthic macrophytes-base model. *Mediterr. Mar. Sci.* **2001**, *2*, 45–65. [CrossRef]

35. Sfriso, A.; Facca, C.; Ghetti, P.F. Rapid Quality Index (R-MaQI), based mainly on macrophyte associations, to assess the ecological status of Mediterranean transitional environments. *Chem. Ecol.* **2007**, *23*, 1–11. [CrossRef]

36. Viaroli, P.; Bartoli, M.; Giordani, G.; Naldi, M.; Orfanidis, S.; Zaldívar, J.M. Community shifts, alternative stable states, biogeochemical controls and feed-backs in eutrophic coastal lagoons: A brief overview. *Aquat. Conserv.* **2008**, *18*, S105–S117. [CrossRef]

37. Mcleod, E.; Chmura, G.L.; Bouillon, S.; Salm, R.; Björk, M.; Duarte, C.M.; Lovelock, C.E.; Schlesinger, W.H.; Silliman, B.R. A blueprint for blue carbon: Toward an improved understanding of the role of vegetated coastal habitats in sequestering CO_2. *Front. Ecol. Environ.* **2011**, *9*, 552–560. [CrossRef]

38. Duarte, C.M.; Sintes, T.S.; Marbà, N. Assessing the CO_2 capture potential of seagrass restoration projects. *J. Appl. Ecol.* **2013**, *50*, 1341–1349. [CrossRef]

39. Scapin, L.; Zucchetta, M.; Facca, C.; Sfriso, A.; Franzoi, P. Using fish assemblage to identify success criteria for seagrass habitat restoration. *Web Ecol.* **2016**, *16*, 33–36. [CrossRef]

40. Scapin, L.; Zucchetta, M.; Sfriso, A.; Franzoi, P. Local habitat and Seascape Structure Influence Seagrass Fish Assemblages in the Venice Lagoon: The Importance of Conservation at Multiple Spatial Scales. *Estuar. Coasts* **2018**, *41*, 2410–2425. [CrossRef]

41. Scapin, L.; Zucchetta, M.; Sfriso, A.; Franzoi, P. Predicting the response of nekton assemblages to seagrass transplantations in the Venice lagoon: An approach to assess ecological restoration. *Aquat. Conserv. Mar. Freshw. Ecosyst.* **2019**, *29*, 849–864. [CrossRef]

Article

Trophic Features, Benthic Recovery, and Dominance of the Invasive *Mytilopsis Sallei* in the Yundang Lagoon (Xiamen, China) Following Long-Term Restoration

Paolo Magni [1,*], Serena Como [1], Maria Flavia Gravina [2,3], Donghui Guo [4], Chao Li [4] and Lingfeng Huang [5,*]

[1]   National Research Council of Italy, Institute for the Study of Anthropogenic Impact and Sustainability in Marine Environment (CNR-IAS), 09170 Oristano, Italy

[2]   Dipartimento di Biologia, Università di Roma "Tor Vergata", 00133 Rome, Italy

[3]   Consorzio Nazionale Interuniversitario per le Scienze del Mare (CoNISMa), Piazzale Flaminio 9, 00196 Rome, Italy

[4]   State Key Laboratory of Marine Environmental Science, College of Ocean and Earth Sciences, Xiamen University, Xiamen 361102, Fujian, China

[5]   Key Laboratory of the Ministry of Education for Coastal and Wetland Ecosystems, College of the Environment and Ecology, Xiamen University, Xiamen 361102, Fujian, China

*   Correspondence: paolo.magni@cnr.it (P.M.); huanglf@xmu.edu.cn (L.H.)

Received: 20 June 2019; Accepted: 12 August 2019; Published: 15 August 2019

Abstract: A comprehensive set of physicochemical variables in near-bottom water and surface sediments, as well as the soft-bottom macrozoobenthic assemblages were investigated at six sites across the Yundang Lagoon (Southeast China) in November 2012. This lagoon was severely damaged in the 1970s due to domestic and industrial pollution and land reclamation and underwent a massive restoration effort over the past 30 years. Our objectives were to: (1) assess the current trophic and environmental condition of the lagoon; (2) investigate the pattern of spatial variation in the macrozoobenthic assemblages; and (3) assess the benthic recovery in relation to the main environmental gradients and the presence of invasive alien species. Nutrient, chlorophyll-*a*, biological oxygen demand (BOD_5), chemical oxygen demand (COD_{Mn}), and total organic carbon (TOC) concentrations were lower than those reported in previous decades, yet organically-enriched conditions occurred at an inner site. From azoic conditions in the 1980s and a few benthic species reported prior to this study, we found a significant increase in benthic diversity with 43 species heterogeneously distributed across the lagoon. The invasive bivalve *Mytilopsis sallei* was the dominant species, which was associated with the richest benthic assemblage. However, *M. sallei* is a pest species, and its spatiotemporal distribution should be carefully monitored. These results highlight the central role of the macrozoobenthos in providing important ecological information on the current status of the Yundang Lagoon and as an effective biological tool to follow the recovery's progress and the future evolution of this highly valued ecosystem.

Keywords: macroinvertebrates; biodiversity; eutrophication; saprobity; organic pollution; spatial variation; sediments; coastal lagoons; invasive alien species

---

1. Introduction

Coastal lagoons are typically human-managed, highly productive ecosystems which provide multiple uses and services [1–3]. However, in the last few decades, engineering interventions have increasingly affected lagoons' morphology, hydrodynamics, and sedimentology, resulting in excessive inputs of nutrients and organic matter (OM) that have led to eutrophication and organic over-enrichment of the sediments [4–7]. These factors have profoundly impaired the biological structure, trophic status, and functioning of coastal lagoon systems worldwide [8–11]. As an example, dystrophic events are the cause of anoxia, sulfide development, mass mortalities, and shifts to alternative ecohydrological states characterized by reducing conditions and lower pH [12–15]. Within this context, the concept of "saprobity", originally developed for rivers and lakes more than a century ago [16,17], applies well to coastal lagoons as a "state descriptor" of the ecosystem's condition resulting from the input and decomposition of OM and the removal of its catabolites [18]. Thus, both the trophic features (i.e., the amount of nutrients and OM in water and sediments) and the degree of saprobity (i.e., the balance between input of OM and other processes such as mineralization, sinking, dilution and export of OM) are instrumental for assessing the natural conditions and the environmental quality of a lagoon. Contrary to estuaries where the salinity gradient is dominant [19], the seawater renewal and hydrodynamics strongly govern the land-sea gradient in coastal lagoons [18,20]. Several field studies have described how species or groups of benthic organisms reduce their abundance in a given order as environmental stress increases, and how the number of taxonomic groups is reduced as stress increases [21–25]. Similarly, the main effect of saprobity on the macrozoobenthic assemblages is that the number of species that can cope with it decrease progressively as saprobity increases, leading to a reduction of species richness and diversity [18,26]. Yet, OM input associated with low saprobity due to high oxygen availability or low by-product (e.g., ammonia and sulfide) concentrations can promote the vitality of the biocoenosis, increasing biomass and abundance.

In China, many coastal areas are experiencing major urban and economic development, which is often coupled with growing anthropogenic pressures such as land reclamation, which often causes loss of coastal wetlands and serious environmental problems [27]. Monitoring and assessment of man-made pollutants and their sources are, thus, crucial for evaluating the environmental quality of these areas, including coastal lagoons, located at the interface between continental and marine ecosystems. Accordingly, a substantial body of legislation exists to address environmental protection in China, such as laws on Water (21 January 1988) and Environmental Protection (26 December 1989), Sea Water Quality GB 3097-1997, Environmental Quality for Surface Water GB 3838-2002, and Provisions for Monitoring of Marine Culture and Propagation Areas (1 April 2002) [28]. The Yundang Lagoon is an urban water body located in the highly populated and industrialized city of Xiamen (Southeast China). This lagoon, a natural bay open to the sea and historically rich in marine species, was severely damaged in the 1970s due to domestic and industrial pollution and land reclamation for agriculture. In particular, the Yundang Lagoon was reduced in size by more than 90% (from 10 km^2 before 1970 to 0.8 km^2 by 1988) and suffered decreased water flow and enhanced siltation, anoxia, loss of habitat and species, and the eventual collapse of the whole ecosystem [29–31]. Thus, an integrated management program aimed at restoring the ecosystem was put in place in the late 1980s, which included clean-up activities, wastewater treatment, flood prevention, and sludge dredging. The lagoon was also opened to the outer bay of the Xiamen Sea through a diversion canal inlet in the early 1990s, coupled with new restoration actions including a mangrove plantation and phytoremediation techniques [29,32]. One of the main current aspirations of residents and the government is to restore the ecosystem services that a healthy Yundang Lagoon could provide [29,33].

Environmental and ecological studies in Chinese coastal lagoons are sparse, but growing steadily, owing to the scientific and socio-economic interest in these highly productive but fragile ecosystems [27,31]. Within this framework, in the present study the environmental and benthic features of the Yundang Lagoon were investigated as a unique case-study of a formerly "dead" lagoon which underwent a massive restoration effort over the past 30 years. We hypothesized that the long-term restoration interventions may have led to improved environmental and biological conditions of the lagoon. To address these issues, we analyzed a comprehensive set of physicochemical variables in water and sediments, as well as the macrozoobenthic assemblages, at six different sites across the Yundang Lagoon. Our objectives were to: (1) evaluate the trophic features of the lagoon and possible areas of major environmental concern;

(2) investigate the pattern of spatial variation in the macrozoobenthic assemblages; and (3) assess the benthic recovery in relation to both the main environmental (e.g., trophic and saprobity) gradients and the presence of invasive alien species.

2. Materials and Methods

2.1. Study Area

The Yundang Lagoon was selected as the study site, a subtropical urban water body in Xiamen City, Southeast China, which connects to the Western Sea of Xiamen Island through a small canal controlled by a sluice (Figure 1). The Yundang Lagoon used to be a natural bay called Yundang Harbor. The harbor covered an area of about 10 km^2 (6.3 km long and 1.6 km wide). A great number of land reclamation projects were carried out in the early 1970s, and the Yundang Harbor gradually became a dead lagoon unable to exchange water with the sea. The lagoon area was reduced from the original 10 km^2 to only 2.2 km^2 (1988 data, including 1.0 km^2 of marshes). At present, the total water area of the Yundang Lagoon is approximately 1.5 km^2, with a maximum depth of around 5 m and a drainage area of 37 km^2, making up 30% of Xiamen Island. The lagoon can be subdivided into different sectors, including a diversion canal, inner and outer sectors, and an innermost canal (Figure 1).

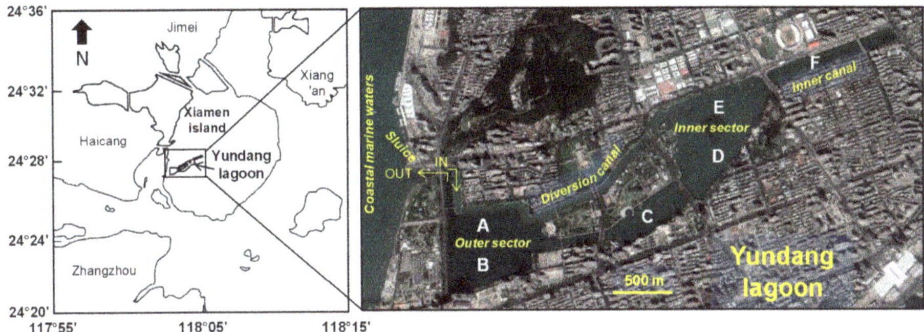

Figure 1. Location of the study area (Xiamen Island, Southeast China) and sampling sites (sites A–F) in the Yundang Lagoon. A schematic representation of water circulation in the Yundang Lagoon is given in [34].

2.2. Experimental Design and Sampling Activities

Samples were collected on November 13–14, 2012 at six different sites (A–F) hundreds of meters apart and covering different sectors of the lagoon (Figure 1). Site selection was based on our extensive knowledge of the lagoon's general features [32,34,35]. All sites were characterized by unvegetated soft-bottom sediments with a depth varying from about 1.5–3 meters. At each site, three stations tens of meters apart were randomly chosen and three replicate sediment samples meters apart were collected at each station using a Van Veen grab (30 × 20 cm, penetration depth 18 cm, approximate volume of 7 L), for a total of 54 samples. Each sediment sample was sieved on a mesh size of 0.5 mm and the residue was fixed in a 5% formalin buffered solution for macrozoobenthos determination. Prior to sieving, a subsample of the top 1 cm of the sediment was collected from each grab to determine water content and grain-size and chemical (total organic carbon and total nitrogen) analysis. These subsamples were kept refrigerated in a cooler-box until further treatment in the laboratory.

At each site and station, water salinity, temperature, and dissolved oxygen (DO) concentrations were measured using a portable CTD cast (YSI 6600) and near-bottom water samples for chemical analysis (nutrients, biological oxygen demand (BOD$_5$), chemical oxygen demand (COD$_{Mn}$), and chlorophyll-*a* (Chl-*a*)) and suspended solid (SS) determination were collected using a two-liter Niskin bottle, for a total of 18 samples.

2.3. Sample Treatment and Analysis

In the laboratory, water samples were filtered using a 0.45 μm filter. Nutrients (ammonia, nitrate, nitrite, and reactive phosphorus (RP)), BOD_5, COD_{Mn}, and SS were measured by applying standard analytical methods (Table 1) according to the national standard GB17378.4 [36]. Chl-*a* was measured by fluorometric analysis using a Turner Designs Fluorometer (Mode 10-AU) (San Jose, USA).

Table 1. Methods used for seawater analysis in Yundang Lagoon. All methods are according to the national standard GB17378.4 [36].

Variable	Method	Lower Detection Limit (mg L^{-1})
Ammonia	Indophenol-blue colorimetry	0.012
Nitrite	N(1-naphty1)-ethylenediamine dihydrochloride spectrophotometric method	0.0005
Nitrate	Cadmium column reduction method	0.012
Phosphate	Phosphomolybdenum blue spectrophotometric method	-
Biological Oxygen Demand (BOD$_5$)	Five-days biochemical culture method	2
Chemical Oxygen Demand (COD$_{Mn}$)	Potassium iodide-alkaline potassium permanganate determination method	0.5
Suspended solids (SS)	Weighting	0.1

For the measurements of sediment particle size, diluted hydrochloric acid and hydrogen peroxide were added to the evenly mixed sample to remove carbonates and organic matter. After being washed to remove the acid to attain neutrality, Na-hexametaphosphate 0.6% solution was added to avoid particle flocculation, and the samples were allowed to rest for 24 h. Subsequently, the median particle size of sediments (Md) was measured with a Malvern Mastersizer 2000 laser particle size analyzer (Malvern, United Kingdom), and the measurement data were outputted at 1/4 Φ intervals. The moment method was used to calculate the grain-size parameters of the sediments [37]. The water content (Wc) of the sediment was obtained after drying a sediment subsample at 70 °C for 24 h. Analysis of total organic carbon (TOC) and total nitrogen (TN) content of sediments was accomplished by freeze-drying and powdering the sediment sample. Carbonate was removed from the sample with 2 N HCl, at which point it was vacuum-dried. The TOC and TN content of the dried sediment sample was measured using an Elemental vario EL-III element analyzer. Replicate analyses of standards of acetanilide yielded a mean precision of about 0.3% for organic carbon and nitrogen.

The macrozoobenthos from each grab sample were sorted, identified to the species level, when possible, counted under a stereo-microscope and preserved in 75% ethanol. After counting, the wet weight was taken for individuals of the same species in each sample.

2.4. Data Analyses

In order to evaluate environmental differences among sites, multivariate analyses were carried out on the environmental data using the principal component analysis (PCA) ordination method and correlation matrices [38]. The components loadings were considered to quantify the correlation between the variables and the principal components scores. Water and sediment variables were analyzed separately in order to investigate the trophic features of the Yundang Lagoon from two different perspectives and to evaluate the consistency of results between the two ecosystem components (water and sediment). Near-bottom water variables included salinity, ammonium, nitrate, nitrite, dissolved inorganic nitrogen (DIN = sum of ammonium, nitrate, and nitrite), RP, N/P ratio, DO, BOD_5, COD_{Mn}, chlorophyll-*a* (Chl-*a*), and suspended solids (SS). Sediment variables included the sand (>63 μm), silt (63-8 μm) and clay (<8 μm) fractions, Md, Wc, and TOC (TN was excluded from the multivariate analyses as it correlated to TOC at $R > 0.90$).

PERMANOVA was used to tests for differences in the arrangement of the sites in terms of biotic parameters (i.e., abundance, number of species, diversity, biomass) and the abundance of individual species [39]. The pairwise post hoc test was also performed for all pairs of site comparisons. The Bonferroni correction was used and 9999 permutations were performed [40]. Differences among sites related to macrozoobenthos data were examined by means of non-metric multidimensional scaling (nMDS) based on the Bray–Curtis similarity matrix. Dominance curves [41] were plotted for each site, and species were ranked in order of importance in terms of percent cumulative abundance. Canonical correspondence analysis (CCA) was performed on macrozoobenthos abundance data (log transformed) and environmental variables (nitrate, ammonium, RP, SS, sand, Md, TOC) at each site, in order to test the response of the macrozoobenthic community to the main environmental (trophic and confinement) gradients to determine what provided the best combination of variables supporting the ordination model [42]. In particular, nitrate, ammonium, and RP for the water and TOC for the sediments were used as trophic variables, while SS for the water and sand and Md for the sediments were used as variables related to the hydrodynamics and confinement of the lagoon.

All statistical analyses were performed by means on the PAST statistics program (version 3.12).

3. Results

3.1. Environmental Variables

3.1.1. Water

Near-bottom water temperature and salinity were homogeneous throughout the lagoon averaging 22.6 ± 0.2 °C and 29.9 ± 0.3 psu (± SE standard error), respectively. Marked spatial variation was found for all the other environmental variables examined (Figure 2). In particular, ammonium, RP, BOD_5, COD_{Mn}, and Chl-*a* concentrations were highest at site F, while nitrate, nitrite and DIN concentrations, and the N/P ratio were lowest at site A. SS concentrations were lowest at both site A and B. Overall, the water variables highlighted the hypertrophic condition of site F, in the innermost sector of the lagoon, as compared to all the other sites. Separation of sites was also clearly revealed by PCA, as site F was separate along the *x*-axis, and sites A and B along the *y*-axis (Figure 3). The PCA loadings showed the highest correlation for ammonium, RP, BOD_5, COD_{Mn}, and Chl-*a* (positive), and DO (negative) on the *x*-axis, and the highest (positive) correlation for nitrate, nitrite DIN, N/P, and SS on the *y*-axis (Table 2).

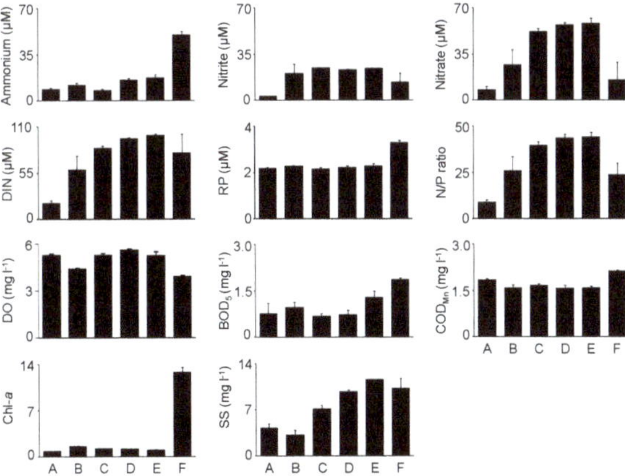

Figure 2. Mean value (n = 3; ±SE standard error) of near-bottom water variables at each site: ammonium, nitrate, nitrite, dissolved inorganic nitrogen (DIN = ammonium, nitrate, and nitrite), reactive phosphorous (RP), N/P, dissolved oxygen (DO), biological oxygen demand (BOD_5), chemical oxygen demand (COD_{Mn}), chlorophyll-*a* (Chl-*a*), and suspended solids (SS).

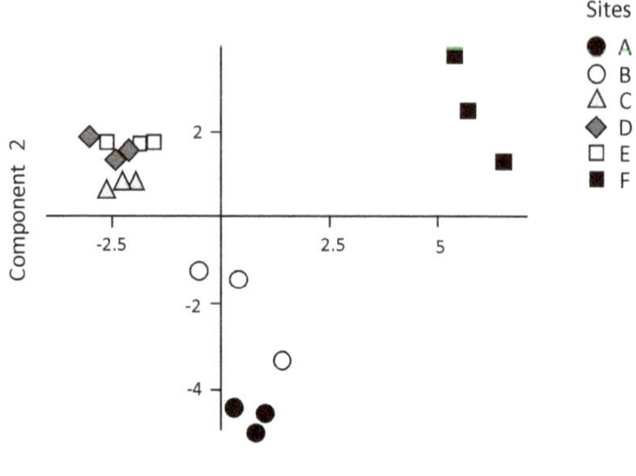

Figure 3. Principal component analysis (PCA) of near-bottom water variables, including salinity, dissolved oxygen (DO), ammonium, nitrate, nitrite, dissolved inorganic nitrogen (DIN = ammonium, nitrate, and nitrite), reactive phosphorus (RP), N/P ratio, suspended solids (SS), biological oxygen demand (BOD$_5$), chemical oxygen demand (COD$_5$), and chlorophyll-*a* (Chl-*a*). PC1 = 46.5% of variance, PC2 = 34.4% of variance.

Table 2. Component correlation coefficients in the principal component analysis (PCA) of the water variables and the first two components (PC1 and PC2 = 46.5% and 34.4% of variance, respectively).

	PC1	PC2
Salinity	0.10	0.25
Dissolved oxygen	**−0.34**	−0.03
Nitrate	−0.29	**0.34**
Nitrite	−0.21	**0.36**
Ammonium	**0.37**	0.24
DIN	−0.11	**0.47**
RP	**0.38**	0.18
N/P	−0.24	**0.40**
Suspended solids	0.04	**0.40**
BOD$_5$	**0.32**	0.19
COD$_{Mn}$	**0.37**	−0.01
Chlorophyll−*a*	**0.39**	0.15

In bold, the highest loadings.

3.1.2. Sediments

Surface sediments were generally muddy, with a prevalence of silt at all sites (Figure 4). However, spatial differences included higher sand content at sites A and B, compared to C, D, and E (Figure 4). The median particle diameter (Md) varied similarly to the variation in the silt content among sites, with the highest Md values at sites C, D, and E. The water content (Wc) varied less markedly among sites, with the largest within−site variation at site A. Finally, the TOC content did not show marked differences among sites, except at site F. In particular, TOC varied from 1.7% to 2.3% at sites A and B, respectively, but was up to 5.0 ± 0.01 at site F, providing an additional evidence of its hypertrophic condition (Figure 4). Similar to the water variables, the ordination model of the PCA carried out on the sediment variables revealed a clear separation of site F along the *y*-axis (Figure 5). Along the *x*-axis, sites A and B were located on the negative pole, opposite to sites C, D, and E. The observed pattern is consistent with the variation in sediment grain-size among sites A–E, with higher loadings for sand (negative values) and silt, clay and Md (positive values) on the *x*-axis. On the contrary, factors related to organic enrichment, found at site F, were mostly correlated with the *y*-axis where Wc (a proxy of mud content) and TOC showed a positive and a negative correlation, respectively (Table 3). Overall, these results were consistent with those obtained from the water variables, indicating that both water and sediment components provide consistent data helpful in assessing the environmental heterogeneity and the trophic features of the Yundang Lagoon.

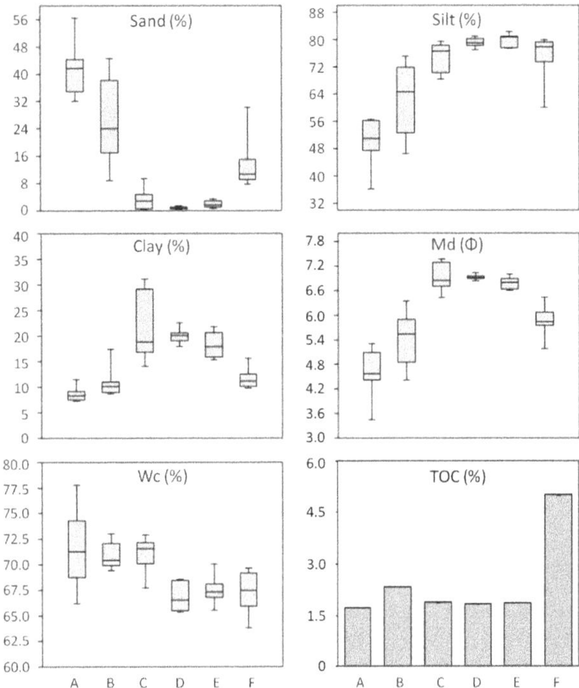

Figure 4. Box-plots of sediment sand, silt and clay content, and median particle diameter (Md) and water content (Wc) at each site (n = 9), as well as the mean value of total organic carbon (TOC) content of sediments at each site (n = 3, ±SE standard error not visible when <0.01).

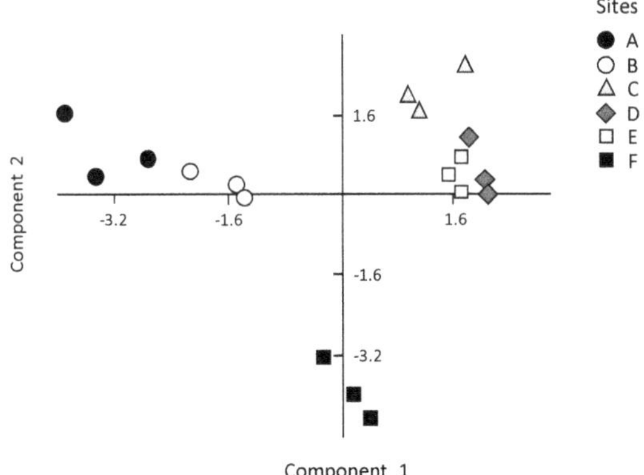

Figure 5. Principal component analysis (PCA) of sediment variables, including sand, silt, and clay content, median particle diameter (Md), and water and total organic carbon content. PC1 = 67.6% of variance, PC2 = 23.9% of variance.

Table 3. Component correlation coefficients in the principal component analysis (PCA) of the sediment variables and the first two components (PC1 and PC2 = 67.6% and 23.9% of variance, respectively).

	PC1	PC2
Sand	**−0.49**	0.00
Silt	**0.48**	−0.17
Clay	**0.43**	0.36
Md	**0.48**	0.16
Wc	−0.33	**0.46**
TOC	0.00	**−0.78**

In bold, the highest loadings.

3.2. Macrozoobenthic Assemblages

In the present study, we found 43 taxa which marked a significant increase in benthic species richness and diversity in the Yundang Lagoon compared to previous years and decades (Table 4). In particular, according to the available published and gray literature, no species were present in the 1980s, a single polychaete, *Neanthes succinea*, was found in the 1990s, and a few more species (n = 9) were reported during the 2000's, with two polychaetes dominating (*Neanthes glandicincta* and *Capitella capitata*), and the first report of the invasive bivalve *Mytilopsis sallei* (Table 4). In our study, *M. sallei* was the most abundant benthic species, accounting for 75% of the total individuals and 88% of the total biomass. Several other less abundant species included the gastropods *Stenothyra glabra* and *Rissolina plicatula*, the bivalve *Pseudopythina tsurumaru*, the crustaceans *Corophium* sp. and *Corophium uenoi*, and the polychaete *Cossurella dimorpha* (Table 4). All these species, except *C. dimorpha*, were found at site A. *M. sallei* was dominant at all stations and replicates of sites A and C, very abundant at a single station of site F, and was not found at sites B, D, and E. Another invasive species, the ascidian *Styela plicata*, was found only in some replicates of site C, but followed *M. sallei* in terms of its contribution to the total biomass (8.4%). *S. glabra*, *P. tsurumaru* and *N. glandicincta* occurred at all sites except B and F, while *C. dimorpha* was dominant in B and absent in A and C. Finally, different subsets of species were exclusive to a single site: nine species in A, four in B, four in C, nine in D and only two in E (Table 4).

The PERMANOVA conducted on the abundance of all species revealed significant differences among sites. Site A was distinct from all the others, while site C differed from B, D, and E, and site F from D and E (Table 5). These differences were similar to those found for *M. sallei* which showed the highest abundance and biomass at site A. Significant differences also emerged from the structural biotic parameters, including total abundance, species richness, diversity, and biomass (Table 6, Figure 6). In the pairwise comparisons, site A was significantly different from most of the other sites in terms of total abundance and species richness, and less so in terms of diversity (≠ from F) and biomass (≠ from B, D, and E) (Table 7). Results suggest that the dominance of *M. sallei* appeared to be one of the discriminating factors responsible for the differentiation of sites and their species richness. In addition, site A had the highest species richness, followed by C, D, and E, and was significantly different in terms of diversity from site F which had the least number of species (Table 7, Figure 6).

Table 4. Changes in species number and composition of macrozoobenthos collected from the bottom sediment in the Yundang Lagoon in the years before the restoration (1982–1987), and during the first (1989–1992) and the third (2001 to present) phase of restoration. Information prior to the present study (2012) was gathered from Xu et al. [43] for 2007–2008, and from Ye et al. [29] and unpublished reports for the other periods. The dominant species in each period/Site are marked by *. For the present study (2012), the abundance (ind. 600 cm^{-2}) per site is also reported as the mean of three stations.

Species/Year	Taxon	Habitat and/or Trait	1982–1987	1992–1993	2001–2002	2007–2008	2012 (This study)	2012 Site A	2012 Site B	2012 Site C	2012 Site D	2012 Site E	2012 Site F
Mytilopsis sallei (Récluz, 1849)	B	F			√	√	√*	230*		67.67	2		63
Rissoina plicatula Gould, 1861	G	F/M					√*	12*		4.33			
Stenothyra glabra Adams, 1861	G	M/S					√*	8.67*	1.33	0.33	4.34	3.33	
Corophium uenoi Stephensen, 1932	C	F/S			√	√	√*	8.67*					
Corophium sp.	C	F/S					√*	8.67*	1.33				
Pseudopythina tsurumaru (Habe, 1959)	B	C					√*	7.33*		2.67	0.67	1	
Molgula manhattensis (De Kay, 1843)	A	F					√	2					
Gammaropsis sp.	C	S					√	2	1				
Cerithidea sp.	G	M/S					√	1.34					
Grandidierella japonica Stephensen, 1938	C	S			√	√	√	1					
Tharyx sp.	P	S					√	0.67					
Diopatra chiliensis Quatrefages, 1865	P	S					√	0.33					
Cephalothrix sp.	N	S					√	0.33					
Cycladicama sp.	B	S					√	0.33					
Neanthes glandicincta (Southern, 1921)	P	S			√	√*	√	0.33		0.33	0.33	0.33	
Melita sp.	C	S					√	0.33					
Spionidae und.	P	S			√	√	√		0.33				
Terebellides stroemii Sars, 1835	P	S/F					√		0.67				
Cossurella dimorpha Hartman, 1976	P	S					√*		1.33*		8	1.34	1.67
Sabellidae und.	P	F			√	√	√		0.67	1		0.33	
Pista sp.	P	S					√		0.33				
Serpulidae und.	P	F					√		0.33				
Paraprionospio pinnata (Ehlers, 1901)	P	M/S					√		0.33	1.67	1		
Praxillella sp.	P	S/F					√			0.33			0.33
Alitta succinea (Leuckart, 1847)	P	S		√									
Neanthes sp.	P	S					√			0.33	0.33	0.33	
Oxyodromus angustifrons (Grube, 1878)	P	M/S					√			0.67			
Mitrella bella (Reeve, 1859)	G	F					√			1			
Styela plicata (Lesueur, 1823)	A	F					√*			4*			
Lumbrinereis sp.	P	M/S					√			0.33			
Tylorrhynchus heterochaetus (Quatrefages, 1866)	P	M/S					√				0.67	1.67	
Nitidotellina minuta (Lischke, 1872)	B	S					√				0.33		
Theora sp.	B	S					√				0.33		
Perinereis sp.	P	S/M					√				0.33	2	0.67
Micronephtys oligobranchia (Southern, 1921)	P	M/S			√		√				0.33		
Eulima bifascialis (Adams, 1864)	G	P					√				0.33		
Capitella capitata (Fabricius, 1780)	P	S			√*	√*	√				0.67		
Poecilochaetus serpens Allen, 1904	P	M/S					√				0.33		
Namalycastis abiuma (Grube, 1872)	P	S					√				1		
Prionospio queenslandica Blake & Kudenov, 1978	P	M/S					√				0.67		
Potamocorbula ustulata (Reeve, 1844)	B	M/S					√				1.34	3.33	1
Capitellidae und.	P	S					√				0.33		
Perinereis nuntia (Lamarck, 1818)	P	S/F					√					1.34	
Potamocorbula laevis (Hinds, 1843)	B	M/S			√	√							
Tellinimactra sp.	B	S					√					0.33	
Typosyllis sp.	P	S				√							
Total number of species			**0**	**1**	**9**	**9**	**43**	**16**	**10**	**13**	**19**	**11**	**5**

Taxon: B = Bivalves; G = Gastropods; C = Crustaceans; P = Polychaetes; A = Ascidians; N = Nemerteans. Habitat or Trait: S = Muddy Sediment; M = Mangrove; F = Fouling; P = Parasite; C = Commensal. Bold numbers indicate the total number of species in each column.

Table 5. Results of PERMANOVA (based on Bray–Curtis similarity index) testing differences among sites (A, B, C, D, E, F) for the abundance of all the taxa. Comparisons of post–hoc test between all pairs of sites are also given. Significant comparisons at $p < 0.05$ are marked by * and at $p < 0.01$ by **. The Bonferroni correction was used. Permutation number = 9999.

Total sum of squares	21.88				
Within–group sum of squares	14.87				
F	4.533				
p (same)	0.0001				
Pairwise between sites	A	B	C	D	E
B	0.003**				
C	0.07	0.02*			
D	0.002**	0.28	0.003**		
E	0.003**	1	0.005**	1	
F	0.03*	0.53	0.11	0.003**	0.02*

Table 6. Results of PERMANOVA (based on Euclidean distance) testing differences among sites for abundance, number of species, Shannon diversity (H′), and biomass. Permutation number = 9999. All values are significant at $p < 0.0005$.

	Abundance	N° Species	Diversity (H′)	Biomass
Total sum of squares	9.08	249	17.37	2.18
Within–group sum of squares	3.69	104.4	10.65	1.15
Pseudo F	14.00	13.29	6.07	8.69
p	0.0001	0.0001	0.0004	0.0001

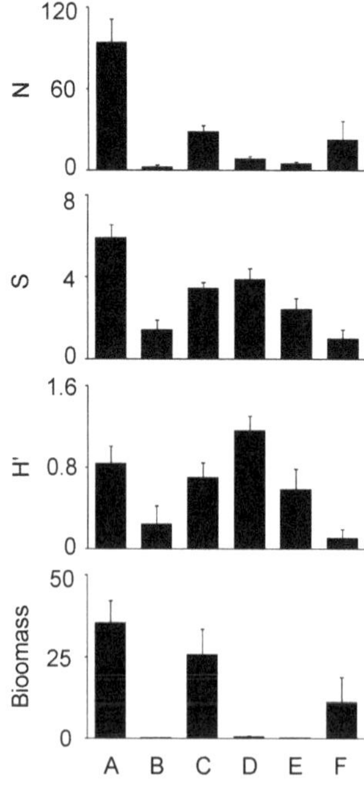

Figure 6. Mean values (n = 9; ±SE standard error) of the total number of individuals (N, ind. 600 cm^{-2}), total number of species (S), Shannon index (H′), and biomass (g.wet weight 600 cm^{-2}) at each site.

Table 7. Results of post–hoc test between all pairs of sites. (A–F). Significant comparisons at $p < 0.05$ are marked by * and at $p < 0.01$ by **. The Bonferroni correction was used.

Abundance	A	B	C	D	E
B	0.002**				
C	0.01**	0.006**			
D	0.02*	0.2	0.05		
E	0.003**	1	0.02*	1	
F	0.09	1	1	1	1
N° Species	A	B	C	D	E
B	0.003**				
C	0.02*	0.05			
D	0.52	0.07	1		
E	0.003**	1	1	0,94	
F	0.002**	1	0.02*	0.04*	1
Diversity (H′)	A	B	C	D	E
B	0.42				
C	1	0.963			
D	1	0.046*	0.628		
E	1	1	1	0.513	
F	0.01**	1	0.04*	0.009**	0.77
Biomass	A	B	C	D	E
B	0.003**				
C	1	0.001**			
D	0.001**	0.016*	0.02*		
E	0.001**	0.7	0.006**	1	
F	0.44	1	1	1	1

Different aspects of diversity, i.e., species richness and the even distribution of individuals over the species, differently affected the benthic distribution patterns of the study sites, as shown by the cumulative dominance curves (Figure 7). The most even distribution in the number of individuals among species was exhibited by site B (low species richness and low diversity), D (high species richness and highest diversity), and E (intermediate between B and D in terms of species richness and diversity). Instead, the highest abundances of *M. sallei* mostly affected the dominance curve of sites A and C notwithstanding their high species richness (highest at site A). Finally, the short and flattened curve for site F reflected both the lowest number of species (n = 5) observed and the high number of *M. sallei* individuals which occurred at one of its three stations.

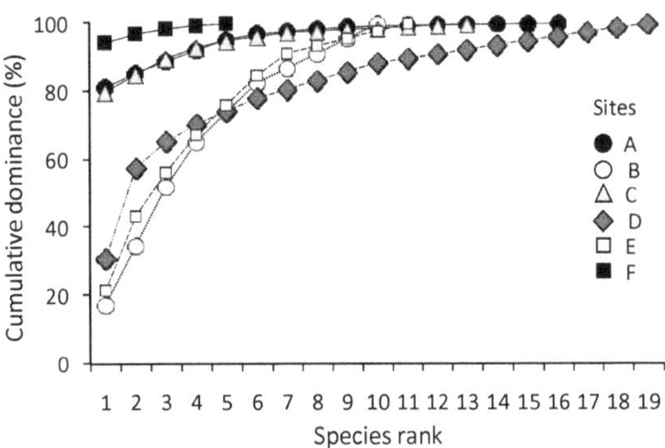

Figure 7. Cumulative dominance curves on macrobenthic abundance data from each sampling site.

The separation of sites based on the macrobenthos structure was revealed by the MDS (Figure 8). The low level of stress (=0.05) confirmed that there was a good representation of the community pattern. Site A was most markedly distinguished from all the other sites, with the exception of one station of site F dominated by *M. sallei*, while both sites A and C were clearly separated from the other sites. The significance of differences among the sites was confirmed by the PERMANOVA (Table 5). Canonical correspondence analysis (CCA) revealed the relationship between the selected environmental variables and the ordination model produced (Figure 9). The first two axes accounted for 69.1% of the explained variability (51.5% and 17.6% for the *x*- and *y*-axes, respectively). The environmental variables were plotted as correlations with site scores and the strongest correlations were found for sand (negative) and for SS, Md, nitrate, and RP (positive); TOC was less correlated with the axes. Overall, the environmental variables mostly affecting the separation of sites were those connected with the trophic and hydrodynamic features of the inner lagoon sites (D, E), with sand vs. Md and SS in opposite directions to indicate, as proxies, higher and lower hydrodynamics, respectively (Figure 9).

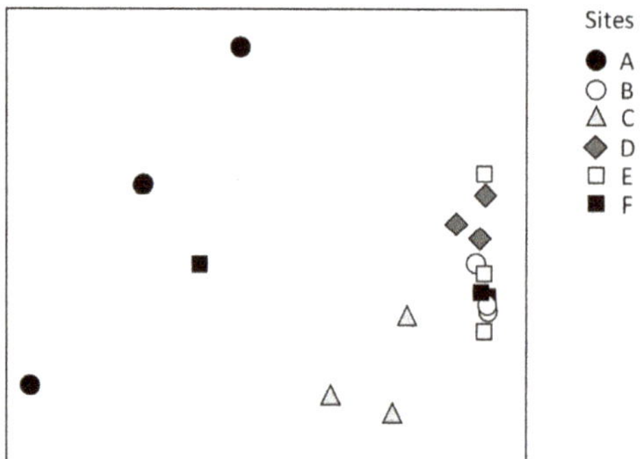

Figure 8. Non-metric multidimensional scaling (nMDS) ordination model of log transformed abundance data (stress: 0.05). Each symbol represents a sampling station at each site obtained as the mean of three replicate samples.

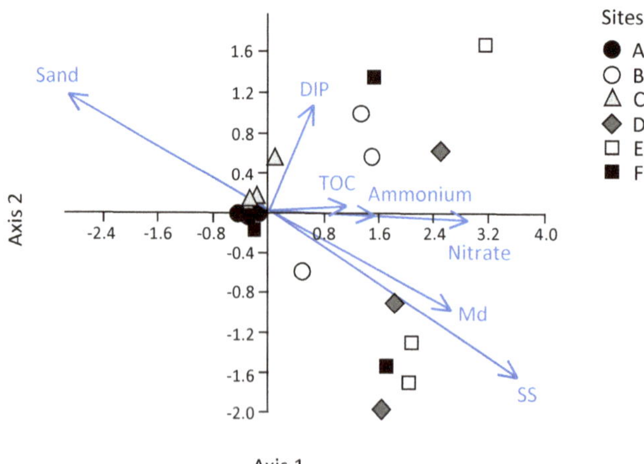

Figure 9. Plot of Canonical correspondence analysis (CCA) performed on species abundance and environmental variables most correlated with two axes and indicative of trophic (i.e., ammonia, nitrate, reactive phosphorus, and TOC) and hydrodynamic (i.e., suspended solids, sand, and size-particle) conditions.

4. Discussion

For the first time, this study assesses the trophic features and benthic diversity of the Yundang Lagoon, an urban coastal lagoon in Southeast China, in a comprehensive and integrated manner. The Yundang Lagoon was referred to as a "dead" lagoon in the 1970s, and has undergone 30-year restoration effort. Microbiological and eco-toxicological studies conducted in the last decade suggest that the recovery of the Yundang Lagoon has not yet been accomplished. In particular, high levels of heavy metals and estrogenic compounds originating mainly from municipal wastewaters have been found in sediments and pore-water of the Yundang Lagoon posing serious toxic risk to the biota including humans [44–46]. Furthermore, evidence has shown that the discharge of surrounding sewage is still a major source of OM pollution in the Yundang Lagoon [34,46].

These results provide a detailed analysis of the trophic features of the Yundang Lagoon and highlight the high spatial heterogeneity of both abiotic (water and sediments) and biotic (macrozoobenthos) components, in spite of its small size and the absence of a salinity gradient. From an abiotic perspective, both water and sediment variables independently point in the same direction. In particular, there was a major separation between the low-nutrient and sandier sites A and B in the outer sector of the lagoon, and the hypertrophic and organically over-enriched site F in the inner canal. Site F departed the most from all the other sites, which was consistent with the fact that this sector of the lagoon is still subjected to numerous sewage discharges [32,47]. The confined location and reduced hydrodynamics of site F are indicated by high levels of suspended solids in the near-bottom water and a high silt content in surface sediments, making the site likely to favor the accumulation of toxic by-products derived from the microbial decomposition of high amounts of OM which were present above established critical thresholds [24,25]. Indeed, higher BOD_5 and COD_{Mn} concentrations measured at this site (where we also observed methane bubbling from the sediment) may be a concurrent evidence of high saprobity levels in this sector of the lagoon which, in turn, will severely affect the biota [18]. On the other hand, all other sites of the lagoon showed significantly lower Chl-*a* (in water) and TOC (in sediment) concentrations than those found at site F, notwithstanding seasonal variation, which was not considered here, during phytoplankton blooms [34]. Interestingly, our results are consistent with those reported by Sun et al. [33] who found spatial variation in Chl-*a* and TOC to be greater in highly urbanized watersheds such as Yundang Lagoon, than in less urbanized watersheds. Overall, our study demonstrates that at the time samples were taken for this study (2012), the Yundang Lagoon had much lower nutrient, BOD_5, and COD_{Mn} concentrations than those found in previous decades [29], supporting the inference that a partial, but significant environmental recovery has occurred.

The overall improvement of the environmental condition of the Yundang Lagoon was reflected in a major recovery and revitalization of the lagoon's soft-bottom benthic assemblage during the last decade. Following anoxic and azoic sediments found when the lagoon was highly polluted and closed to the adjacent coastal waters until 1987 [29], a single polychaete species, *Neanthes succinea*, typical of organically enriched sediments with high H_2S content [14,48], was reported during the first phase of the restoration program (1989–1992). Between 2001 and 2008, nine species started to colonize the lagoon's sediments [43]. Among them, the opportunistic polychaetes *Capitella capitata* and *Neanthes glandicincta*, typical of muddy, organically enriched sediments were dominant, while they occurred with few individuals in our study. These results show a significant recovery of the macrozoobenthic assemblages for the first time, with a marked increase in species richness and diversity, yet a heterogeneous distribution across the lagoon. Of the 43 species found, eight were dominant (accounting for 93% of the total abundance), mainly represented by mollusks and amphipods, and including the polychaete *Cossurella dimorpha* which was not reported previously. Overall, the most dominant species was the bivalve Dreissenidae *Mytilopsis sallei*, an invasive alien species native to the Caribbean and a massive component of fouling events [49,50]. *M. sallei* was transported to China via ballast waters, and was first recorded in Xiamen waters (Maluan Bay) in 1990 and in the Yundang Lagoon in 2000 [51–53]. In a basin as small as the Yundang Lagoon, one may expect little spatial variation in the soft-bottom benthic assemblages. However, our extensive sampling effort allowed us to identify significant variation in the macrozoobenthic assemblages within relatively short distances of a few hundred meters. A number of environmental and biological factors may help explaining the high spatial variability and the recovery of the macrozoobenthic assemblages in the Yundang Lagoon. These are: (1) the heterogeneous morphology and the peculiar hydrology of the lagoon, including the daily water exchange with the adjacent coastal sea, (2) the high environmental variability in water and sediments across the lagoon, (3) the presence of nonpoint sources of pollution and other anthropogenic pressures, and (4) the dominance of the invasive *M. sallei* and the biotic interactions within the benthic community. Thus, the present work represents a unique case-study to evaluate the complex dynamics of macrozoobenthic assemblages in a formerly dead urban coastal lagoon following long-term restoration.

As an example, sites A and B in the outer sector of the lagoon, although similar in several environmental features and, thus, significantly different from the other sites, were the least similar to one another in terms of benthic species composition, richness and diversity. In particular, site A was the most abundant, biomass-rich, and species-rich site in the lagoon, while the opposite was true for site B, which had a species richness and diversity as low as that found at the most degraded site, F. We infer that such biotic differences between sites A and B may be related to their position within the outer sector of the lagoon. In particular, site B was located near the heavily modified shoreline made by a concrete wall, following earlier dredging and reclamation, which may have left little of the historic ecohydrological system on this site of the lagoon. This may have contributed to determine a poor macrozoobenthic assemblage which included few species typical of muddy sediments, such as *Cossurella dimorpha*, *Corophium* sp. and *Gammaropsis* sp. On the other hand, site A was located closer to the shoreline where mangrove species, such as *Kandelia candel*, *Avicennia marina*, *Rhizophora stylosa*, and *Heritiera littoralis*, had been successfully planted from 2001 to 2006 [54], contributing to increases of birds and fish species in the lagoon [29]. It is likely that the presence of mangroves along this shore has favored the colonization of the invasive fouling species *M. sallei*. This bivalve was associated with the richest benthic assemblage of the lagoon, correlating with the highly diverse assemblages found at site A which also included species typical of mangrove habitats such as *R. plicatula* and *S. glabra*. The diversity of the site was also seen in *Corophium uenoi*, which lives within muddy tubes on soft bottom sediments, but is also present in the fouling community, and *P. tsurumaru*, which is typical of muddy sediments and possibly shares the burrow of holothurian or crabs (commensal). On the contrary, the innermost site F had the most impoverished assemblage, consistent with its degraded condition in both water and sediments. Yet, *M. sallei* was also found at one station at this site, providing evidence of its high resistance and resilience, and its ability to settle when a substrate is available, owing to the characteristic traits of opportunistic *r*-strategists [49,50]. The apparently random distribution of *M. sallei*, heavily colonizing some sectors of the lagoon, was also indicated by its dominance at site C, where environmental conditions were intermediate between the outer (sites A and B) and the innermost (site F) sectors. This indicates that, in addition to the environmental factors, species-specific biological interactions should be considered to evaluate the complex dynamics that occur within a benthic assemblage.

Our results suggest that *M. sallei*, known as an invasive ecosystem engineer that creates new habitats and favors the settlement of other species [55], may support a rich and diversified community. In particular, in the present study *M. sallei* was associated with the highest number of species having different biological traits (e.g., fouling, parasite and commensal) and typical of different habitats (e.g., muddy sediments, mangroves). Among them, *S. glabra*, an IUCN threatened species [56] in need of protection measures, and *P. tsurumaru* which is known to establish symbiotic relations with other species [57]. Nevertheless, *M. sallei* is also well known as a pest species characterized by wide temperature, salinity, and oxygen tolerances, as well as a fast growth-rate and a high fecundity which may favor competition with native species and inhibit the growth of other species [49]. As an example, an experimental field study on the fouling macrofauna associated with *M. sallei* conducted in the

Yundang Lagoon showed a reduction in species diversity in summer when environmental conditions worsened, but the density and biomass of both *M. sallei* and fouling macrofauna were highest [52]. Furthermore, *M. sallei* is known to erode fishing facilities and artificial structures leading to ecosystem damage, economic loss, and inconvenience to locals [53], thus, its spatial and temporal variation should be carefully monitored.

With regards to the spatiotemporal evolution of the macrozoobenthic assemblages in the Yundang Lagoon, one possibility is an even more massive colonization of the lagoon by *M. sallei* as the most dominant species. The successful establishment of *M. sallei* in the Yundang Lagoon may prefigure a wider, yet probably overlooked colonization of this species as a structural component of the benthic assemblages in other Chinese coastal waters and lagoons for which information is very scant [53]. A similar colonization process occurred in the Mediterranean Sea during the first half of the last century by Serpulidae *Ficopomatus enigmaticus* which is now a typical bioengineer species in Mediterranean lagoons [58–60]. However, after a period of dominance, *M. sallei* may decrease in dominance or even be replaced by other species typical of sessile fouling, such as *S. plicata*. A similar alternation of species has been found between the mussel *Mytilus edulis* and the polychaete *Sabellaria spinulosa* along the eastern Atlantic coast, and in the Mediterranean Sea between *M. galloprovincialis* and the extensive bioconstructions of *S. spinulosa* along the Italian Puglia coast [61,62].

5. Conclusions

The present study highlighted the central role of soft-bottom macrozoobenthic assemblages and associated water and sediment variables in assessing the environmental status of the Yundang Lagoon following the massive restoration effort conducted over the past 30 years. We believe that the macrozoobenthos, which greatly contributed to the lagoon's recovery, will also have a future role in the evolution of the lagoon's biodiversity and functioning, especially considering the dominance of invasive alien species. In particular, the presence of *M. sallei* as a dominant species may be of concern for its possible role as a pest species, thus its spatiotemporal evolution should be carefully monitored. Our results provide valuable, data–based knowledge of the environmental quality, as well as early warning, detection, and prevention of any invasion process in the Yundang Lagoon, an area of special relevance for current legislation. Overall, the environmental and biological information gathered in the present study provides insights relevant to both the activities carried out within the long-term integrated management program, and the forecast of possible future ecological scenarios for the lagoon. As such, these results can support the local administration as they strive towards proper management of the lagoon and the restoration of the ecosystem services that this urban water body may provide.

Author Contributions: Supervision, data curation. and funding acquisition: P.M. and H.L.; investigation: P.M.; H.L., D.G. and C.L.; formal analysis: P.M., S.C. and M.F.G.; visualization: P.M. and M.F.G.; writing—original draft: P.M. and M.F.G.; writing—review and editing: P.M., S.C., M.F.G., D.G., C.L. and H.L.

Funding: The present study was carried out under the Executive Program of Scientific and Technological Cooperation between Italy and China (2010–2012) of the Ministry of Foreign Affairs of Italy granted to PM and LH, and the Short Term Mobility Program (STM–2015) of the National Research Council of Italy granted to PM and LH. The study was also partially funded by the National Key R and D Program of China (grant no. 2016YFC0502904).

Acknowledgments: We gratefully thank Shiqiang Zhou for help in macrozoobenthos identification, the staff of the Xiamen Yundang Lagoon Management Office for logistic and technical support, and Marcello Giorgi, University of Rome "Tor Vergata", for technical support in data analysis procedures. We gratefully acknowledge three anonymous reviewers whose comments greatly contributed to an improved version of the original manuscript.

Conflicts of Interest: The authors declare no conflict of interest.

References

1. Basset, A.; Elliott, M.; West, R.J.; Wilson, J.G. Estuarine and lagoon biodiversity and their natural goods and services. *Estuar. Coast. Shelf Sci.* **2013**, *132*, 1–4. [CrossRef]
2. Marcos, C.; Torres, I.; Lopez–Capel, A.; Perez–Ruzafa, A. Long term evolution of fisheries in a coastal lagoon related to changes in lagoon ecology and human pressures. *Rev. Fish Biol. Fish.* **2015**, *25*, 689–713. [CrossRef]

3. Newton, A.; Brito, A.C.; Icely, J.D.; Derolez, V.; Clara, I.; Angus, S.; Béjaoui, B.; Sousa, A.I.; Solidoro, C.; Yamamuro, M.; et al. Assessing, quantifying and valuing the ecosystem services of coastal lagoons. *J. Nat. Conserv.* **2018**, *44*, 50–65. [CrossRef]

4. da Cunha, L.; Wasserman, J.C. Relationships between nutrients and macroalgal biomass in a Brazilian coastal lagoon: The impact of a lock construction. *Chem. Ecol.* **2003**, *19*, 283–298. [CrossRef]

5. Como, S.; Magni, P.; Casu, D.; Floris, A.; Giordani, G.; Natale, S.; Fenzi, G.A.; Signa, G.; De Falco, G. Sediment characteristics and macrofauna distribution along a human−modified inlet in the Gulf of Oristano (Sardinia, Italy). *Mar. Poll. Bull.* **2007**, *54*, 733–744. [CrossRef] [PubMed]

6. Gong, W.; Shen, J.; Jia, J. The impact of human activities on the flushing properties of a semi−enclosed lagoon: Xiaohai, Hainan, China. *Mar. Environ. Res.* **2008**, *65*, 62–76. [CrossRef] [PubMed]

7. Ayache, F.; Thompson, J.R.; Flower, R.J.; Boujarra, A.; Rouatbi, F.; Makina, H. Environmental characteristics, landscape history and pressures on three coastal lagoons in the Southern Mediterranean Region: Merja Zerga (Morocco), Ghar El Melh (Tunisia) and Lake Manzala (Egypt). *Hydrobiologia* **2009**, *622*, 15–43. [CrossRef]

8. Mendes, C.L.T.; Soares−Gomes, A. Macrobenthic community structure in a Brazilian chocked lagoon system under environmental stress. *Zoologia* **2011**, *28*, 365–378. [CrossRef]

9. Hung, J.J.; Huang, W.C.; Yu, C.S. Environmental and biogeochemical changes following a decade's reclamation in the Dapeng (Tapong) Bay, southwestern Taiwan. *Estuar. Coast. Shelf Sci.* **2013**, *130*, 9–20. [CrossRef]

10. Magni, P.; Draredja, B.; Melouah, K.; Como, S. Patterns of seasonal variation in lagoonal macrozoobenthic assemblages (Mellah lagoon, Algeria). *Mar. Environ. Res.* **2015**, *109*, 168–176. [CrossRef]

11. Magni, P.; Como, S.; Kamijo, A.; Montani, S. Effects of *Zostera marina* on the patterns of spatial distribution of sediments and macrozoobenthos in the boreal lagoon of Furen (Hokkaido, Japan). *Mar. Environ. Res.* **2017**, *131*, 90–102. [CrossRef] [PubMed]

12. Viaroli, P.; Bartoli, M.; Giordani, G.; Naldi, M.; Orfanidis, S.; Zaldivar, J.M. Community shifts, alternative stable states, biogeochemical controls and feedbacks in eutrophic coastal lagoons: A brief overview. *Aquat. Conserv. Mar. Freshw. Ecosyst.* **2008**, *18*, S105–S117. [CrossRef]

13. Magni, P.; Rajagopal, S.; van der Velde, G.; Fenzi, G.; Kassenberg, J.; Vizzini, S.; Mazzola, A.; Giordani, G. Sediment features, macrozoobenthic assemblages and trophic relationships (δ^{13}C and δ^{15}N analysis) following a dystrophic event with anoxia and sulphide development in the Santa Giusta lagoon (western Sardinia, Italy). *Mar. Poll. Bull.* **2008**, *57*, 125–136. [CrossRef] [PubMed]

14. Kanaya, G.; Nakamura, Y.; Koizumi, T.; Yamada, K. Seasonal changes in infaunal community structure in a hypertrophic brackish canal: Effects of hypoxia, sulfide, and predator−prey interaction. *Mar. Environ. Res.* **2015**, *108*, 14–23. [CrossRef] [PubMed]

15. Kanaya, G.; Uehara, T.; Kikuchi, E. Effects of sedimentary sulfide on community structure, population dynamics, and colonization depth of macrozoobenthos in organic−rich estuarine sediments. *Mar. Poll. Bull.* **2016**, *109*, 393–401. [CrossRef] [PubMed]

16. Kolkwitz, R.; Marsson, M. Ökologie der tierischen Saprobien. *Int. Rev. der Gesamten Hydrobiol. Hydrogr.* **1909**, *2*, 126–152. (In German) [CrossRef]

17. Sládeček, V. The future of the saprobity system. *Hydrobiologia* **1965**, *25*, 518–537. [CrossRef]

18. Tagliapietra, D.; Sigovini, M.; Magni, P. Saprobity: A unified view of benthic succession models for coastal lagoons. *Hydrobiologia* **2012**, *686*, 15–28. [CrossRef]

19. Remane, A. Ecology of brackish water. In *Biology of Brackish Water*, 2nd ed.; Remane, A., Schlieper, C., Eds.; Schweizer-bart'sche: Stuttgart, Germany, 1971; pp. 1–210.

20. Guelorget, O.; Perthuisot, J.P. Paralic ecosystems. Biological organization and functioning. *Vie Milieu* **1992**, *42*, 215–251.

21. Magni, P.; Micheletti, S.; Casu, D.; Floris, A.; Giordani, G.; Petrov, A.N.; De Falco, G.; Castelli, A. Relationship between chemical characteristics of sediments and macrofaunal communities in the Cabras lagoon (Western Mediterranean, Italy). *Hydrobiologia* **2005**, *550*, 105–119. [CrossRef]

22. Semprucci, F.; Gravina, M.F.; Magni, P. Meiofaunal Dynamics and Heterogeneity along Salinity and Trophic Gradients in a Mediterranean Transitional System. *Water* **2019**, *11*, 1488. [CrossRef]

23. Magni, P. Biological benthic tools as indicators of coastal marine ecosystems health. *Chem. Ecol.* **2003**, *19*, 363–372. [CrossRef]

24. Hyland, J.; Balthis, L.; Karakassis, I.; Magni, P.; Petrov, A.; Shine, J.; Warwick, R.; Vestergaard, O. Organic carbon content of sediments as an indicator of stress in the marine benthos. *Mar. Ecol. Prog. Ser.* **2005**, *295*, 91–103. [CrossRef]

25. Magni, P.; Tagliapietra, D.; Lardicci, C.; Balthis, L.; Castelli, A.; Como, S.; Pessa, G.; Rismondo, A.; Viaroli, P.; Tataranni, M.; et al. Animal–sediment relationships: Evaluating the 'Pearson–Rosenberg paradigm' in Mediterranean coastal lagoons. *Mar. Poll. Bull.* **2009**, *58*, 478–486. [CrossRef] [PubMed]

26. Foti, A.; Fenzi, G.; Di Pippo, F.; Gravina, M.F.; Magni, P. Testing the saprobity hypothesis in a Mediterranean lagoon: Effects of confinement and organic enrichment on benthic communities. *Mar. Environ. Res.* **2014**, *99*, 85–94. [CrossRef] [PubMed]

27. Chen, L.; Ren, C.; Zhang, B.; Li, L.; Wang, Z.; Song, K. Spatiotemporal dynamics of coastal wetlands and reclamation in the Yangtze estuary during past 50 years (1960s–2015). *Chin. Geogr. Sci.* **2018**, *28*, 386–399. [CrossRef]

28. Borja, A.; Bricker, S.B.; Dauer, D.M.; Demetriades, N.T.; Ferreira, J.G.; Forbes, A.T.; Zhu, C.; Hutchings, P.; Marques, J.C.; Kenchington, R. Overview of integrative tools and methods in assessing ecological integrity in estuarine and coastal systems worldwide. *Mar. Poll. Bull.* **2008**, *56*, 1519–1537. [CrossRef]

29. Ye, G.; Chou, L.M.; Hu, W. The role of an integrated coastal management framework in the long–term restoration of Yundang Lagoon, Xiamen, China. *J. Environ. Plan. Man.* **2014**, *57*, 1704–1723. [CrossRef]

30. Lu, Z.B.; Du, Q.; Huang, Y.J. Ecological effects of Integrated Management on Yundang Lake in Xiamen. *J. Oceanogr. Taiwan Strait.* **1997**, *16*, 306–310. (In Chinese)

31. Zhang, X.; Xue, X. Analysis of marine environmental problems in a rapidly urbanising coastal area using the DPSIR framework: A case study in Xiamen, China. *J. Environ. Plan. Manag.* **2013**, *56*, 720–742. [CrossRef]

32. Huang, L.F.; Zhuo, J.F.; Guo, W.D.; Spencer, R.G.M.; Zhang, Z.Y.; Xu, J. Tracing organic matter removal in polluted coastal waters via floating bed phytoremediation. *Mar. Pollut. Bull.* **2013**, *71*, 74–82. [CrossRef] [PubMed]

33. Sun, Y.W.; Guo, Q.H.; Liu, J.; Wang, R. Scale Effects on Spatially Varying Relationships Between Urban Landscape Patterns and Water Quality. *J. Environ. Manag.* **2014**, *54*, 272–287. [CrossRef] [PubMed]

34. Zheng, X.; Como, S.; Magni, P.; Huang, L. Spatiotemporal variation in environmental features and elemental/isotopic composition of organic matter and primary producers in the Yundang lagoon (Xiamen, China). *Environ. Sci. Pollut. Res.* **2019**, *26*, 13126–13137. [CrossRef] [PubMed]

35. Guo, W.; Xu, J.; Wang, J.; Wen, Y.; Zhuo, J.; Yan, Y. Characterization of dissolved organic matter in urban sewage using excitation emission matrix fluorescence spectroscopy and parallel factor analysis. *J. Environ. Sci. China* **2010**, *22*, 1728–1734. [CrossRef]

36. *National standard GB17378.4-2007. The Specification for Marine Monitoring, Part 4: Seawater Analysis*; Standards Press of China: Beijing, China, 2007.

37. McManus, J. Grain size determination and interpretation. In *Techniques in Sedimentology*; Tucker, M., Ed.; Black-Well: Oxford, UK, 1988; pp. 63–85.

38. Davis, J.C. *Statistics and Data Analysis in Geology*; John Wiley & Sons: New York, NY, USA, 1986.

39. Anderson, M.J. A new method for non–parametric multivariate analysis of variance. *Austral. Ecol.* **2001**, *26*, 32–46.

40. Rice, W.R. Analyzing tables of statistical tests. *Evolution* **1989**, *43*, 223–225. [CrossRef] [PubMed]

41. Clarke, K.R. Comparisons of dominance curves. *J. Exp. Mar. Biol. Ecol.* **1990**, *138*, 143–157. [CrossRef]

42. Clarke, K.R.; Ainsworth, M. A method of linking multivariate community structure to environmental variables. *Mar. Ecol. Prog. Ser.* **1993**, *92*, 205–219. [CrossRef]

43. Xu, Y.B.; Li, X.Z.; Wu, L.F.; Qian, X.M.; Zheng, H.D.; Cai, J.D. Investigation and assessment on the macrobenthos of Yundang Lake in Xiamen. *J. Fujian Fish.* **2009**, *2*, 17–23. (In Chinese)

44. Yan, C.; Li, Q.; Zhang, X.; Li, G. Mobility and ecological risk assessment of heavy metals in surface sediments of Xiamen Bay and its adjacent areas, China. *Environ. Earth Sci.* **2009**, *60*, 1469–1479. [CrossRef]

45. Chen, C.; Lu, Y.; Hong, J.; Ye, M.; Wang, Y.; Lu, H. Metal and metalloid contaminant availability in Yundang Lagoon sediments, Xiamen Bay, China, after 20 years continuous rehabilitation. *J. Hazard. Mater.* **2010**, *175*, 1048–1055. [CrossRef] [PubMed]

46. Zhang, X.; Gao, Y.J.; Li, Q.Z.; Li, G.X.; Guo, Q.H.; Yan, C.Z. Estrogenic compounds and estrogenicity in surface water, sediments, and organisms from Yundang Lagoon in Xiamen, China. *Arch. Environ. Contam. Toxicol.* **2011**, *61*, 93–100. [CrossRef] [PubMed]

47. Zhuo, J.; Guo, W.; Deng, X.; Zhang, Z.; Xu, J.; Huang, L. Fluorescence excitation–emission matrix spectroscopy of CDOM from Yundang Lagoon and its indication for organic pollution. *Spectrosc. Spect. Anal.* **2010**, *30*, 1539–1544.

48. Como, S.; Magni, P. Temporal changes of a macrobenthic assemblage in harsh lagoon sediments. *Estuar. Coast. Shelf Sci.* **2009**, *83*, 638–646. [CrossRef]

49. Morton, B. The biology and functional morphology of *Mytilopsis sallei* (Recluz) (Bivalvia: Dreissenacea) fouling Visakhapatnam harbour, Andhra Pradesh, India. *J. Molluscan Stud.* **1981**, *47*, 25–42. [CrossRef]

50. Morton, B. Life–history characteristics and sexual strategy of *Mytilopsis sallei* (Bivalvia: Dreissenacea), introduced into Hong Kong. *J. Zool.* **1989**, *219*, 469–485. [CrossRef]

51. Wang, J.; Huang, Z.; Zheng, C.; Lin, N. Population dynamics and structure of alien species *Mytilopsis sallei* in Fujian, China. *J. Oceanogr. Taiwan Strait* **1999**, *18*, 372–377. (In Chinese)

52. Cai, L.Z.; Hwang, J.S.; Dahms, H.U.; Fu, S.J.; Zhuo, Y.; Guo, T. Effect of the invasive bivalve *Mytilopsis sallei* on the macrofaunal fouling community and the environment of Yundang Lagoon, Xiamen, China. *Hydrobiologia* **2014**, *741*, 101–111. [CrossRef]

53. Song, J.-W.; Dong, Y.-H.; Li, H.T.; Zhao, Y. Invasive shellfish and their impacts on Chinese coastal waters. *J. Biosaf.* **2015**, *24*, 177–183. (In Chinese)

54. Zheng, F.Z.; Lu, C.Y.; Zhang, Y.K.; Zeng, Q.F. Study on the ecological engineering of mangrove landscape in Yundang Lagoon. In *Yundang: From a Harbor to a Lagoon*; Lu, C.Y., Ed.; Xiamen University Press: Xiamen, China, 2003; pp. 174–179. (In Chinese)

55. Guy-Haim, T.; Lyons, D.A.; Kotta, J.; Ojaveer, H.; Queirós, A.M.; Chatzinikolaou, E.; Orav-Kotta, H.; Magni, P.; Blight, A.J.; Como, S.; et al. Diverse effects of invasive ecosystem engineers on marine biodiversity and ecosystem functions: A global review and meta–analysis. *Glob. Chang. Biol.* **2018**, *24*, 906–924. [CrossRef]

56. The IUCN Red List of Threatened Species. Available online: www.iucnredlist.org (accessed on 15 August 2019).

57. Lutzen, J.; Jespersen, A.; Takahashi, T.; Kai, T. Morphology, structure of dimorphic sperm, and reproduction in the hermaphroditic commensal bivalve *Pseudopythina Tsurumaru*. *J. Morphol.* **2004**, *262*, 407–420. [CrossRef] [PubMed]

58. Magni, P.; Micheletti, S.; Casu, D.; Floris, A.; De Falco, G.; Castelli, A. Macrofaunal community structure and distribution in a muddy coastal lagoon. *Chem. Ecol.* **2004**, *20*, S397–S407. [CrossRef]

59. Nonnis Marzano, C.; Baldacconi, R.; Fianchini, A.; Gravina, M.F.; Corriero, G. Settlement seasonality and temporal changes in hard substrate macrozoobenthic communities of Lesina Lagoon (Apulia, Southern Adriatic Sea). *Chem. Ecol.* **2007**, *23*, 479–491. [CrossRef]

60. Cardone, F.; Corriero, G.; Fianchini, A.; Gravina, M.F.; Nonnis Marzano, C. Biodiversity of transitional waters: Species composition and comparative analysis of hard bottom communities from the south–eastern Italian coast. *J. Mar. Biol. Assoc. UK* **2014**, *94*, 25–34. [CrossRef]

61. Gravina, M.F.; Cardone, F.; Bonifazi, A.; Bertrandino, M.S.; Chimienti, G.; Longo, C.; Nonnis Marzano, C.; Moretti, M.; Lisco, S.; Moretti, V.; et al. *Sabellaria spinulosa* (Polychaeta, Annelida) reefs in the Mediterranean Sea: Habitat mapping, dynamics and associated fauna for conservation management. *Estuar. Coast. Shelf Sci.* **2018**, *200*, 248–257. [CrossRef]

62. Ingrosso, G.; Abbiati, M.; Badalamenti, F.; Bavestrello, G.; Belmonte, G.; Cannas, R.; Benedetti–Cecchi, L.; Bertolino, M.; Cerrano, C.; Chemello, R.; et al. Mediterranean Bioconstructions Along the Italian Coast. *Adv. Mar. Biol.* **2018**, *79*, 63–134.

Article

Balance between the Reliability of Classification and Sampling Effort: A Multi-Approach for the Water Framework Directive (WFD) Ecological Status Applied to the Venice Lagoon (Italy)

Federica Cacciatore [1,*], **Andrea Bonometto** [1], **Elisa Paganini** [2], **Adriano Sfriso** [3], **Marta Novello** [4], **Paolo Parati** [4], **Massimo Gabellini** [2] and **Rossella Boscolo Brusà** [1]

1   ISPRA (Institute for Environmental Protection and Research), Loc. Brondolo, 30015 Chioggia (Venice), Italy
2   ISPRA (Institute for Environmental Protection and Research), via Vitaliano Brancati 60, 00144 Rome, Italy
3   Department of Environmental Sciences, Informatics & Statistics, University of Venice Ca' Foscari, via Torino 155, 30170 Ve-Mestre (Venice), Italy
4   ARPAV (Environmental Prevention and Protection Agency of Veneto Region), via Ospedale Civile 24, 35121 Padova, Italy
*   Correspondence: federica.cacciatore@isprambiente.it; Tel.: +39-06-50074990

Received: 8 July 2019; Accepted: 26 July 2019; Published: 29 July 2019

Abstract: The Water Framework Directive (WFD) requires Member States to assess the ecological status of water bodies and provide an estimation of the classification confidence and precision. This study tackles the issue of the uncertainty in the classification, due to the spatial variability within each water body, proposing an analysis of the reliability of classification, using the results of macrophyte WFD monitoring in the Venice Lagoon as case study. The level of classification confidence, assessed for each water body, was also used as reference to optimize the sampling effort for the subsequent monitorings. The ecological status of macrophytes was calculated by the Macrophyte Quality Index at 114 stations located in 11 water bodies. At water body scale, the level of classification confidence ranges from 54% to 100%. After application of the multi-approach (inferential statistics, spatial analyses, and expert judgment), the optimization of the sampling effort resulted in a reduction of the number of stations from 114 to 84. The decrease of sampling effort was validated by assessing the reliability of classification after the optimization process (54–99%) and by spatial interpolation of data (Kernel standard error of 22.75%). The multi-approach proposed in this study could be easily applied to any other water body and biological quality element.

Keywords: Macrophyte Quality Index (MaQI), transitional waters; uncertainty analysis; confidence interval; Kernel standard error

---

1. Introduction

The Water Framework Directive (WFD) [1] is a framework by European Commission that requires Member States to monitor each relevant biological quality element (BQE) in order to assess the ecological status of each water body (WB), which represents the classification and management unit of the WFD [2]. Macroalgae, phanerogams, macroinvertebrates, and fish faunal are the BQEs to be evaluated for the assessment of the ecological status of European transitional WBs under the WFD. In addition to BQEs, the physicochemical and hydromorphological supporting elements contribute to the ecological classification, confirming or not the classification provided by the BQEs. Biological monitoring results have to be expressed as ecological quality ratios (EQRs), by comparing sampled data with those equivalents from undisturbed or minimally disturbed reference sites. Depending on the scale of deviation from reference conditions, the EQR range is divided into five status classes: High, good, moderate, poor, and bad. The most important distinction is between good and moderate status: the first is defined as a "slight deviation from reference condition", whilst the latter is a "moderate deviation from reference conditions". When the quality status is less than good, Member States/River Basin District Authorities must adopt a plan of measures to improve a WB until good status is achieved [1,2].

The WFD also requires all Member States to provide information about the estimation of the confidence interval and precision attained by the monitoring system used in their classification [1] (Annex V). As the critical good/moderate boundary can require expensive remedial measures to be included in the River Basin Management Plan, it is important that water body monitoring and management organizations estimate the confidence to which an individual WB can be assigned to an ecological status class [3]. The biological, physical, and chemical data from current monitoring programs underlie large sources of uncertainty, which can be partitioned into spatial variations, temporal variations, sampling, and analytical errors, in addition to the uncertainty introduced by the classification system, especially by the reference conditions [4]. Some authors proposed assessment systems based on statistical principles to provide information on the level of confidence and precision of monitoring program results, focusing on field sampling variability and laboratory procedures and protocols [3–7]. Mascarò et al. [8] investigated the sources of variability associated with sampling design of a selection of macrophyte-based (both macroalgae and seagrasses) monitoring programs and observed that uncertainty is mainly affected by spatial variation. The authors consequently suggested that spatial uncertainty analyses be performed to optimize sampling strategy and to improve the reliability of classification of ecological status [8].

The MaQI (Macrophyte Quality Index) is the macrophyte index adopted by the Italian Law (Italian Ministry decree 260/2010) in agreement with the WFD requirements for ecological classification in transitional waters, and it was successfully intercalibrated in the framework of the WFD intercalibration exercise [9]. The MaQI was applied to assess the ecological status by macrophyte assemblages in the Venice Lagoon, Italy's largest wetland and one of the most important coastal ecosystems in the whole Mediterranean basin [10–12]. During the first cycle of the WFD operational monitoring, 114 stations were monitored in the Venice Lagoon. The ecological status classification of the Lagoon, resulting from the above monitoring, has been adopted by Veneto Region as published as Decision n. 140/2014 in the first revision of the Management Plan of the hydrographic district of the Eastern Alps [10]. However, for subsequent assessments, due to limitations of budget, optimization of the monitoring strategy was needed. Reducing monitoring efforts is a common consequence of cuts in government spending, adopted by many countries to save money in order to face the current global crisis [13]. Therefore, the balance between sampling effort and classification confidence becomes a critical issue in implementing the WFD [14,15].

In this context, the aim of this study was first to assess the reliability of ecological status classification in transitional waters, using results from monitoring of macrophyte assemblages in the Venice Lagoon, as a case study. Then, a multi-approach method to optimize the monitoring strategies minimizing both sampling effort and risk of misclassification was proposed, based on inferential statistical principles, geostatistical analyses, and expert judgment.

2. Materials and Methods

2.1. Study Area

The Venice Lagoon is a shallow water body of about 550 km^2, located in the north-eastern part of Italy (Figure 1). It has an average depth of approximately 1.1 m, and it is characterized by a variety of aquatic and terrestrial habitats, such as canals, shoals, salt-marshes and islands [16]. According to the WFD, the Venice Lagoon is a large coastal microtidal lagoon, divided into 11 natural WBs and 3 heavily modified WBs [10]. In 2011, during the first cycle of the WFD Operational Monitoring, 114 stations, spread within the 11 natural WBs of the Venice Lagoon, were sampled to assess the ecological status by macrophyte assemblages (Figure 1). The dataset analyzed during the current study is available online at Environmental prevention and protection agency of Veneto Region (ARPAV) web portal [17]. For the classification of the BQE macrophytes, the MaQI [9] was used.

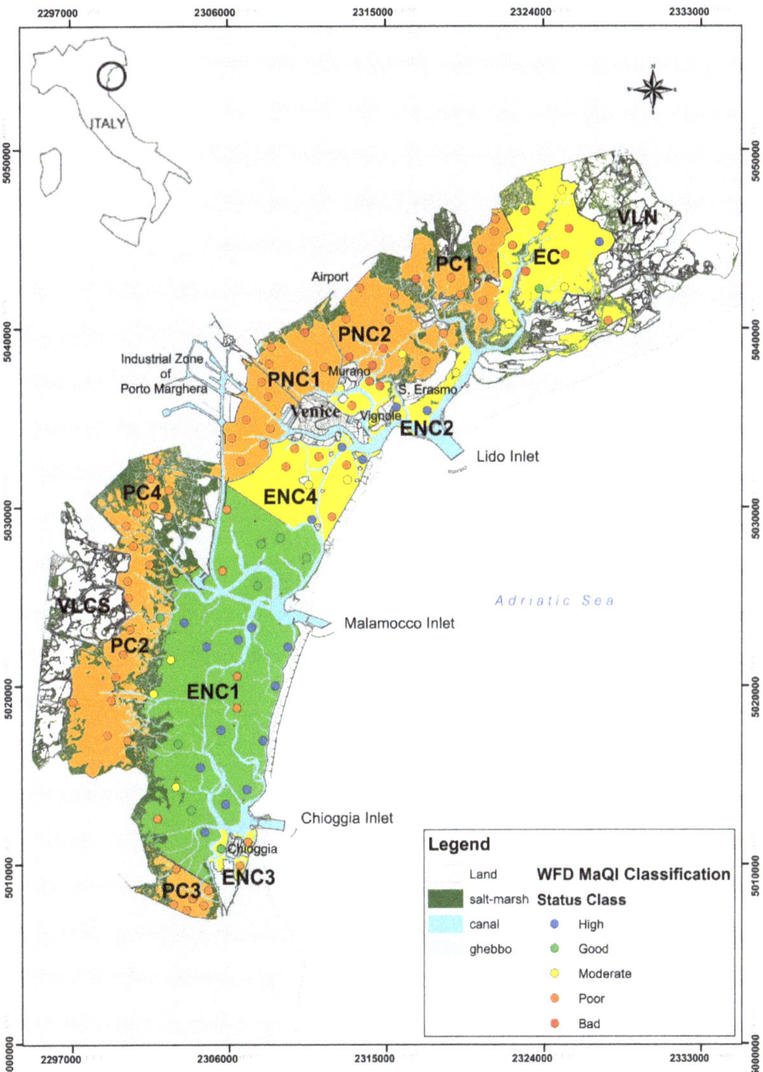

Figure 1. The Venice Lagoon map with its morphological and hydrological characteristics, and the WFD (Water Framework Directive) MaQI (Macrophyte Quality Index [8]) classification in 2011. Names of water bodies (WBs) are indicated in uppercase bold characters; place names are in lowercase. Final assessment of each WB is labeled like for stations: blue for high, green for good, yellow for moderate, orange for poor, red for bad status classes.

2.2. Reliability of Classification at Water Body Scale

The classification of the macrophyte assemblage at WB scale was obtained by averaging the MaQI EQRs of all stations within each WB (j).

First, the confidence interval (L) was also calculated for each WB (j) by the following formula:

$$L_j = t_{N_j-1,\alpha/2} \frac{S_j}{\sqrt{N_j}} \tag{1}$$

where $t_{N_j-1,\alpha/2}$ is the critical value of the t-student distribution for the confidence level $1-\alpha$ (two tiled distribution); S_j is the standard deviation of the EQRs within each WB; N_j is the number of stations within each WB. In this study, the confidence intervals were stated at the 95% confidence level ($\alpha = 0.05$).

The L value provides an overall view of the confidence of the mean EQRs, but it does not take into account the closeness of the face value to the class boundaries, being independent from the WFD classification system. Therefore, the reliability of the classification was assessed in terms of probabilities that the observed ecological status classification (sample mean) lies within the right class. The cumulative probability that the actual (population) mean value of MaQI fell in each status class was assessed by t-student distribution. Data normality was tested by the Kolmogorov–Smirnov test whilst the absence of autocorrelations was verified by variogram analyses and Moran's test of autocorrelation. Classes were identified by the WFD boundaries reported by the Italian Ministry decree 260/2010 (high/good = 0.8, good/moderate = 0.6, moderate/poor = 0.4, poor/bad = 0.2). The confidences related to the critical boundary good/moderate were also calculated by the sum of probability of classes being lower/higher than good.

Statistics were carried out using the R software [18].

2.3. Optimization of Sampling Effort

2.3.1. Statistical Approach

The relationship between the sampling effort (N of stations) and the classification confidence (confidence interval "L" and level "α") was investigated by the following equation applied to each WB of the Venice Lagoon:

$$N = t_{N_j-1,\alpha/2}^2 \frac{S_j^2}{L^2}. \tag{2}$$

Three scenarios with different L values were investigated. The first two L values were selected considering the relationship between the width of the confidence interval, the distance between the class boundaries, and the maximum error of classification. Italian normative (Italian Ministry Decree 260/2010) provides equidistant boundaries for MaQI EQR (bad/poor = 0.2; poor/moderate = 0.4; moderate/good = 0.6; good/high = 0.8), therefore the width of the classes is 0.2. As a consequence, $L = 0.1$ ($L_{0.1}$) entails a maximum error of one ecological class at only one direction (1st scenario), whilst $L = 0.2$ ($L_{0.2}$) entails a maximum error of one ecological class at both directions (2nd scenario).

Finally, the mean confidence interval (L_{mean}) was calculated as average L of all WBs and applied to each WB (3rd scenario), in order to obtain a more consistent (homogeneous) classification confidence between WBs within the Venice Lagoon.

2.3.2. Expert Judgment Criteria

Starting from the three above-mentioned statistical scenarios, a two-step expert judgment analysis was applied to define the final optimization of monitoring effort (number of stations), based on level of confidence and hydromorphological features.

First of all, a suitable L value (L_{opt}) was defined in order to ensure higher homogeneity between WBs with an acceptable minimum level of reliability. Accordingly, to calculate N by Equation (2) for each WB, the following rules were adopted:

(a) If $L_{mean} < L_j < 0.2$ (i.e., maximum error of one ecological class at both directions), then $N_{new} = N_{old}$;
(b) If $L_j > 0.2$, then Equation (2) is calculated with $L_{opt} = 0.2$;
(c) If $L_j < L_{mean}$, then Equation (2) is calculated with $L_{opt} = L_{mean}$.

The risk of misclassification in relation to the critical boundary G/M was also considered.

Finally, the whole dimension and the hydrological and morphological heterogeneity within each WB were also taken into account, to avoid an oversized number of stations in small WBs or, vice versa, an excessive reduction in large WBs. Table 1 summarizes the elements considered for each WB of the Venice Lagoon.

Table 1. Dimension and hydrological and morphological characteristics considered for each water body (WB) during the optimization of the monitoring sampling effort of the Venice Lagoon.

WB	Whole Dimension (km^2)	Hydrological and Morphological Characteristics (dimension, km^2)
EC	49.10	salt-marshes (4.17), canals (0.54)
ENC1	134.71	salt-marshes (6.08), canals (20.62)
ENC2	20.15	salt-marshes (0.34), canals (10.14)
ENC3	4.48	canals (2.09)
ENC4	26.75	canals (2.64)
PC1	27.64	salt-marshes (6.44), canals (3.41)
PC2	50.44	salt-marshes (0.87), canals (3.95)
PC3	10.04	salt-marshes (2.28), canals (1.08)
PC4	21.05	salt-marshes (0.81), canals (2.98)
PNC1	32.93	salt-marshes (1.48), canals (3.17)
PNC2	31.53	salt-marshes (3.44), canals (2.68)

2.4. Validation Method

2.4.1. Reliability Assessment

First, the reliability of classification related to the critical boundary good/moderate was recalculated for each WB using the dataset obtained from the optimization process. The analyses were carried out as reported in Section 2.2 by calculating the cumulative probability of Student's *t*-distribution.

2.4.2. Spatial Analyses

To estimate the error in the sampling effort reduction, spatial interpolations of the two (original and reduced) monitoring networks of the whole Venice Lagoon were performed by Kernel interpolation with barrier available in the ArcMap toolbox of ArcGIS Desktop 10.1 (http://esri.com). Barriers were represented by islands and salt-marshes (Figure 1). Exponential equations were used during the regression analysis as Kernel function, while the other parameters were optimized by default in ArcMap. Mean prediction errors and root-mean-square errors (cross-validation) were calculated to test the interpolation maps obtained. The relative error (RE) reduction was computed by the modified formula reported by [19]:

$$RE(\%) = \frac{KBSE_{original} - KBSE_{reduced}}{KBSE_{original}} \times 100 \tag{3}$$

where $KBSE_{original}$ and $KBSE_{reduced}$ are the Kernel standard error of the MaQI network of 2011 (original) and the new one (reduced), respectively.

Finally, for each WB, the ecological classes resulted by the original network and the reduced one were also calculated by averaging the interpolated EQR values of all Kernel interpolation grid cells. The significance of the differences between the two networks were tested using Student's *t*-test.

3. Results

3.1. Reliability Assessment

The probability that the actual mean value of MaQI EQRs fell within each of the five WFD classes, assessed by the Student's *t*-distributions; MaQI EQRs and status classifications; confidence interval

(L_j), computed by Equation (1); and the cumulative probability related to the critical boundary G/M are shown for each WB in Table 2.

The Venice Lagoon mean L value (L_{mean}) used for the following analyses was 0.109.

The WB ENC3 was left out from analyses because of its very small surface, its urban features, and the low number of stations ($N = 3$) that affected the meaning of statistical elaborations.

Table 2. MaQI (Macrophyte Quality Index) status class results, confidence, and reliability of classification of each WB (water body) of the Venice Lagoon.

WB	MaQI EQRs (mean ± st. dev.)	MaQI Classification	Classification Confidence (L_j)	Classification Reliability (%)					Probability to be Less than Good (%)
				Bad	Poor	Moderate	Good	High	
EC	0.408 ± 0.236	Moderate	0.143	0.45	44.99	53.89	0.67	0	99.33
ENC1	0.715 ± 0.280	Good	0.113	0	0	2.28	90.95	6.77	2.28
ENC2	0.479 ± 0.293	Moderate	0.271	2.27	22.94	59.05	14.37	1.37	84.26
ENC4	0.520 ± 0.250	Moderate	0.179	0.14	8.00	74.99	16.56	0.31	83.13
PC1	0.317 ± 0.071	Poor	0.054	0.06	99.56	0.38	0	0	100
PC2	0.350 ± 0.104	Poor	0.066	0.02	93.71	6.27	0	0	100
PC3	0.300 ± 0.055	Poor	0.057	0.33	99.34	0.33	0	0	100
PC4	0.307 ± 0.053	Poor	0.049	0.09	99.72	0.19	0	0	100
PNC1	0.314 ± 0.067	Poor	0.045	0.01	99.9	0.09	0	0	100
PNC2	0.270 ± 0.160	Poor	0.115	10.02	88.46	1.52	0	0	100

3.2. Optimization of Sampling Effort

The number of stations (N) resulted by the application of the statistical approach (Equation (2)) with the three scenarios (i.e., considering $L = 0.1$, $L = 0.2$, and $L = L_{mean}$) changed from 111 (2011 sampling) to 169 ($L = 0.1$), 54 ($L = 0.2$), and 147 ($L = L_{mean}$), respectively (Table 3). After emendation by the two steps of expert judgment, following criteria described in Section 2.3.2, 81 stations were proposed for future monitoring.

Table 3. Results of the application of the statistical approach and expert judgement criteria to optimize the sampling effort in the MaQI (Macrophyte Quality Index) monitoring program of the Venice Lagoon.

WB	$N_{original}$	Statistical Approach			Expert Judgment Criteria	
		$N_L = 0.1$	$N_L = 0.2$	$N_L = L_{mean}$	$N_L = L_{opt}$	N_{new}
EC	13	26	7	22	13 [a]	13
ENC1	26	33	8	28	26 [a]	21
ENC2	7	51	13	43	13 [b]	7
ENC4	10	32	8	27	10 [a]	10
PC1	9	3	3*	3*	3 [c]	6
PC2	12	5	3*	4	4 [c]	4
PC3	6	2	3*	3*	3 [c]	3
PC4	7	2	3*	3*	3 [c]	3
PNC1	11	2	3*	3*	3 [c]	5
PNC2	10	13	3	11	10 [a]	9
Total	111	169	54	147	88	81

Keys: '$N_{original}$' = original number of stations sampled in 2011, '$N_L = L_{mean}$' = number of stations resulted by Equation (2) with L_{mean} (0.109), '$N_L = L_{opt}$' = number of stations resulted by Equation (2) with a suitable L value (L_{opt}) deriving from rules (a), (b) and (c), 'N_{new}' = number of stations considering also elements of Table 1, '3*' = a default N of 3 was given, when the application of Equation (2) resulted as $N < 2$, since standard deviation tends to be zero.

Briefly, no modifications of the number of stations were made to EC and ENC4 as their L_j values (0.143 and 0.179, respectively) were higher than L_{mean} but lower than 0.2 (rule (a)).

Rule (c) was strictly adopted at PC2, PC3, and PC4 since their L values (0.066, 0.057, and 0.049, respectively) were lower than L_{mean}. Accordingly, the four stations to be maintained at PC2 were selected on the basis of the presence of salt-marshes and the small canals, which characterize the WB; the three stations deleted at PC3 were chosen from those on the most southern part of the WB, where several restoration activities, such as the construction of artificial salt-marshes, are underway to replace shallow waters available just before; and finally, the three stations maintained at PC4 were those mostly located in the middle of the WB.

The other WBs needed the expert judgment as following. A small or no reduction of monitoring effort was done for ENC1, ENC2, PC1, PNC1, and PNC2 due to their dimensions (Table 1), habitat heterogeneity (Table 1 and Figure 2), and their reliability of the classification (Table 3). Accordingly, five stations with spatially redundant information were deleted at ENC1, the largest WB of the Venice Lagoon, with several canals and two Lagoon inlets (Malamocco and Chioggia) and 97.7% of probability to be good or more. No modifications were done at ENC2, due to its very small dimension and high reliability of classification (84.3% of probability to be less than good). Regarding PC1, three stations were not considered to be sufficient to correctly represent, since it consists of large expanses of salt-marshes, the Dese river mouth, and several small canals that divide it mainly into four areas. No modifications were made for PNC1, which is located between the industrial area of Porto Marghera and the city of Venice; it is quite homogenous ($L_j = 0.045$), but it is separated into five sections by the presence of six canals. The morphological characteristics of PNC2 are quite similar to PNC1 with the airport of Venice on the north-west side and Murano and Sant'Erasmo islands on the south-east side. Considering its high reliability of the status classification (100% less than good), one station, located in the middle of the WB, was deleted. Finally, a special approach was adopted for ENC3, which was excluded from the optimization process because no more than three stations could be sampled there, due to its very small dimension, the presence of Chioggia island in the middle of the WB, and the proximity to the Lagoon inlet of Chioggia.

After application of the optimization process, the final number of stations established for the subsequent monitoring programs of the 11 natural WBs (including 3 stations of ENC3) of the Venice Lagoon was 84 (Figure 2).

3.3. Validation Process

The classification reliabilities resulted with the reduced network and comparisons with the MaQI network of 2011 (original network) are shown for each WB in Table 4. The change in probability that the quality status calculated with the reduced network is less than good and differences with the 2011 MaQI network are also reported.

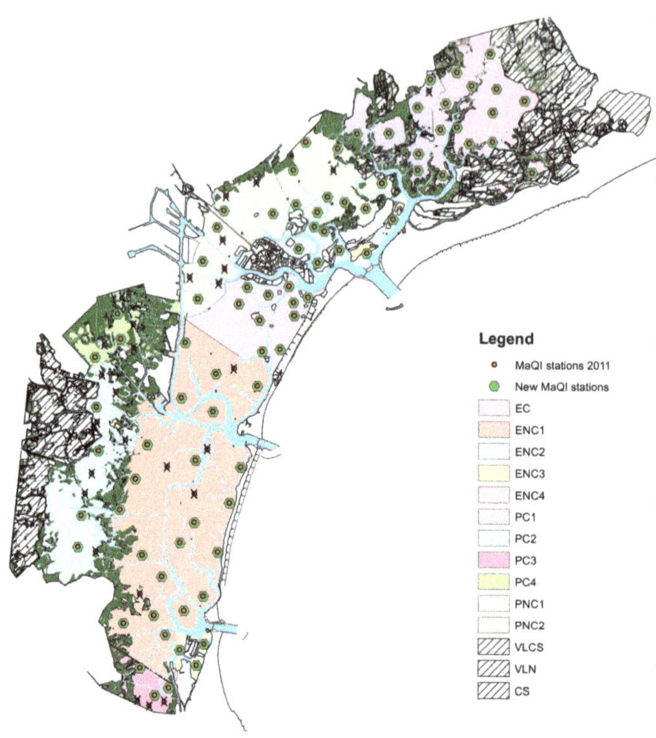

Figure 2. MaQI (Macrophyte Quality Index) network of 2011 (MaQI stations 2011) and the new network (new MaQI stations) with reduced number of stations by the multi-approach application. Crossed stations are those deleted.

Table 4. Reliability of classification of each water body (WB) of the Venice Lagoon after the optimization process (new network) and comparison with the original monitoring network results.

| WB | Number of Stations | | Classification Reliability (%) | | Probability to be Less than Good (%) | |
	Reduced Network	Differences from the Original Network	Reduced Network	Differences from the Original Network	Reduced Network	Differences from the Original Network
EC	13	0	53.89% Moderate	0	99.33	0
ENC1	21	−5	86.59% Good	−4.36	2.94	0.66
ENC2	7	0	59.05% Moderate	0	84.26	0
ENC4	10	0	74.99% Moderate	0	83.13	0
PC1	6	−3	96.41% Poor	−3.15	99.98	−0.02
PC2	4	−8	96.35% Poor	2.64	99.92	−0.08
PC3	3	−3	89.9% Poor	−9.44	99.34	−0.66
PC4	3	−4	89.9% Poor	−9.82	99.34	−0.66
PNC1	5	−6	98.61% Poor	−1.29	99.99	−0.01
PNC2	9	−1	82.88% Poor	−5.58	99.99	−0.01

Figure 3 shows the two maps produced by Kernel interpolation with barrier for the MaQI network of 2011 and the new one, respectively. Prediction errors are reported in Table 5. From Equation (3), an RE of 22.75% resulted.

Comparisons (Student's *t*-test) between status classifications of each WB before and after the reduction of sampling effort are reported in Figure 4. To test the differences, values were extrapolated from each Kernel interpolation map.

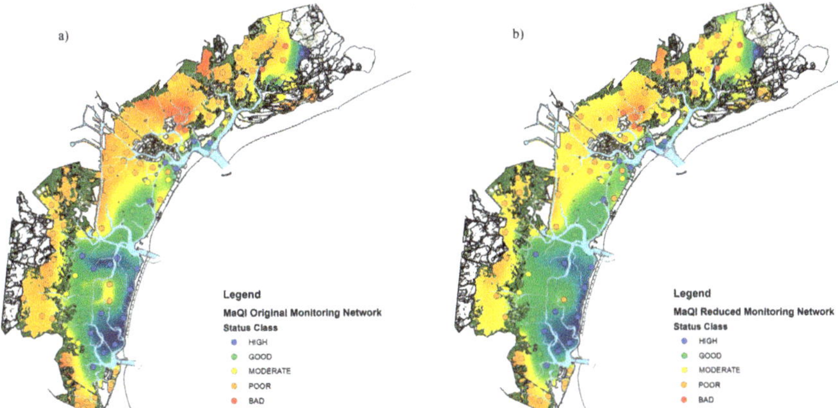

Figure 3. Kernel interpolation maps for (**a**) the MaQI (Macrophyte Quality Index) network of 2011 and (**b**) the new one with reduced station of 2011 from 114 to 84.

Table 5. Regression functions and prediction errors resulted by Kernel interpolation with barrier for the MaQI network of 2011 (original network) and the new one (reduced network).

	Original Network ($N = 114$)	Reduced Network ($N = 84$)
Mean prediction error	0.000047	0.005881
Root-mean-square error	0.213	0.189
Average standard error	0.234	0.190

Figure 4. Box plots of water body values resulted by the two Kernel interpolation maps of the original (Orig.) and reduced (Reduc.) networks. Results of Student's *t* tests are also given (n.s. = not significance, $p > 0.05$; * = $p < 0.05$; ** = $0.5 < p < 0.01$; *** = $0.01 < p < 0.001$). Note: blue line (first from the top) is the high/good boundary, green line (second from the top) is the good/moderate boundary, orange line (third from the top) is the moderate/poor boundary, red line (fourth from the top) is the poor/bad boundary.

4. Discussion

In this study, an analysis of reliability of the ecological status classification resulted by monitoring macrophyte assemblages according to the WFD in the Venice Lagoon was proposed and discussed in relation to monitoring effort review.

Previous theoretical studies proposed mixed models to estimate uncertainties in monitoring data, considering numerous different sources of variation that could affect an indicator, from the uncertainty related to sampling and analysis, to spatial and temporal variations and their interactions. However, when real data were considered, if some sources of uncertainty were small, they were disregarded by analysis [20–26]. In the current study, temporal variations were not considered, as the application of the MaQI index foresees to merge both spring and autumnal data collected in the year of monitoring, avoiding contribution from intra-annual variations to uncertainty [9]. Uncertainties associated with sampling and analysis methodology were excluded as the same laboratory staff was involved in the whole campaign at all activities, strictly following national protocols. Accordingly, only spatial variations were considered, also taking into account the practical need to reduce monitoring effort. However, when uncertainty is estimated from a larger dataset, pooling observations from multiple ecosystems with similar characteristics, with several spatial, temporal, and analytical method variations, it might be desirable to quantify every component at different scales [3,22]. In addition, as soon as data from subsequent monitoring cycles are available, inter-annual variability may also be assessed [7,24–26].

In the current study, the reliability of classification was assessed in terms of probabilities that the observed ecological status classification (mean value at WB scale) lies inside the right class, i.e., inside the class assigned from the index. Spatial variations were estimated by the confidence interval: higher values resulted at ENC2, a small WB characterized by high hydromorphological, and pressure gradient from Lido inlet to Venice island. Lower values were mostly observed at polyhaline WBs (annual mean of salinity < 30), such as PC1, PC2, and PC3, which are mainly characterized by lower internal ecological variability. By Student's t-distribution, the probability that the actual mean value of MaQI EQRs of each WB fell within each one of the five WFD classes ranged between 53.9% and 99.6%. However, it should be considered that the confidence interval itself is not sufficient to determine the risk of misclassification. The uncertainty is in fact determined both by the width of the confidence interval and by the proximity of the mean to the class boundary and in particular to the critical threshold good/moderate. Anyway, considering the critical good/moderate boundary, which is important for making decision about measures by governments, results highlighted a satisfactory reliability of the WFD MaQI classification of 2011 (83–100%).

From results of the estimation of confidence, it was possible to investigate where a reduction of the monitoring network effort could be allowed, avoiding excessively increasing the risk of misclassification. Recent studies proposed, for monitoring programs assessing status of WBs, to identify the optimal allocation of samples in time and space, through the quantification of the different uncertainty components affecting monitoring data [22,23]. As reported above, in this study, the main factors affecting the reliability of MaQI results were spatial variations, therefore changes focused only on the number and location of sampling stations.

Again, statistical principles offer relatively simple and suitable tools to address the optimization of sampling effort, providing a quantitative and objective assessment of the impact of sampling strategy on the risk of misclassification. On the other hand, results of statistical analysis require a careful analysis before their application and the operative choice is likely to benefit from including a final revision by expert judgment. Indeed, this study highlighted as the purely statistical approach, based on the amplitude of the interval of confidence, could lead to not-applicable or meaningless results. For instance, according to the first scenario ($L = 0.1$), to reduce the L value of the WB ENC2 from 0.27 to 0.1, the number of stations should increase from 7 to 51. Considering that the sampleable area of ENC2 is about 10 km^2, it means five stations per km^2, an effort unachievable in the framework of Institutional monitoring and far away from the concept of "optimization". Under scenario $L = 0.1$, more than one station per km^2 would be also required within WB ENC4, and similar issues of incoherence, in case of the strictly application of statistical approach, are observed in the scenario $L = L_{mean}$. Conversely, the application of the second scenario ($L = 0.2$) resulted in very little restriction for some large WBs, especially those with lower standard deviations. For instance, under this scenario, the number of stations within the WB ENC1 decreased from 26 to 8 (1 station for 13.5 km^2). All these evaluations are obviously linked to real data. However, there could be contexts where statistical results are confirmed even after expert judgment.

Accordingly, a revision process by expert judgment is essential. Intrinsically, the expert judgment is difficult to standardize and it introduces subjectivity into the evaluation. Therefore, it is crucial to guarantee maximum transparency on the followed criteria and later provide an objective estimation (validation) of the impact of the choices. In this study, the expert judgment followed the criteria described in Sections 2.3.2 and 3.2, and it aims (i) to homogenize the reliability of classification between WBs, (ii) to ensure for all WBs a minimum efficient reliability, (iii) to ensure a high reliability of the status classification regarding the critical boundary good/moderate, and (iv) to consider other particular elements of each WB such as dimensions, and hydrological and morphological characteristics.

To ensure the validation process, spatial interpolation of data was performed. Geostatistical techniques are based on the hypothesis that nearer observations are more similar to one another than to distant observations, and therefore allow you not only to insert new points where knowledge is more approximate, but also to eliminate others in those where they are redundant [27]. Furthermore, mathematical variogram models could assess the reductions of sampling effort by interpolations. Previous attempts to validate optimal locations of monitoring networks were performed adopting specific variogram models, such as in ordinary Kriging or Bayesian maximum entropy, depending on characteristics of their study areas [19,28]. The Venice Lagoon is a complex transitional area characterized by a composite mosaic of canals, salt-marshes, mud and sand intertidal flats, shoals, man-made structures, and islands [29,30]. Accordingly, spatial interpolations of the MaQI EQRs were performed by applying Kernel interpolation with barrier. The choice of this model, rather than others, was made in order to obtain interpolations of data which reflect the morphology of the Venice Lagoon, therefore with the presence of breaklines. Cross-validation results of both interpolations confirmed the right choice of the model, as mean prediction errors tended to be zero and quite small root-mean-square errors were observed. Moreover, both mean prediction errors resulted positive, which indicated that both models slightly overestimated the data [31], with relatively better results in the MaQI network of 2011 rather than in that reduced. To quantify the relative error of the reduced MaQI network, standard errors of the two interpolations were related by the formula of [19] modified using Kernel standard errors instead of Kriging.

In summary, results of the monitoring effort review of the WFD MaQI network of the Venice Lagoon showed a relative error of 22.7%, which can be considered acceptable taking into account the total reduction of stations of 26.3%. To evaluate the optimization process, Adhikary et al. [19] reported values below 30% as acceptable as considering the zero trend as a criterion. In order to obtain a more objective assessment, we performed a further check to validate the optimization process. As the aim of this study was also to avoid risk of misclassification, we extrapolated, both from the Kernel interpolation map of the original and the reduced networks, the EQR values of each WB. Then, results of each WB were tested by Student's *t*-test to verify that the reduction had no significantly modified from the original network. All comparisons did not significantly differ, except for PC1, PC3, and PNC1. Moreover, the classification results were not affected by the optimization process at all WBs.

5. Conclusions

From the statistical analyses, it appears that the first results of the WFD status classes of the Venice Lagoon obtained by macrophyte assemblages produced a satisfactory reliability and provided useful information about estimation of the confidence, which can be useful to management organizations to optimize resources for remedial measure adoption. Results also allowed proposing a reduction of the monitoring sampling effort, which definitely affects river basin management costs. For this purpose, a suitable multi-approach method based on inferential statistics, spatial analyses, and expert judgment was applied in this study, in order to review the sampling effort with the aim at ensuring and keeping on a high reliability of the status classification. Following the approaches described in this study, an increasing role of expert judgment could be observed: from a clear and strictly statistical approach based on L values *a priori* defined by ecological class amplitude, to a mixed method where L_{opt} was *a posteriori* defined for each WB, and resulting stations further emendated taking into account the site-specific characteristics. However, this study showed that a strictly statistical approach is obviously objective and standardized, but it could be unachievable with realistic data, since some results could lead to a waste of resource or vice versa to an excessive reduction of information. For these reasons, to optimize the monitoring effort, the inferential statistics were applied as a guideline, with the implementation of criteria arisen from expert knowledge of the area and the problem. Moreover, a robust process of validation was also proposed and adopted which definitively ensure on spatial reductions and reliability of information.

Apart from the complex nature of the Venice Lagoon, the multi-approach proposed in this study could also be applied to any other water body to assess the WFD classification reliability, and to optimize the monitoring effort to any other area.

Author Contributions: Conceptualization, R.B.B., P.P., F.C., and A.B.; methodology, F.C., A.B.; formal analysis, F.C., A.B., and E.P.; investigation, A.S.; resources, R.B.B., A.S., P.P. and M.G.; data curation, A.S., F.C., A.B., and M.N.; writing—original draft preparation, F.C.; writing—review and editing, A.B., R.B.B., A.S., and P.P.; visualization, F.C.; supervision, R.B.B.; project administration, R.B.B., M.G., and P.P.; funding acquisition, M.G., P.P., and R.B.B.

Funding: This work was supported by ARPAV in the framework of the Mo.V.Eco Project, funded by Veneto Region; contract of 31/01/2013 from the general agreement between ISPRA and ARPAV of 27/04/2012.

Conflicts of Interest: The authors declare no conflict of interest.

References

1. Water Framework Directive, European Union (WFD E.U.). *Establishing a Framework for Community Action in the Field of Water Policy; Directive 2000/60/EC of the European Parliament and of the Council of 23 October 2000*; EU: Brussels, Belgium, 2000.

2. Arle, J.; Mohaupt, V.; Kirst, I. Monitoring of Surface Waters in Germany under the Water Framework Directive—A Review of Approaches, Methods and Results. *Water* **2016**, *8*, 217. [CrossRef]

3. Clarke, R.T. Estimating confidence of European WFD ecological status class and WISER Bioassessment Uncertainty Guidance Software (WISERBUGS). *Hydrobiologia* **2013**, *704*, 39–56. [CrossRef]

4. Carstensen, J. Statistical principles for ecological status classification of Water Framework Directive monitoring data. *Mar. Pollut. Bull.* **2007**, *55*, 3–15. [CrossRef] [PubMed]

5. Clarke, R.T.; Hering, D. Errors and Uncertainty in Bioassessment Methods—Major Results and Conclusions from the STAR Project and their Application Using STARBUGS. In *The Ecological Status of European Rivers: Evaluation and Intercalibration of Assessment Methods*; Springer: Berlin/Heidelberg, Germany, 2006; Volume 188, pp. 433–439.

6. Sundermann, A.; Pauls, S.U.; Clarke, R.T.; Haase, P. Within-stream variability of benthic invertebrate samples and EU Water Framework Directive assessment results. *Fund Appl. Limnol.* **2010**, *173*, 21–34. [CrossRef]

7. Pasquaud, S.; Brind'Amour, A.; Berthelé, O.; Girardin, M.; Elie, P.; Boët, P.; Lepage, M. Impact of the sampling protocol in assessing ecological trends in an estuarine ecosystem: The empirical example of the Gironde estuary. *Ecol. Indic.* **2012**, *15*, 18–29. [CrossRef]

8. Mascaró, O.; Alcoverro, T.; Dencheva, K.; Díez, I.; Gorostiaga, J.M.; Krause-Jensen, D.; Balsby, T.J.S.; Marbà, N.; Muxika, I.; Neto, J.M.; et al. Exploring the robustness of macrophyte-based classification methods to assess the ecological status of coastal and transitional ecosystems under the Water Framework Directive. *Hydrobiologia* **2013**, *704*, 279–291. [CrossRef]

9. Sfriso, A.; Facca, C.; Bonometto, A.; Boscolo, R. Compliance of the macrophyte quality index (MaQI) with the WFD (2000/60/EC) and ecological status assessment in transitional areas: The Venice lagoon as study case. *Ecol. Indic.* **2014**, *46*, 536–547. [CrossRef]

10. Progetto di Aggiornamento del Piano di Gestione del Distretto Idrografico delle Alpi Orientali—Secondo Ciclo di Pianificazione (2015–2021). Available online: http://www.alpiorientali.it/ (accessed on 29 July 2019).

11. Maggi, C.; Ausili, A.; Boscolo, R.; Cacciatore, F.; Bonometto, A.; Cornello, M.; Berto, D. Sediment and biota to assess the trend monitoring of contaminants of transitional waters in the context of the Water Framework Directive: The Lagoon of Venice as a case study. *TrAC Trends Anal. Chem.* **2012**, *36*, 82–91. [CrossRef]

12. Cacciatore, F.; Noventa, S.; Antonini, C.; Formalewicz, M.; Gion, C.; Berto, D.; Gabellini, M.; Brusà, R.B. Imposex in *Nassarius nitidus* (Jeffreys, 1867) as a possible investigative tool to monitor butyltin contamination according to the Water Framework Directive: A case study in the Venice Lagoon (Italy). *Ecotoxicol. Environ. Saf.* **2018**, *148*, 1078–1089. [CrossRef]

13. Borja, Á.; Elliott, M. Marine monitoring during an economic crisis: The cure is worse than the disease. *Mar. Pollut. Bull.* **2013**, *68*, 1–3. [CrossRef]

14. De Jonge, V.; Elliott, M.; Brauer, V. Marine monitoring: Its shortcomings and mismatch with the EU Water Framework Directive's objectives. *Mar. Pollut. Bull.* **2006**, *53*, 5–19. [CrossRef] [PubMed]

15. Jaanus, A.; Kuprijanov, I.; Kaljurand, K.; Lehtinen, S.; Enke, A. Optimization of phytoplankton monitoring in the Baltic Sea. *J. Mar. Syst.* **2017**, *171*, 65–72. [CrossRef]

16. Umgiesser, G. Modelling the Venice Lagoon. *Int. J. Salt Lake Res.* **1997**, *6*, 175–199. [CrossRef]

17. Environmental Prevention and Protection Agency of Veneto Region (ARPAV). Laguna di Venezia—Dati Macrofite—Anno 2011. Available online: http://www.webcitation.org/78gb8RPtp (accessed on 27 May 2019).

18. R Core Team. R: A Language and Environment for Statistical Computing. Available online: https://www.R-project.org/ (accessed on 29 July 2019).

19. Adhikary, S.K.; Yilmaz, A.G.; Muttil, N. Optimal design of rain gauge network in the Middle Yarra River catchment, Australia. *Hydrol. Process.* **2014**, *29*, 2582–2599. [CrossRef]

20. Balsby, T.J.S.; Carstensen, J.; Krause-Jensen, D. Sources of uncertainty in estimation of eelgrass depth limits. *Hydrobiologia* **2013**, *704*, 311–323. [CrossRef]

21. Dromph, K.M.; Agusti, S.; Basset, A.; Franco, J.; Henriksen, P.; Icely, J.; Lehtinen, S.; Moncheva, S.; Revilla, M.; Roselli, L.; et al. Sources of uncertainty in assessment of marine phytoplankton communities. *Hydrobiologia* **2013**, *704*, 253–264. [CrossRef]

22. Carstensen, J.; Lindegarth, M. Confidence in ecological indicators: A framework for quantifying uncertainty components from monitoring data. *Ecol. Indic.* **2016**, *67*, 306–317. [CrossRef]

23. Clarke, R.T.; Davy-Bowker, J.; Sandin, L.; Friberg, N.; Johnson, R.K.; Bis, B. Estimates and comparisons of the effects of sampling variation using 'national' macroinvertebrate sampling protocols on the precision of metrics used to assess ecological status. *Hydrobiologia* **2006**, *566*, 477–503. [CrossRef]

24. De La Fuente, G.; Chiantore, M.; Gaino, F.; Asnaghi, V. Ecological status improvement over a decade along the Ligurian coast according to a macroalgae based index (CARLIT). *PLoS ONE* **2018**, *13*, e0206826. [CrossRef]

25. Bennett, S.; Roca, G.; Romero, J.; Alcoverro, T. Ecological status of seagrass ecosystems: An uncertainty analysis of the meadow classification based on the *Posidonia oceanica* multivariate index (POMI). *Mar. Pollut. Bull.* **2011**, *62*, 1616–1621. [CrossRef]

26. Torras, X.; Pinedo, S.; García, M.; Weitzmann, B.; Ballesteros, E. Environmental Quality of Catalan Coastal Waters Based on Macroalgae: The Interannual Variability of CARLIT Index and Its Ability to Detect Changes in Anthropogenic Pressures over Time. In *The Handbook of Environmental Chemistry*; Springer: Cham, Switzerland, 2015; pp. 183–199.

27. Goovaerts, P. *Geostatistics for Natural Resources Evaluation, Applied Geostatistics Series, 1st Edition*; Oxford University Press: Oxford, UK, 1997; p. 483.

28. Bayat, B.; Hosseini, K.; Nasseri, M.; Karami, H. Challenge of rainfall network design considering spatial versus spatiotemporal variations. *J. Hydrol.* **2019**, *574*, 990–1002. [CrossRef]

29. Ferrarin, C.; Ghezzo, M.; Umgiesser, G.; Tagliapietra, D.; Camatti, E.; Zaggia, L.; Sarretta, A. Assessing hydrological effects of human interventions on coastal systems: numerical applications to the Venice Lagoon. *Hydrol. Earth Syst. Sci. Discuss.* **2012**, *9*, 13839–13878. [CrossRef]

30. Scapin, L.; Zucchetta, M.; Bonometto, A.; Feola, A.; Brusà, R.B.; Sfriso, A.; Franzoi, P. Expected Shifts in Nekton Community Following Salinity Reduction: Insights into Restoration and Management of Transitional Water Habitats. *Water* **2019**, *11*, 1354. [CrossRef]
31. Lessio, F.; Tota, F.; Alma, A. Tracking the dispersion of *Scaphoideus titanus* Ball (Hemiptera: Cicadellidae) from wild to cultivated grapevine: use of a novel mark–capture technique. *Bull. Èntomol. Res.* **2014**, *104*, 432–443. [CrossRef] [PubMed]

Article

Meiofaunal Dynamics and Heterogeneity along Salinity and Trophic Gradients in a Mediterranean Transitional System

Federica Semprucci [1,*]**, Maria Flavia Gravina** [2,3] **and Paolo Magni** [4,*]

[1]   DiSB, Università degli Studi di Urbino Carlo Bo, Campus Scientifico Enrico Mattei, Località Crocicchia, 61029 Urbino, Italy

[2]   Dipartimento di Biologia, Università di Roma Tor Vergata, Via della Ricerca Scientifica, 1, 00133 Roma, Italy

[3]   Consorzio Nazionale Interuniversitario per le Scienze del Mare (CoNISMa), Piazzale Flaminio 9, 00196 Rome, Italy

[4]   Consiglio Nazionale delle Ricerche, Istituto per lo studio degli impatti Antropici e Sostenibilità in ambiente marino (CNR-IAS), Località Sa Mardini, Torregrande, 09170 Oristano, Italy

*   Correspondence: federica.semprucci@uniurb.it (F.S.); paolo.magni@cnr.it (P.M.)

Received: 10 May 2019; Accepted: 13 July 2019; Published: 18 July 2019

Abstract: The spatiotemporal variation in meiofaunal assemblages were investigated for the first time in the Cabras Lagoon, the largest transitional system in the Sardinian Island (W-Mediterranean Sea). Two main environmental (salinity and trophic) gradients highlighted a significant separation of the three study sites across the lagoon, which were consistent through time. The environmental variability and habitat heterogeneity of the Cabras Lagoon influenced the meiofauna. In particular, salinity and dissolved oxygen, primarily, shaped the meiofaunal assemblage structure at the seaward site which was significantly different from both the riverine and the organically enriched sites. On the other hand, the trophic components (e.g., organic matter, Chlorophyll-*a*, and phaeopigments) and the different degrees of confinement and saprobity among sites were the secondary factors contributing mostly to the separation between the latter two sites. The lack of significant differences in the temporal comparison of the meiofaunal assemblage structure along with the very low contribution of temperature to the meiofaunal ordination indicated that this assemblage was more affected by spatial rather than by temporal variation. This pattern was also supported by significant differences between the three sites in several univariate measures, including total number of individuals, number of taxa, Pielou's evenness, and the ratio between nematodes and copepods. Thus, the present study corroborates the hypothesis that meiofaunal organisms are good indicators of the spatial heterogeneity in transitional waters (TWs) and could have a greater species richness than that expected. Indeed, the Cabras Lagoon overall showed one of the highest meiofaunal richness values found from both Mediterranean and European TWs.

Keywords: biodiversity; spatial variation; sediments; confinement; saprobity; organic enrichment; coastal lagoons; Mediterranean sea

---

1. Introduction

Transitional waters, being a continuum between continental and marine ecosystems, represent areas with high environmental heterogeneity. As such, there is a complex association between abiotic and biotic components that makes these water bodies ideal to study the distribution and dynamics of the benthic assemblages with the aim to further our understanding of the ecosystem functioning [1,2]. Lagoons have a historically relevant "social" value because they offer a high biological productivity [3]. For this reason, they host many human activities (i.e., fisheries, aquaculture, agriculture, industry, and tourism [4]) that, on the other side, have endangered their integrity and ecological quality status as well [5,6]. Therefore, there is general agreement among the scientific community, which is also recognized by legislations worldwide (e.g., the US Clean Water Act, European Water Framework Directive, Marine Strategy Framework Directive, and the National Water Act in South Africa), about the need to assess their health status and ensure proper management of their resources [4,7].

Meiofauna are small benthic invertebrates that have a well-recognized role in the food webs of lagoon systems connecting microbial components to higher trophic levels that contributes to the overall carbon fluxes and organic matter mineralization [8,9]. Because of their high taxonomic diversity, rapid generation times, lack of larval stages, and various life strategies meiofaunal organisms are considered excellent bioindicators of natural or anthropogenic stressful conditions [10–12]. However, their role in ecosystems tends to be overlooked, mainly due to the lack of taxonomists and the small size of meiofauna, which require time and the appropriate techniques for their study [13].

In the Mediterranean basin, there are more than 100 coastal lagoons, half of which have available physico-chemical or ecological data in the scientific literature [4]. Among them, the largest amount of information on meiofaunal spatial pattern is available for the upper Adriatic Sea, including the Venice lagoon [14–20]. Meiofaunal diversity and assemblage structure are also well-documented in the southern part of the Adriatic Sea, including the Lesina and Varano lagoons [6,8,17,21–23]. Instead, meiofaunal studies in transitional environments along the Tyrrhenian coast, with the exception of the Stagnone of Marsala (Western Sicily), are largely lacking [17,24]. Furthermore, most of the available literature on meiofauna from coastal lagoons takes into consideration the spatial pattern of the assemblages, while only in a few cases their temporal dynamics is reported [2,8,24–26]. Finally, little is known on spatiotemporal dynamics of meiofauna in Mediterranean transitional systems characterized by different physico-chemical gradients related to the riverine inflow, the connection to the sea, and the organic matter (OM) enrichment of sediments.

Within the Tyrrhenian coast, the Sardinian Island is one of the richest Italian regions in number and extension of lagoons [27], yet knowledge on meiofaunal composition and distribution in these systems is absent. In the present study, we describe for the first time the spatiotemporal variation in meiofaunal assemblages in the Cabras Lagoon, the largest and most complex transitional system in the Sardinia Island. This lagoon is characterized by a large environmental heterogeneity, with an increasing salinity along its main longitudinal axis and varying degrees of trophic condition across the basin [28–30]. For these reasons, it represents a valuable case-study in which to test the general hypotheses on the meiofaunal dynamics in these highly variable systems. Our main objectives were to investigate the pattern of spatial variation in meiofaunal diversity and community structure in relation to the main environmental gradients, and to assess whether this pattern was consistent through time. In particular, we tested whether spatiotemporal variation could be identified in: (1) the whole meiofaunal assemblage of the three sites in terms of (i) total number of individuals, (ii) total number of taxa, and (iii) Shannon diversity (H′) and Pielou's evenness (J) indices; and (2) the abundance of dominant taxa, including the ratio between nematodes and copepods. We anticipate that the response of meiofaunal assemblages to the environmental drivers (both in water and sediments) identified in the present study will provide one of the few evidences of the importance of meiofaunal studies to further our understanding of the functioning of Mediterranean lagoons.

2. Materials and Methods

2.1. Study Area and Sampling Sites

The Cabras Lagoon (central-western Sardinia; Figure 1) is the largest lagoon in the Sardinia island, with a surface area of 22 km^2 and a watershed of ~430 km^2 inhabited by approximately 38,000 people. Its main freshwater riverine source is the Rio Mare e Foghe located in the northern sector of the lagoon, with a minor contribution from the Rio Tanui, southward. The lagoon is connected to the adjacent Gulf of Oristano only via three narrow creeks that flow into a large channel ("scolmatore") built in the late 70's, closed in proximity of the lagoon by a 30 cm high dam. In the last two decades, the Cabras Lagoon has been extensively investigated from various perspectives and using different approaches, including physical/modeling [31,32], biogeochemical [33,34], biological [35–38], and ecological [5,39,40]. However, while several studies have been conducted in the Cabras Lagoon on the macrozoobenthos [28–30,41,42], nothing is known regarding the spatiotemporal variation in meiofaunal assemblages. In fact, no such studies are available for transitional waters in the Sardinian Island, one of the richest Italian regions in number and extension of lagoons [27], with only few examples conducted in fully marine coastal waters [43,44].

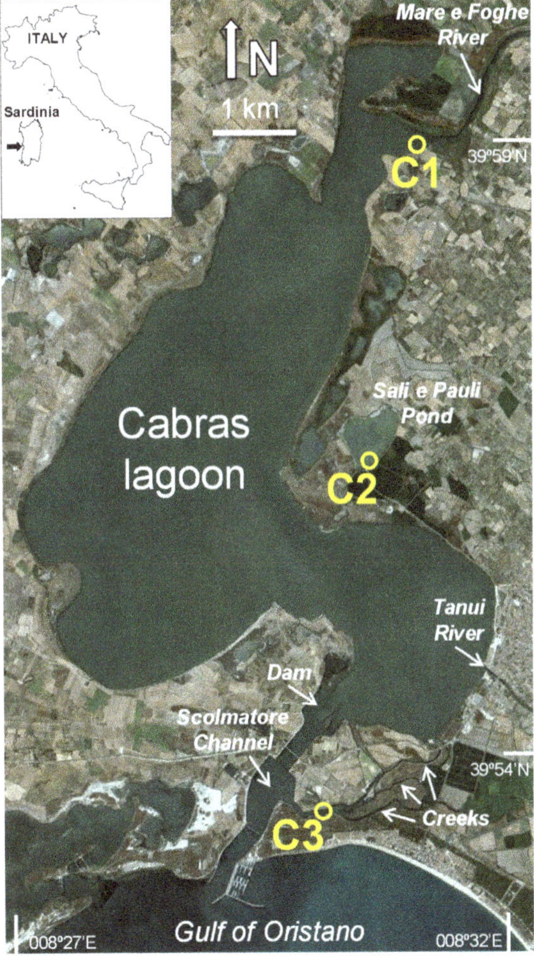

Figure 1. Location of the study area (Cabras lagoon, western Sardinia, Italy) and sampling sites (C1, C2, and C3). Image source: Google Earth.

For the present study, three sites (C1, C2, and C3; Figure 1) were selected along the longitudinal axis of the Cabras Lagoon, being representative of different environmental (e.g., salinity, confinement, and sediment grain-size) and trophic (e.g., sediment OM and phytopigments) conditions. Site C1 was located in the northern sector of the lagoon, connected to the main freshwater tributary the Rio Mare e Foghe. This site was characterized by sandy sediments, low OM content of sediments, and the presence of halophytic vegetation (*Phragmites* sp.) along the shore. Site C2, was located in the satellite pond of Sali e Pauli and surrounded by halophytic vegetation (*Salicornia* sp.). This site was highly confined and characterized by a high OM content of sediments [33]. Biofilm-forming cyanobacterial strains with extremely growth rates were also found here [45,46]. Site C3 was located in the southern sector of the lagoon, at the confluence of the three creeks connecting the Cabras Lagoon to the main channel. This site was characterized by muddy-sandy sediments, limited OM enrichment of sediments, abundant submerged vegetation (e.g., *Ruppia*), and a significantly higher hydrodynamics than at the other sites [47].

2.2. Field Surveysand Sample Treatment

The field surveys were carried out at sites C1, C2, and C3 on 6 July 2010 and 2 February 2011. At each site and date, water temperature, salinity, and dissolved oxygen (DO) were measured using portable probes (WTW LF 197 and WTWOxil 197, respectively). Subsequently, sediment samples for the determination of the water content (Wc) and chemical analysis (OM, chlorophyll-a, and phaeopigment content) were collected using a manual core (40 cm long, 5.5 cm diameter) gently pushed by hand into the sediments. Procedural details of sediment collection and chemical analysis are given in the companion paper by [30].

For the analyses of the meiofauna, six replicates were collected at each site by means of plexiglas corers (diameter: 3.6 cm) inserted 5 cm in the sediment. These samples were pre-filtered with magnesium chloride ($MgCl_2$; 80 g L^{-1}) to allow organisms to relax before fixation and facilitate subsequent taxonomic identification [48]. This treatment appears important because the "soft-bodied" taxa (e.g., Gastrotricha, Plathelminthes, and Nemertina) usually undergo the major morphological alterations after fixation and they can remain in good conditions with magnesium chloride treatment. The sediment samples were then fixed in a solution of pre-filtered seawater containing formalin buffered with sodium tetraborate $Na_2B_4O_7$ to reach a pH of ca. 8.2 [49]. The amount of formalin to be added to the sample to obtain a final concentration of 4% was calculated based on the total volume of sediment and water present in the sample. A few drops of a Rose Bengal solution (0.5 g L^{-1}) were added to the sample in order to facilitate the identification of organisms in the sorting phase [50].

2.3. Meiofaunal Analysis

The samples were rinsed with a gentle jet of fresh water through a 0.5 mm sieve to separate the macrofauna from the meiofauna [48]. They were then decanted, sieved 10 times through a 42 μm mesh and centrifuged three times with Ludox HS30 (specific density 1.18 g/cm^3) [51]. The obtained animals were then transferred to a "Delfuss" Petri dish with a checkered bottom (200 squares, to make counting easier), sorted into their major taxa under a Leica G26 stereomicroscope, and counted.

All the values obtained was recalculated as abundance per 10 cm^2. The richness (number of major taxa), Shannon's diversity, Pielou's evenness indices (log_2) were calculated to describe the structure of the meiofaunal assemblage. The possible occurrence of anthropogenic impact on the meiofaunal community was also assessed by the total number of nematodes (Ne) and copepods (Co) computed in the Ne:Co ratio. This index was proposed by Raffaelli and Mason [52] for the pollution monitoring with meiofauna. The hypothesis was that the divergent auto-ecological characteristics of these two abundant and frequent meiofaunal components (the extreme tolerance of nematodes and the high sensitivity of copepods) might detect the occurrence of human stress in marine sediments.

2.4. Statistical Analysis

Both abiotic and biotic data were used for the data analysis. Water variables were temperature, salinity, and dissolved oxygen; sediment variables included water content, OM, chlorophyll-*a*, and phaeopigments. Biotic data consisted in the abundance of the meiofauna and were used to construct a taxa-by-site and period matrix. The environmental data variation was represented by means of box-plots for each variable and each site. The biotic parameters computed were the number of taxa (S, taxon richness), the number of individuals per taxa (A, abundance), and the Shannon (H', diversity) and Pielou (J, evenness) indices, as well as Ne/Co ratio. These biotic variables were computed for the three sampling sites C1, C2, and C3, for each replicate and date.

As for multivariate analysis, the non-parametric permutational analysis of variance (two way-PERMANOVA), based on Bray-Curtis (dis)similarity measures [53] was carried out to test significant differences of the structure of community among sites (three levels: C1, C2, and C3), periods (two levels: July and February), and site × period interactions as fixed factors. The data were log(x+1) transformed before the analysis. The PERMANOVA, based on Euclidean distance, was also used to test the significant differences of all the biotic univariate measures (i.e., total meiofaunal abundance, number of taxa, Shannon-diversity, Pielou evenness, and Ne/Co ratio). A log (x+1) transformation of data was applied only for the total meiofaunal abundance. The significance was computed by permutation with 9999 replicates. The pairwise comparisons between all pairs of sites were computed as post-hoc test and the Bonferroni correction procedure was followed to account for multiple simultaneous correlations [54]. The principal component analysis (PCA), based on the correlation matrix, was used to explore the faunal variations within the lagoon and periods. The environmental variables were used to understand the key environmental variables accounting for the much % of variance affecting the meiofaunal distribution. The multivariate procedure non-metric multidimensional scaling (nMDS) was used to investigate the differences between the sites; the more informative environmental variables were added in the analysis to best explain the meiobenthos structure and they were superimposed in the graph [55]. The meiofaunal major taxa contributing most to (dis)similarities among the sites were identified using the similarity percentages (SIMPER) test.

3. Results

3.1. Environmental Variables

Among the water variables, both salinity and DO showed increasingly higher values from C1 to C3 (Figure 2), most variation found in salinity at C3 due to large differences between the summer (27 psu) and winter (3 psu) dates [30]. Differently from salinity, the median of DO was significantly higher at C3 than at both C1 and C2. Sediment variables had large within- and between-site variation. In particular, while C1 had low and homogeneous Wc, OM, chlorophyll-*a* (Chl-*a*), and phaeopigment values, C2 was most variable and had the highest peaks and outliers in OM, Chl-*a*, and phaeopigments among the three sites. C3 was in an intermediate position, with the medians of all sediment variables between C1 and C2 (Figure 2).

The PCA showed a clear separation of both sites and dates (Figure 3). C1 data-points were located on the left-had side of the ordination model (second and third quadrants), while those of C2 were located on the right-hand side (first and fourth quadrants). C3 was in intermediate position, yet clearly separated from C1 and C2. The two sampling dates were also clearly distinguishable for each site, summer and winter data-point being positioned on the upper and lower portion of the model, respectively. Among the environmental variables, sediment Wc, OM, Chl-*a*, and phaeopigments were mostly correlated with PC1, while water temperature and salinity showed the highest correlation with PC2 (Table 1). This indicates that sites were discriminated on the *x*-axis by the sediment variables mostly affected by the trophic features, while on the *y*-axis they were distributed in relation to the confinement gradient as indicated by the water variables. The variance explained by the model was 46.1% and 23.5% for PC1 and PC2, respectively.

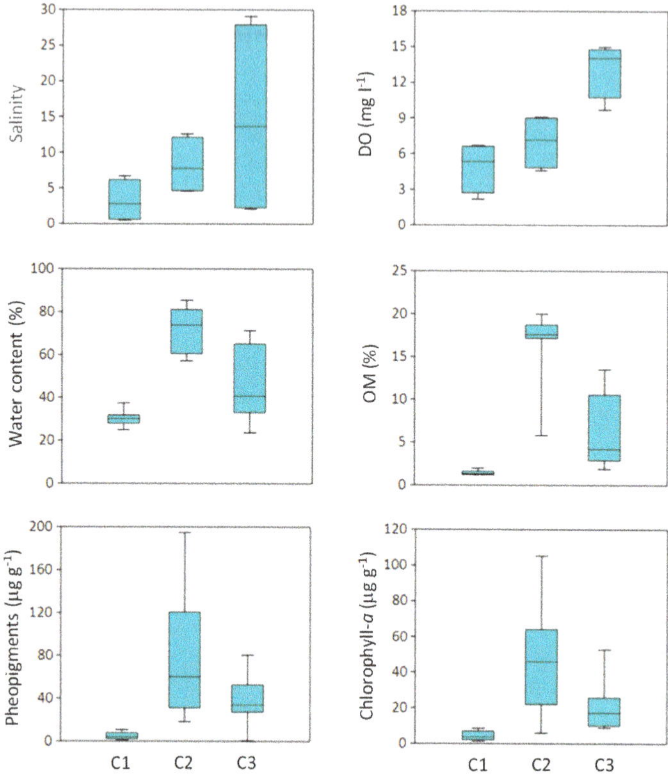

Figure 2. Mean values (n = 4, ±SE) of water salinity and dissolved oxygen (DO) concentration, and mean values (n = 12, ±SE) of sediment water (Wc), phaeopigment (Pha), organic matter (OM), and chlorophyll-*a* (Chl-*a*) contents at the sampling sites C1, C2, and C3 during the study period.

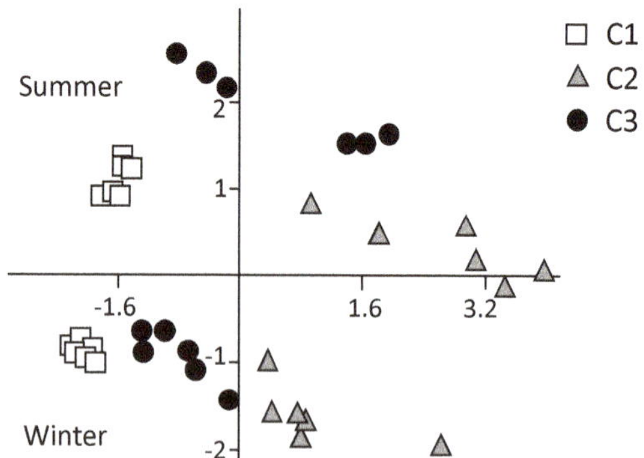

Figure 3. Principal component analysis (PCA) on environmental variables.

Table 1. Component correlation coefficients in the PCA of the environmental variables and the first two components. 69.6% of variance explained by Principal Components 1 (46.1%) and 2 (23.5%). In bold, the highest loadings.

Variable	PC 1	PC 2
Temperature	0.17	**0.71**
Salinity	0.24	**0.59**
Dissolved oxygen	0.004	−0.12
Water content	**0.49**	−0.27
OM	**0.50**	−0.22
Chlorophyll-*a*	**0.47**	0.06
Phaeopigments	**0.45**	−0.12

3.2. Meiofauna

A total of 16 meiofaunal taxa were found: Plathelminthes, Nemertina, Nematoda, Kinorhyncha, Bivalvia, Polychaeta (adults and nectochaetes), Oligochaeta, Copepoda (adults and juveniles), Ostracoda, Amphipoda, Cladocera, Isopoda, Tanaidacea, Insecta, Halacaroidea, and Pycnogonida. The dominant taxa were generally nematodes (from 41% at C2 in winter 2011 to 86% at C3 in summer 2010), copepods (from 4% at C3 in summer 2010 to 33% C3 in winter 2011), and ostracods (from 0% at C2 in summer 2010 to 12% at C3 in winter 2011).

The PERMANOVA carried out on the structure of the meiofaunal assemblage indicated highly significant differences only among sites ($p < 0.001$), while no significant differences emerged among periods or site × period interactions (Table 2). In particular, pairwise comparisons highlighted significant differences between C3 and the other two sites (C1 and C2). The aforementioned variations were tested by means of the SIMPER analysis, which highlighted that five main groups, i.e., Nematoda, Copepoda, Ostracoda, Nauplii, and Halacaroidea, contributed up to the 95% of the cumulative similarity, with an individual contribution varying from 58% for Nematoda to 2.3% for Halacaroidea (Table 3; Figure 4). All the other taxa contributed to less than 2%.

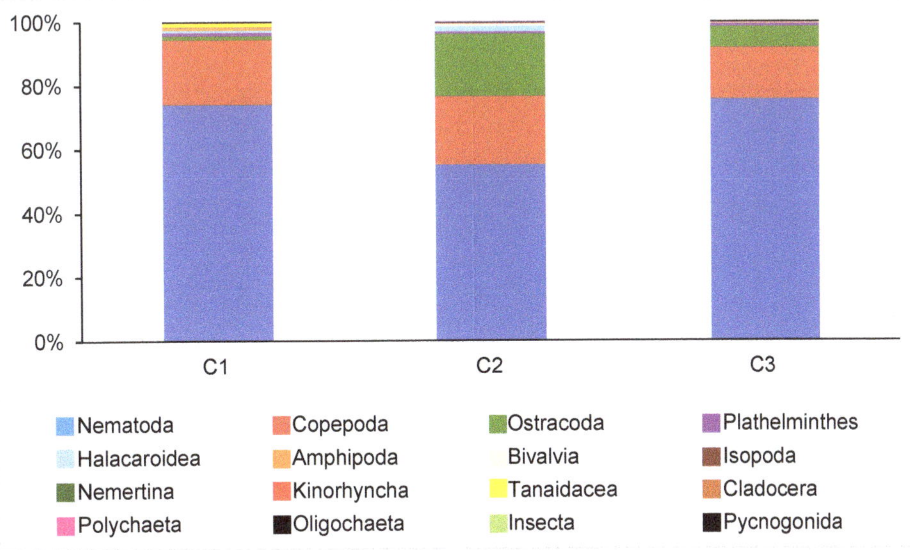

Figure 4. Meiofauna community structure in the sampling sites C1, C2, and C3.

Table 2. Results of two-way PERMANOVA (Bray Curtis (dis)similarity based) testing the differences among sites, periods and their interactions for the meiofaunal assemblage structure. Df = degrees of freedom. Significant p values are marked, ** for $p < 0.001$.

Factors	Total Sum of Squares	Df	Mean Square	Pseudo-F	p	Pairwise
Sites	2.72	2	1.36	6.36	0.0001 **	C1 vs. C3 $p = 0.0003$ **; C2 vs. C3 $p = 0.0006$ **
Periods	0.20	1	0.20	0.94	0.44	
Sites × Periods	0.66	2	0.33	1.55	0.12	
Residual	6.41	30	0.21			
Total	9.99	35				

Table 3. Results of the Similarity Percentage (SIMPER) analysis showing the average dissimilarity (Av. Dissim.), % contribution of each taxon (Cont. %), % cumulative for the pair comparisons (Cum. %), and average abundances of each sites (Av. ab.). The overall average dissimilarity is 70.9, 76.0, and 82.6 for C1 vs. C2, C1 vs. C3, and C2 vs. C3, respectively.

C1 vs. C2	Av. Dissim.	Cont. %	Cum. %	Av.ab. C1	Av.ab. C2	C1 vs. C3	Av. Dissim.	Cont. %	Cum. %	Av.ab. C1	Av.ab. C3	C2 vs. C3	Av. Dissim.	Cont. %	Cum. %	Av.ab. C2	Av.ab. C3
Nematoda	41.2	58.1	58.1	73.1	54.8	Nematoda	52.8	69.4	69.4	73.1	619	Nematoda	58.0	70.2	70.2	54.8	619
Copepoda	13.4	18.9	77.0	17.5	20.3	Copepoda	12.5	16.4	85.8	17.5	89.8	Copepoda	13.0	15.8	85.9	20.3	89.8
Ostracoda	9.4	13.2	90.3	1.3	19.4	Ostracoda	4.9	6.5	92.3	1.3	53.4	Ostracoda	5.6	6.8	92.8	19.4	53.4
Nauplii	2.4	3.4	93.6	2.5	0.9	Nauplii	3.7	4.8	97.1	2.5	43.3	Nauplii	3.8	4.6	97.3	0.9	43.3
Halacaroidea	1.7	2.3	96.0	0.8	1.8	Plathelminthes	0.9	1.2	98.3	1.0	6.3	Plathelminthes	0.9	1.0	98.4	0.6	6.3
Tanaidacea	0.8	1.1	97.1	1.2	0.0	Oligochaeta	0.3	0.4	98.7	0.3	4.3	Halacaroidea	0.3	0.4	98.8	1.8	0.6
Plathelminthes	0.5	0.8	97.9	1.0	0.6	Tanaidacea	0.2	0.3	99.0	1.2	0.1	Oligochaeta	0.3	0.4	99.2	0.3	4.3
Amphipoda	0.3	0.5	98.3	0.7	0.0	Amphipoda	0.2	0.2	99.2	0.7	1.2	Polychaeta	0.2	0.2	99.4	0.4	2.1
Bivalvia	0.3	0.4	98.8	0.1	0.7	Polychaeta	0.2	0.2	99.5	0.0	2.1	Bivalvia	0.2	0.2	99.6	0.7	0.5
Cladocera	0.3	0.4	99.2	0.1	0.1	Halacaroidea	0.1	0.2	99.6	0.8	0.6	Amphipoda	0.1	0.1	99.8	0.0	1.2
Pycnogonida	0.2	0.2	99.4	0.0	0.3	Bivalvia	0.1	0.2	99.8	0.1	0.5	Insecta	0.1	0.1	99.9	0.0	0.4
Oligochaeta	0.2	0.2	99.6	0.8	0.3	Insecta	0.1	0.1	99.9	0.0	0.4	Pycnogonida	0.0	0.1	99.9	0.3	0.0
Kinorhyncha	0.1	0.2	99.8	0.1	0.0	nectochaetes	0.0	0.0	99.9	0.0	0.3	nectochaetes	0.0	0.0	100.0	0.0	0.3
Isopoda	0.1	0.1	99.9	0.2	0.0	Isopoda	0.0	0.0	99.9	0.2	0.0	Cladocera	0.0	0.0	100.0	0.1	0.0
Polychaeta	0.1	0.1	100.0	0.0	0.4	Cladocera	0.0	0.0	100.0	0.1	0.0	Nemertina	0.0	0.0	100.0	0.0	0.1
Nemertina	0.0	0.0	100.0	0.0	0.0	Kinorhyncha	0.0	0.0	100.0	0.1	0.0	Tanaidacea	0.0	0.0	100.0	0.0	0.1
Insecta	0.0	0.0	100.0	0.0	0.0	Nemertina	0.0	0.0	100.0	0.0	0.1	Kinorhyncha	0.0	0.0	100.0	0.0	0.0
nectochaetes	0.0	0.0	100.0	0.0	0.0	Pycnogonida	0.0	0.0	100.0	0.0	0.0	Isopoda	0.0	0.0	100.0	0.0	0.0

Variation in the meiofaunal abundance, number of taxa, diversity, evenness, and the Ne/Co ratio are shown in Figure 5. While differences between sites in abundance, evenness, and the Ne/Co ratio were high, those in taxon number and diversity were not. In particular, abundance was the highest at C3, which also showed the most marked standard error, and consequently the lowest evenness. The Ne/Co ratio was higher in summer, but always lower in winter both at C2 and C3, which coincided, with the drop in salinity. Finally, the most confined and organically enriched site C2 showed the most variation in the Ne/Co ratio up to various orders of magnitude, particularly in summer. The 2-way PERMANOVA conducted individually on each biotic measure showed significant differences among sites in the total abundance, taxon number, evenness and the Ne/Co ratio, while no difference was found for the diversity index (Table 4). The pairwise comparisons among the three sites are showed in Table 4 and highlighted higher and significantly differences especially of C3. PERMANOVA did not reveal significant differences of meiofaunal abundance, taxon richness, Shannon diversity, and Ne/Co ratio among the periods investigated, while some differences of evenness ($p < 0.05$) were found (Table 4).

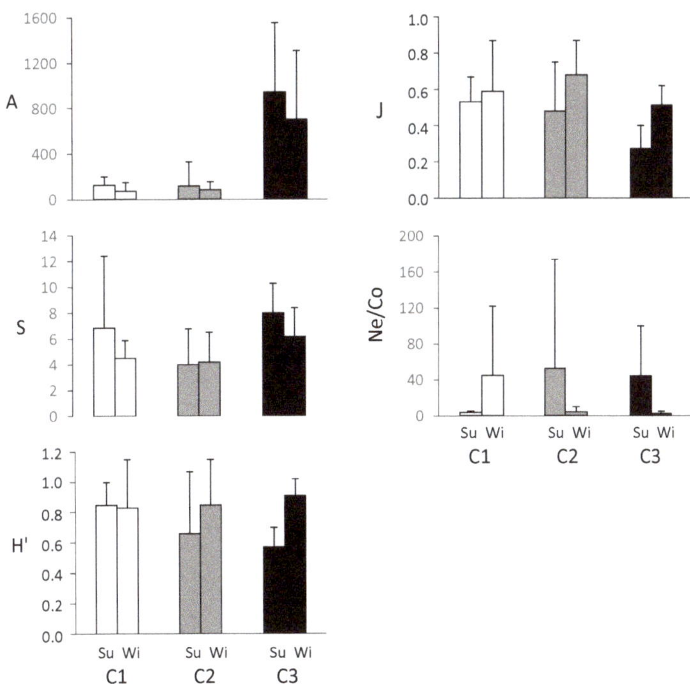

Figure 5. Mean values (n = 6, ±SE standard error) of meiofaunal community synthetic measures at the sampling sites C1, C2, and C3 in summer (Su) and winter (Wi).

The nMDS on the structure of assemblage similarly highlighted three major groups corresponding to the three study sites, with a partial overlap between C1 and C2, and a major separation of C3 from both C1 and C2 (Figure 6). In this analysis, the environmental variables superimposed to the biotic data showed that salinity and DO were the main responsible factors for the separation of C3 from C1 and C2 (on the right-hand side) and that Wc, OM, Chl-*a*, and phaeopigments contributed mostly to the separation between C1 and C2 (Figure 6). These results demonstrated two main gradients influencing the spatiotemporal variation in the meiofaunal assemblages in the Cabras Lagoon. Temperature, on the contrary, did not make a relevant contribution to the ordination of the sites as indicated by its short segment.

Table 4. Results of two-way PERMANOVA testing differences among sites for the total meiofaunal abundance, number of taxa, diversity, evenness, and Ne/Co ratio (Euclidean distance based). Significant p values are marked, * for $p < 0.05$; ** for $p < 0.01$.

Variable	Factors	Total Sum of Squares	df	Mean Square	Pseudo-F	p	Pairwise
Total abundance	Sites	0.54	2	0.27	10.74	0.0003 **	C1 vs. C3 $p = 0.01$ *; C2 vs. C3 $p < 0.01$ **
	Periods	0.01	1	0.01	0.46	0.57	
	Sites × Periods	0.11	2	0.06	2.2	0.11	
	Residual	0.75	30	0.03			
	Total	1.41	35				
number of taxa	Sites	0.49	2	0.24	6.15	0.0023 **	C2 vs. C3 $p < 0.05$ *
	Periods	0.03	1	0.03	0.76	0.42	
	Sites × Periods	0.02	2	0.01	0.21	0.76	
	Residual	1.19	30	0.04			
	Total	1.73	35				
Shannon-diversity	Sites	0.08	2	0.04	0.64	0.8	-
	Periods	0.07	1	0.07	1.11	0.31	
	Sites × Periods	0.16	2	0.08	1.29	0.22	
	Residual	1.9	30	0.06			
	Total	2.22	35				
Pielou-evenness	Sites	0.31	2	0.16	2.52	0.0166 *	C1 vs. C3 $p < 0.05$ *
	Periods	0.17	1	0.17	2.71	0.0356 *	
	Sites × Periods	0.19	2	0.1	1.56	0.13	
	Residual	1.86	30	0.06			
	Total	2.53	35				
Ne/Co ratio	Sites	1.16	2	0.58	3	0.02 *	C1 vs. C3 $p < 0.01$ **
	Periods	0.24	1	0.24	1.25	0.28	
	Sites × Periods	0.71	2	0.36	1.83	0.13	
	Residual	5.81	30	0.19			
	Total	7.92	35				

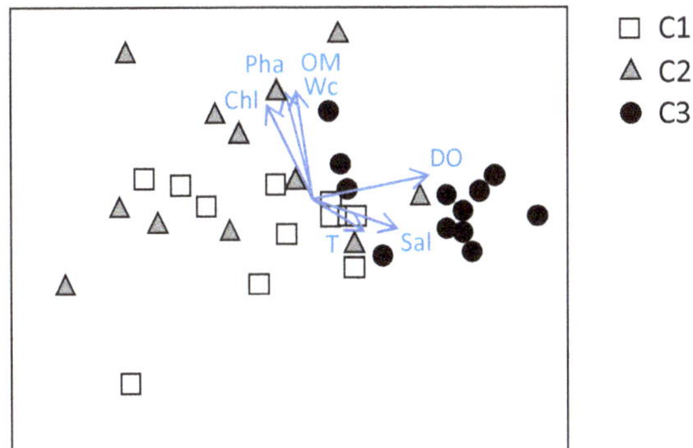

Figure 6. MDS on meiofaunal abundance (stress = 0.09) with superimposed environmental.

4. Discussion

The present study provides the first information of the spatiotemporal variation in lagoonal meiofaunal assemblages in one of the richest Italian regions in number and extension of lagoons within the Tyrrhenian coast, i.e., the Sardinian Island. We demonstrated significant changes in meiofaunal diversity and community structure in relation to the main environmental gradients which, in the studied lagoon, are driven by salinity and dissolved oxygen concentration in water, sediment organic enrichment, and different degrees of confinement and saprobity. Our results highlight the importance of meiofaunal studies often neglected, by ecologists and policy makers, to further our understanding of the functioning of Mediterranean lagoons.

From an abiotic point of view, the heterogeneity of the Cabras Lagoon can be related to two main interlinked environmental and ecological gradients. The first gradient is due to the marine influence seaward, the second one is related to spatial differences in the trophic status and saprobity condition across the lagoon. In particular, the most confined site C2 differed from the other two investigated sites as being characterized by muddy, organically enriched sediments, and due to its lack of connection and exchange with both continental and marine waters, mostly influencing C1 and C3, respectively. Differently, the northern site C1 was most directly affected by the riverine input of freshwater and had sandy sediments with a very low phytopigment and OM content. This is consistent with the high energy and the low water residence time found at C1, which helps explaining the resuspension and export of fine sediment particles from this site to the central sector of the lagoon [56,57]. Finally, the seaward site C3 located in one of the three creeks connecting the lagoon to the Gulf of Oristano was in terms of sediment features intermediate between C1 and C2. It is worthwhile noting that in this sector of the lagoon man-made structures constructed in proximity of the inlet have caused modifications in the sedimentary regime of the lagoon and the water exchange with the adjacent Gulf [41,56]. Yet, the largest variation in salinity was found at C3 due to a marked drop in salinity in the whole lagoon in the winter of the present study, nullifying temporarily the salinity gradient and further highlighting the high seasonal and interannual environmental variability of the Cabras Lagoon [35,36].

Overall, these results show that the environmental (dis)similarities between the three study sites were constant in the two periods investigated, although summer and winter data-points were most clumped at C1 and more widespread, but still clearly separated one another, at C2 and C3. Thus, temporal variation did not alter significantly the location of the site-points in the multivariate model ordination. This highlighted that the most relevant source of variation in the study area was not due to differences between summer and winter dates, but to differences between sampling sites.

Regarding the meiofaunal richness, our data can be compared only with a few studies because the complete list of meiofaunal taxa is rarely reported. However, an overall high number of richness (i.e., 16 taxa) was recorded in the present study showing levels higher than those documented from both Mediterranean and European transitional water bodies [17–19,22,58,59]. Nematodes were the dominant taxon as frequently reported in lagoon systems worldwide [2,17,22,60,61]. This is likely related to their capacity to colonize the fine (suboxic or anoxic) sediments that generally characterize lagoons [19]. As reported in literature, the second most abundant group is represented by copepods [14,22,58,59]. Copepods are one of the most sensitive taxa to oxygen limitation and therefore they are confined to the oxic sediments [20,48], but they seem to take advantage of the high abundance of the microphytobenthos occurring in lagoon sediments and that are a primary food source for numerous copepod species [24].

The meiofaunal structure assemblage seemed to be very sensitive to the spatial environmental heterogeneity found in the Cabras Lagoon appearing a promising indicator of biotic changes in transitional water bodies. Indeed, the multivariate analysis (nMDS and PERMANOVA) applied on the assemblage structure clearly distinguished the three different sites in line with their environmental features with a partial overlap of the meiofaunal structure of C1 and C2, and a greater separation of C3 (Table 2, Figure 6). The latter site, characterized by "marine conditions", had the highest abundance of all the meiofaunal taxa and in particular, it was distinguished by the greater abundances of Plathelminthes and Oligochaeta. These two taxa are often closely associated with each other and are among the primary components of transitional environment sediments [20,24]. Oligochaeta are regarded as taxa able to adapt to numerous environmental stress [18,48]. Plathelminthes are effective predators of many meiobenthic organisms such as copepods that could explain their higher abundance at C3 where the higher meiofaunal densities and copepods was found [14,62].

Instead, the high degree of confinement and consequent trophic load existing at site C2 was marked by a higher presence of Halacaroidea and Ostracoda. Ostracoda are generally recognized as sensitive taxon to environmental perturbations, but adaptive behaviors to numerous natural and anthropogenic environmental (e.g., organic load and trace element contamination) changes have been documented in several species [22,24,63]. In that site, there was also the only record of Pycnogonida that is generally recognized as a marine taxon. Pycnogonida have a few representatives in the meiofauna, but the *Anoplodactylus* genus has some species that ranges in the meiofaunal body size and are also tolerant to salinity variations until values comparable to those found in C2 (11 PSU) [64].

In C1, the site with the lowest salinity, Tanaidacea was one of the discriminating taxon. Noteworthy is that although many euryhaline species from Tanaidacea are found in transitional habitats, most occur only temporarily in these environments, appearing unable to form stable populations there [65]. Furthermore, Ateş et al. [66] reported that some species are particularly related to coarse grain size and low content of organic matter that were the conditions that distinguish C1.

The environmental variables that appeared to mainly affect the meiofaunal assemblage of the study area were salinity and DO that were the main responsible parameters for the separation of C3 and the other two sites. Salinity gradient is one of the primary factors that influence meiofauna in transitional environments [8,18] along with the oxygen availability that seems to influence all the meiofaunal taxa and not only the oxygen sensitive ones such as copepods (see above references). The quantity and quality of the organic matter (OM, Chl-*a*, and phaeopigments) and Wc, which is an indirect indication of the grain size of the substrates, were also important for the meiofaunal distribution as suggested by many authors [8,17,22] and contributed mostly to the separation between the other two sites (i.e., C1 and C2). Instead, temperature did not show a relevant contribution to the ordination of the meiofauna. This issue as well as the lack of significant differences in the comparison of the meiofaunal assemblage structure suggest that meiofauna was more affected by spatial than temporal variations. This pattern resembled the distribution patterns of the macrozoobenthos observed at the same study sites where spatial differences were greater than significant seasonal changes [30].

When the spatiotemporal variation in the univariate measures were statistically studied, the greatest differences were observed between the three sites further supporting the results revealed by the structure of the meiofaunal assemblage (Table 4). In particular, PERMANOVA showed significant differences of the total meiofaunal abundance, number of taxa, evenness and the Ne/Co ratio. Among them, the Ne/Co ratio showed a very temporal variable trend (i.e., it was higher in C1 in the winter and lower in C2 and C3 in the summer) which likely is why PERMANOVA did not reveal significant differences between periods. However, the variation in the Ne/Co ratio did not highlight the presence of anthropogenic stress being the values lower than the thresholds reported by Raffaelli and Mason [52] for stressful conditions. Instead, the highest evenness at C1 was likely related to the coarser grain size of the sediments that generally host a more diversified meiofaunal assemblage [67,68].

5. Conclusions

The present study is one of the few investigations on the meiofaunal community structure, composition, and diversity conducted in transitional waters (TWs) in the Western Mediterranean Sea. The high environmental variability and habitat heterogeneity of the Cabras Lagoon, the largest TW in Sardinia, was reflected in significant differences in meiofauna among the study sites. Spatial differences in several faunal parameters (i.e., community structure, richness, Pielou-evenness, and Ne/co ratio) were stronger than temporal variation, suggesting that meiofaunal organisms are good indicators of the physical-chemical variation in TWs. Furthermore, the Cabras Lagoon showed high values of meiofaunal species richness further supporting the idea that TWs may be biodiversity hotspots and meiofauna is an important biotic component to understand their functioning.

Author Contributions: Conceptualization, F.S., M.F.G. and P.M.; methodology, F.S., M.F.G. and P.M.; formal analysis, F.S., M.F.G. and P.M.; data curation, F.S., M.F.G. and P.M.; writing-original draft preparation, F.S., M.F.G. and P.M.; writing-review and editing, F.S., M.F.G. and P.M.; visualization, F.S., M.F.G. and P.M.

Funding: We would like to thank Marcello Giorgi, University of Rome "Tor Vergata", for technical support in data analysis procedures.

Acknowledgments: We gratefully acknowledge the two reviewers whose comments greatly contributed to an improved version of the original manuscript.

References

1. Cardone, F.; Corriero, G.; Fianchini, A.; Gravina, M.F.; Nonnis Marzano, C. Biodiversity of transitional waters: Species composition and comparative analysis of hard bottom communities from south-eastern Italian coast. *J. Mar. Biol. Assoc. UK* **2014**, *94*, 25–34. [CrossRef]

2. Kandratavicius, N.; Muniz, P.; Venturini, N.; Giménez, L. Meiobenthic communities in permanently open estuaries and open/closed coastal lagoons of Uruguay (Atlantic coast of South America). *Estuar. Coast. Shelf Sci.* **2015**, *163*, 44–53. [CrossRef]

3. Barnes, R.S.K. *Coastal Lagoons: The Natural History of a Neglected Habitat*; Cambridge University Press: Cambridge, UK, 1980; pp. 1–180.

4. Pérez-Ruzafa, A.; Marcos, C.; Pérez-Ruzafa, I.M. Mediterranean coastal lagoons in an ecosystem and aquatic resources management context. *Phys. Chem. Earth* **2011**, *36*, 160–166.

5. Magni, P.; Tagliapietra, D.; Lardicci, C.; Balthis, L.; Castelli, A.; Como, S.; Frangipane, G.; Giordani, G.; Hyland, J.; Maltagliati, F.; et al. Animal-sediment relationships: Evaluating the 'Pearson-Rosenberg paradigm' in Mediterranean coastal lagoons. *Mar. Pollut. Bull.* **2009**, *58*, 478–486. [CrossRef] [PubMed]

6. Armynot du Châtelet, E.; Bout-Roumazeilles, V.; Coccioni, R.; Frontalini, F.; Francescangeli, F.; Margaritelli, G.; Rettori, R.; Spagnoli, F.; Semprucci, F.; Trentesaux, A.; et al. Environmental control on a land-sea transitional setting–Integrated microfaunal, sedimentological, and geochemical approaches. *Environ. Earth Sci.* **2016**, *75*, 123. [CrossRef]

7. Bouchet, V.M.P.; Goberville, E.; Frontalini, F. Benthic foraminifera to assess Ecological Quality Status in Italian transitional waters. *Ecol. Indic.* **2018**, *84*, 130–139. [CrossRef]

8. Gambi, C.; Totti, C.; Manini, E. Impact of organic loads and environmental gradients on microphytobenthos and meiofaunal distribution in a coastal lagoon. *Chem. Ecol.* **2003**, *19*, 207–223. [CrossRef]

9. Moens, T.; Braeckman, U.; Derycke, S.; Fonseca, G.; Gallucci, F.; Ingels, J.; Leduc, D.; Vanaverbeke, J.; Van Colen, C.; Vanreusel, A.; et al. Ecology of free-living marine nematodes. In *Handbook of Zoology: Gastrotricha, Cycloneuralia and Gnathifera Nematoda*; Schmidt-Rhaesa, A., Ed.; deGruyter: Berlin, Germany, 2013; Volume 2, pp. 109–152.

10. Boufahja, F.; Semprucci, F.; Beyrem, H. An experimental protocol to select nematode species from an entire community using progressive sedimentary enrichment. *Ecol. Indic.* **2016**, *60*, 292–309. [CrossRef]

11. Baldrighi, E.; Semprucci, F.; Franzo, A.; Cvitkovic, I.; Bogner, D.; Despalatovic, M.; Berto, D.; MalgorzataFormalewicz, M.; Scarpato, A.; Frapiccini, E.; et al. Meiofaunal communities in four Adriatic ports: Baseline data for risk assessment in ballast water management. *Mar. Pollut. Bull.* **2018**. [CrossRef]

12. Yang, X.; Lin, C.; Song, X.; Xu, M.; Yang, H. Effects of artificial reefs on the meiofaunal community and benthic environment—A case study in Bohai Sea, China. *Mar. Pollut. Bull.* **2019**, *140*, 179–187. [CrossRef]

13. Semprucci, F.; Balsamo, M. Free-living Marine Nematodes as Bioindicators: Past, Present and Future Perspectives. *Trends Environ. Sci.* **2014**, *6*, 17–36.

14. Colangelo, M.A.; Ceccherelli, V.U. Meiofaunal recolonization of azoicsediment in a Po Delta lagoon (Sacca di Goro). *Ital. J. Zool.* **1994**, *61*, 335–342.

15. Villano, N.; Warwick, R.M. Meiobenthic communities associated with the seasonal cycle of growth and decay of Ulva rigidaArardh in the Palude Della Rosa, Lagoon of Venice. *Estuar. Coast. Shelf. Sci.* **1995**, *4*, 181–194. [CrossRef]

16. Guerrini, A.; Colangelo, M.A.; Ceccherelli, V.U. Recolonization patterns of meiobenthic communities in brackish vegetated and unvegetated habitats after induced hypoxia/anoxia. *Hydrobiologia* **1998**, *375–376*, 73–87. [CrossRef]

17. Pusceddu, A.; Gambi, C.; Manini, E.; Danovaro, R. Trophic state, ecosystem efficiency and biodiversity of transitional aquatic ecosystems: Analysis of environmental quality based on different benthic indicators. *Chem. Ecol.* **2007**, *23*, 1–11. [CrossRef]

18. Cibic, T.; Franzo, A.; Celussi, M.; Fabbro, C.; Del Negro, P. Benthic ecosystem functioning in hydrocarbon and heavy-metal contaminated sediments of an Adriatic lagoon. *Mar. Ecol. Prog. Ser.* **2012**, *458*, 69–87. [CrossRef]

19. Semprucci, F.; Balsamo, M.; Sandulli, R. Assessment of the Ecological quality (EcoQ) of the Venice lagoon using the structure and biodiversity of the meiofaunal assemblages. *Ecol. Indic.* **2016**, *67*, 451–457. [CrossRef]

20. Semprucci, F.; Facca, C.; Ferrigno, F.; Balsamo, M.; Sfriso, A.; Sandulli, R. Biotic and abiotic factors affecting seasonal and spatial distribution of meiofauna and macrophytobenthos in transitional coastal waters. *Estuar. Coast. Shelf Sci.* **2019**, *219*, 328–340. [CrossRef]

21. Fabbrocini, A.; Guarino, A.; Scirocco, T.; Franchi, M.; D'Adamo, R. Integrated biomonitoring assessment of the Lesina Lagoon (Southern Adriatic Coast, Italy): Preliminary results. *Chem. Ecol.* **2005**, *21*, 479–489. [CrossRef]

22. Frontalini, F.; Semprucci, F.; Armynot du Châtelet, E.; Francescangeli, F.; Margaritelli, G.; Rettori, R.; Spagnoli, F.; Balsamo, M.; Coccioni, R. Biodiversity trends of the meiofauna and foraminifera assemblages of Lake Varano (southern Italy). *Proc. Biol. Soc. Wash.* **2014**, *127*, 7–22. [CrossRef]

23. Semprucci, F.; Balsamo, M.; Frontalini, F. The nematode assemblage of a coastal lagoon (Lake Varano, Southern Italy): Ecology and biodiversity patterns. *Sci. Mar.* **2014**, *78*, 579–588. [CrossRef]

24. Mirto, S.; La Rosa, T.; Morcciaro, G.; Costa, K.; Sara, G.; Mazzola, A. Meiofauna and benthic microbial biomass in a semi-enclosed Mediterranean marine system (Stagnone of Marsala, Italy). *Chem. Ecol.* **2004**, *20*, S387–S396. [CrossRef]

25. Manini, E.; Fiordelmondo, C.; Gambi, M.C.; Pusceddu, A.; Danovaro, R. Benthic microbial loop functioning in coastal lagoons: A comparative approach. *Oceanol. Acta* **2003**, *26*, 27–38. [CrossRef]

26. Barnes, N.; Bamber, R.; Moncrieff, C.; Sheader, M.; Ferrero, T. Meiofauna in closed coastal saline lagoons in the United Kingdom: Structure and biodiversity of nematode assemblage. *Estuar. Coast. Shelf Sci.* **2008**, *79*, 328–340. [CrossRef]

27. Abbiati, M.; Mistri, M.; Bartoli, M.; Ceccherelli, V.U.; Colangelo, M.A.; Ferrari, C.R.; Giordani, G.; Munari, C.; Nizzoli, D.; Ponti, M.; et al. Trade-off betweenconservation and exploitation of the transitional water ecosystems of the northernAdriatic Sea. *Chem. Ecol.* **2010**, *26*, 105–119. [CrossRef]

28. Magni, P.; Micheletti, S.; Casu, D.; Floris, A.; Giordani, G.; Petrov, A.; De Falco, G.; Castelli, A. Relationships between chemical characteristics of sediments and macrofaunal communities in the Cabras lagoon (western Mediterranean, Italy). *Hydrobiologia* **2005**, *550*, 109–115. [CrossRef]

29. Magni, P.; Rajagopal, S.; van der Velde, G.; Fenzi, G.; Kassenberg, J.; Vizzini, S.; Mazzola, A.; Giordani, G. Sediment features, macrozoobenthic assemblages and trophic relationships (δ^{13}C and δ^{15}N analysis) following a dystrophic event with anoxia and sulphide development in the Santa Giusta lagoon (western Sardinia, Italy). *Mar. Pollut. Bull.* **2008**, *57*, 125–136. [CrossRef] [PubMed]

30. Foti, A.; Fenzi, G.; Di Pippo, F.; Gravina, M.F.; Magni, P. Testing the saprobity hypothesis in a Mediterranean lagoon: Effects of confinement and organic enrichment on benthic communities. *Mar. Environ. Res.* **2014**, *99*, 85–94. [CrossRef] [PubMed]

31. Ferrarin, C.; Umgiesser, G. Hydrodynamic modeling of a coastal lagoon: The Cabras lagoon in Sardinia, Italy. *Ecol. Model.* **2005**, *188*, 340–357. [CrossRef]

32. Molinaroli, E.; Guerzoni, S.; De Falco, G.; Sarretta, A.; Cucco, A.; Como, S.; Simeone, S.; Perilli, A.; Magni, P. Relationships between hydrodynamic parameters and grain size in two contrasting transitional environments: The lagoons of Venice and Cabras, Italy. *Sediment. Geol.* **2009**, *219*, 196–207. [CrossRef]

33. Bartoli, M.; Longhi, D.; Nizzoli, D.; Como, S.; Magni, P.; Viaroli, P. Short term effects of hypoxia and bioturbation on solute fluxes, denitrification and buffering capacity in a shallow dystrophic pond. *J. Exp. Mar. Biol. Ecol.* **2009**, *381*, 105–113. [CrossRef]

34. Specchiulli, A.; Cilenti, L.; D'Adamo, R.; Fabbrocini, A.; Guo, W.; Huang, L.; Lugliè, A.; Padedda, B.M.; Scirocco, T.; Magni, P. Dissolved organic matter dynamics in Mediterranean lagoons: The relationship between DOC and CDOM. *Mar. Chem.* **2018**, *202*, 37–48. [CrossRef]

35. Pulina, S.; Padedda, B.M.; Satta, C.T.; Sechi, N.; Lugliè, A. Long-term phytoplankton dynamics in a Mediterranean Eutrophic lagoon (Cabras Lagoon, Italy). *Plant Biosyst.* **2011**, *146*, 259–272. [CrossRef]

36. Pulina, S.; Padedda, B.M.; Sechi, N.; Lugliè, A. The dominance of cyanobacteria in Mediterranean hypereutrophic lagoons: A case study of Cabras lagoon (Sardinia, Italy). *Sci. Mar.* **2012**, *75*, 111–120. [CrossRef]

37. Magni, P.; Micheletti, S.; Casu, D.; Floris, A.; De Falco, G.; Castelli, A. Macrofaunal community structure and distribution in a muddy coastal lagoon. *Chem. Ecol.* **2004**, *20*, S397–S407. [CrossRef]

38. Satta, C.T.; Anglès, S.; Garcés, E.; Sechi, N.; Pulina, S.; Padedda, B.M.; Stacca, D.; Lugliè, A. Dinoflagellate cyst assemblages in surface sediments from three shallow Mediterranean lagoons (Sardinia, North Western Mediterranean Sea). *Estuaries Coasts* **2014**, *37*, 646–663. [CrossRef]

39. Como, S.; Magni, P.; Van Der Velde, G.; Blok, F.S.; Van De Steeg, M.F.M. Spatial variations in δ^{13}C and δ^{15}N values of primary consumers in a coastal lagoon. *Estuar. Coast. Shelf Sci.* **2012**, *115*, 300–308. [CrossRef]

40. Cucco, A.; Sinerchia, M.; Le Francois, C.; Magni, P.; Ghezzo, M.; Umgiesser, G.; Perilli, A.; Domenici, P. Coupled empirical and numerical model of fish response to environmental changes. *Ecol. Model.* **2012**, *237–238*, 132–141. [CrossRef]

41. Como, S.; Magni, P.; Casu, D.; Floris, A.; Giordani, G.; Natale, S.; Fenzi, G.A.; Signa, G.; De Falco, G. Sediment characteristics and macrofauna distribution along a human-modified inlet in the Gulf of Oristano (Sardinia, Italy). *Mar. Pollut. Bull.* **2007**, *54*, 733–744. [CrossRef]

42. Como, S.; Magni, P. Temporal changes of a macrobenthic assemblage in harsh lagoon sediments. *Estuar. Cost. Shelf Sci.* **2009**, *83*, 638–646. [CrossRef]

43. Sandulli, R.; De Leonardis, C.; Vanaverbeke, J. Meiobenthic communities in the shallow subtidal of three Italian Marine Protected Areas. *Ital. J. Zool.* **2010**, *77*, 186–196. [CrossRef]

44. Curini-Galletti, M.; Artois, T.; Delogu, V.; De Smet, W.H.; Fontaneto, D.; Jondelius, U.; Leasi, F.; Martinez, A.; Meyer-Wachsmuth, I.; Nilsson, K.S.; et al. Patterns of Diversity in Soft-Bodied Meiofauna: Dispersal Ability and Body Size Matter. *PLOS ONE* **2012**, *7*, e33801. [CrossRef] [PubMed]

45. Di Pippo, F.; Ellwood, N.T.W.; Gismondi, A.; Bruno, L.; Rossi, F.; Magni, P.; De Philippis, P. Characterization of biofilm-forming cyanobacteria for exopolysaccharide production and potential biotechnological applications. *J. Appl. Phycol.* **2013**, *25*, 1697–1708. [CrossRef]

46. Di Pippo, F.; Magni, P.; Congestri, R. Microphytobenthic biomass, diversity and exopolymeric substances in a shallow dystrophic coastal lagoon. *J. Mar. Microbiol.* **2018**, *2*, 6–12.

47. Magni, P.; Como, S.; Cucco, A.; De Falco, G.; Domenici, P.; Ghezzo, M.; Lefrançois, C.; Simeone, S.; Perilli, A. A Multidisciplinary and Ecosystemic Approach in the Oristano Lagoon-Gulf System (Sardinia, Italy) as a Tool in Management Plans. *Transit. Waters Bull.* **2008**, *2*, 41–62.

48. Danovaro, R.; Gambi, M.C.; Mirto, S.; Sandulli, R.; Ceccherelli, V.U. Meiofauna. *Biol. Mar. Mediterr.* **2004**, *11*, 55–97.

49. Higgins, R.P.; Thiel, H. *Introduction to the Study of Meiofauna*; Smithsonian Institution Press: Washington, DC, USA, 1988; pp. 1–488.

50. Semprucci, F.; Sbrocca, C.; Rocchi, M.; Balsamo, M. Temporal changes of the meiofaunal assemblage as a tool for the assessment of the ecological quality status. *J. Mar. Biol. Assoc. UK* **2015**, *95*, 247–254. [CrossRef]

51. Pfannkuche, O.; Thiel, H. Sample processing. In *Introduction to the Study of Meiofauna*; Higgins, R.P., Thiel, H., Eds.; Smithsonian Institute: Washington, DC, USA, 1988; pp. 134–145.

52. Raffaelli, D.G.; Mason, D.F. Pollution monitoring with meiofauna, using the ratio of nematodes to copepods. *Mar. Pollut. Bull.* **1981**, *12*, 158–163. [CrossRef]

53. Anderson, M.J. A new method for non-parametric multivariate analysis of variance. *Aust. Ecol.* **2001**, *26*, 32–46.

54. Rice, W.R. Analyzing tables of statistical tests. *Evolution* **1989**, *43*, 223–225. [CrossRef]

55. Clarke, K.R.; Warwick, R.M. *Changes in Marine Communities: An Approach to Statistical Analysis and Interpretation*; Plymouth Marine Laboratory: Plymouth, UK, 2001.

56. De Falco, G.; Magni, P.; Teräsvuori, L.; Matteucci, G. Sediment grain size and organic carbon distribution in the Cabras lagoon (Sardinia, west Mediterranean). *Chem. Ecol.* **2004**, *20*, S367–S377. [CrossRef]

57. Magni, P.; De Falco, G.; Como, S.; Casu, D.; Floris, A.; Petrov, A.N.; Castelli, A.; Perilli, A. Distribution and ecological relevance of fine sediments in organic-enriched lagoons: The case study of the Cabras lagoon (Sardinia, Italy). *Mar. Pollut. Bull.* **2008**, *56*, 549–564. [CrossRef] [PubMed]

58. Smol, K.A.; Willems, J.C.; Govaere, R.; Sandee, A.J.J. Composition, distributionand biomass of meiobenthos in the Oosterschelde estuary (SW Netherlands). *Hydrobiologia* **1994**, *282–283*, 197–217. [CrossRef]

59. Jouili, S.; Essid, N.; Semprucci, F.; Boufahja, F.; Nasri, A.; Beyrem, H. Environmental quality assessment of El Bibane lagoon (Tunisia) using taxonomical and functional diversity of meiofauna and nematodes. *J. Mar. Biol. Assoc. UK* **2017**, *97*, 1593–1603. [CrossRef]

60. McArthur, V.E.; Koutsoubas, D.; Lampadariou, N.; Dounas, C. The meiofaunal community structure of a Mediterranean lagoon (Gialova lagoon, Ionian Sea). *Helgol. Mar. Res.* **2000**, *54*, 7–17. [CrossRef]

61. Alves, A.S.; Adão, H.; Patrício, J.; Magalhães Neto, J.; Costa, M.J.; Marques, J.C. Spatial distribution of subtidal meiobenthos along estuarine gradients in two southern European estuaries (Portugal). *J. Mar. Biol. Assoc. UK* **2009**, *89*, 1529–1540. [CrossRef]

62. Watzin, M.C. Interactions among temporary and permanent meiofauna: Observations on the feeding and behaviour of selected taxa. *Biol. Bull.* **1985**, *169*, 397–416. [CrossRef]

63. Vandekerkhove, J.; Martens, K.; Rossetti, G.; Mesquita-Joanes, F.; Namiotko, T. Extreme tolerance to environmental stress of sexual and parthenogenetic resting eggs of Eucypris virens (Crustacea, Ostracoda). *Freshw. Biol.* **2013**, *58*, 237–247. [CrossRef]

64. Alvarez, F.; Ojeda, M. First record of a sea spider (Pycnogonida) from an anchialine habitat. *Lat. Am. J. Aquat. Res.* **2018**, *46*, 219–224. [CrossRef]

65. Jaume, D.; Boxshall, G.A. Global diversity of cumaceans & tanaidaceans (Crustacea: Cumacea & Tanaidacea) in freshwater. *Hydrobiologia* **2008**, *595*, 225–230.

66. Ateş, A.S.; Katağan, T.; Sezgin, M.; Acar, S. The Response of *Apseudopsis latreillii* (Milne-Edwards, 1828) (Crustacea, Tanaidacea) to Environmental Variables in the Dardanelles. *Turk. J. Fish. Aquat. Sci.* **2014**, *14*, 113–124. [CrossRef]

67. Vanaverbeke, J.; Gheskiere, T.; Steyaert, M.; Vincx, M. Nematode assemblages from subtidal sandbanks in theSouthern Bight of the North Sea: Effect of smallsedimentological differences. *J. Sea Res.* **2002**, *48*, 197–207. [CrossRef]

68. Semprucci, F.; Balsamo, M.; Appolloni, L.; Sandulli, R. Assessment of ecological quality status along the Apulian coasts (Eastern Mediterranean Sea) based on meiobenthic and nematode assemblages. *Mar. Biodivers.* **2018**, *48*, 105–115. [CrossRef]

Article

Expected Shifts in Nekton Community Following Salinity Reduction: Insights into Restoration and Management of Transitional Water Habitats

Luca Scapin [1,*], Matteo Zucchetta [1], Andrea Bonometto [2], Alessandra Feola [2], Rossella Boscolo Brusà [2], Adriano Sfriso [1] and Piero Franzoi [1]

[1]   Dipartimento di Scienze Ambientali, Informatica e Statistica (DAIS), Università Ca' Foscari Venezia, Via Torino 155, 30170 Venezia, Italy
[2]   Istituto Superiore per la Protezione e la Ricerca Ambientale (ISPRA), Loc. Brondolo, 30015 Chioggia, Italy
*   Correspondence: luca.scapin@unive.it; Tel.: +39-041-2347752

Received: 20 May 2019; Accepted: 25 June 2019; Published: 29 June 2019

Abstract: A restoration project is planned to take place in the northern Venice lagoon (northern Adriatic Sea, Italy), aiming at introducing freshwater into a confined shallow water lagoon area and recreating transitional water habitats. This work describes the shifts in the nekton (fish and decapods) community structure to be expected following the future salinity decrease in the restoration area. Nekton was sampled at a series of natural shallow water sites located along salinity gradients in the Venice lagoon. A multivariate GLM approach was followed in order to predict species biomass under the salinity and environmental conditions expected after restoration. Biomass of commercially important species, as well as species of conservation interest, is predicted to increase following salinity reduction and habitat changes. From a functional perspective, an increase in biomass of hyperbenthivores-zooplanctivores, hyperbenthivores-piscivores and detritivores is also expected. This study emphasises the efficacy of a predictive approach for both ecological restoration and ecosystem management in transitional waters. By providing scenarios of community structure, the outcomes of this work could be employed in future evaluations of restoration success in the Venice lagoon, as well as to develop management tools to forecast the effects of alterations of salinity regimes in coastal lagoons due to climate change.

Keywords: nekton; transitional waters; restoration; salinity; predictive models

---

1. Introduction

Transitional waters, including estuaries and coastal lagoons, are highly heterogeneous ecosystems, being characterised by the presence of strong gradients in water and sediment properties and composed of a diverse mosaic of morphologies and biogenic structures [1,2]. This makes them highly valuable ecosystems supporting unique biological communities. Nekton fauna (fish and swimming invertebrates) play a central role in transitional waters, mediating multiple ecological processes and including species of commercial and conservation interest [1,3–5]. The distribution of biological communities in transitional waters is driven by multiple environmental factors, among which salinity is crucial in determining the organism responses at the physiological level [3,6]. The nekton community structure, in particular, may be affected by the different species tolerances and preferences to salinity, with migratory, marine and freshwater straggler taxa being especially influenced by spatial and inter-annual variations in salinity levels within estuarine ecosystems [3,7–11].

A variety of anthropogenic pressures affect transitional water ecosystems, which may lead to habitat degradation, alterations of ecological processes and depletion of biological communities, ultimately impairing the ecosystem status and functionality [12,13]. The restoration of both abiotic and biotic components of transitional water ecosystems is recognised as a strategic approach to both enhance the ecological status of degraded transitional water ecosystems (sensu Dir. 2000/60/EC Water Framework Directive, WFD) and tackle the loss of biodiversity [14–16]. In recent years, a variety of schemes have been carried out in estuaries and coastal lagoons all over the world, aiming at re-establishing specific habitats previously degraded or lost [17,18] and enhancing the status of faunal communities [19–22]. Due to the central role of salinity in transitional water ecosystems, its management by means of hydrological and morphological improvement have been of particular interest for scientists and practitioners in these environments [16,23,24], and in many instances, the control of salinity has been proposed as a measure to achieve ecological restoration. For example, increasing and stabilising salinity levels in enclosed lagoon basins by enhancing connections with the sea would improve trophic status and conservation of vulnerable habitats such as saltmarshes and seagrass meadows [25], as well as fishery yields and faunal diversity [26]. On the other hand, recreating the salinity gradient in estuaries and lagoons subjected to previous river diversions and the unsustainable use of freshwater could significantly contribute to the restoration of key ecological components [27], as well as influence the structure and functioning of biological communities and the provision of ecosystem services [28,29].

Defining the targets of restoration, i.e., which measurable ecological goals are expected to be met once the restoration scheme is completed, is a critical step in the evaluation of the success of interventions [30]. Traditionally, targets are defined in terms of abiotic conditions of soil, water and sediments, or vegetation structure and composition that characterise habitats subjected to restoration [31]. However, in recent years, a wider approach emerged in restoration ecology, emphasising the importance of including functional attributes of re-created habitats (e.g., their trophic role for faunal communities and their ability to support overall biodiversity) among the measurable targets of restoration [32–34]. Furthermore, from a methodological perspective, restoration ecology is rapidly starting to incorporate predictive approaches for the evaluation of project performances [35]. Employing forecasting techniques would indeed contribute to provide robust and less uncertain assessments of restoration success, for instance, by predicting in advance the structure of biological communities prior to the end of habitat re-creation actions, [36] hence allowing for an adaptive management of conservation efforts [37,38].

Defining realistic targets and evaluating restoration success can be particularly difficult in transitional water ecosystems. Here, the high levels of natural stress and the hysteresis exhibited by many ecological components after restoration often mask the response of the ecosystem to management and conservation measures [16,24,39,40]. Due to their central role in ecosystem functioning, nekton fauna in transitional waters is widely employed as indicator of ecological health, being also included among the biological elements to be evaluated for the assessment of the ecological status of European transitional water bodies under the Water Framework Directive [41–44]. Estuarine and lagoon nekton communities are also considered in studies investigating the effects of restoration measures, including those involving the management of salinity regimes in hydrologically impaired water bodies [19,20,29,45–48].

In the Mediterranean basin, the intense human development occurred in the last centuries and the subsequent alterations of hydrodynamics and sediment balances determined major losses of transitional water habitats in estuaries and coastal lagoons [49–54]. The Venice lagoon (northern Adriatic Sea, Italy) in particular, experienced centuries of land claim and decreased freshwater and sediment inputs from the mainland due to historical diversion of major rivers [55,56]. During the last century, the higher hydrodynamic energy caused by jetty construction at the sea inlets and channel dredging further enhanced the morphological and hydrological alterations [57,58]. Overall, this determined the loss of extensive portions of saltmarshes and intertidal flats, and the almost complete disappearing of reedbed and oligo and mesohaline conditions that originally dominated the interface between the lagoon and the drainage basin [49,50,59,60].

In the Venice lagoon, a restoration project (LIFE Lagoon ReFresh; www.lifelagoonrefresh.eu) is planned to establish a salinity gradient in a currently euhaline shallow water area, by connecting the basin to an adjacent river course and creating a freshwater inflow up to 1000 L s^{-1}. In addition, a series of morphological interventions are planned in the area. These include the arrangement of biodegradable modular elements and the transplantation of reed (*Phragmites australis* (Cav.) Trin. ex Steud.) and submerged angiosperms (*Ruppia cirrhosa* (Petagna) Grande and *Zostera noltei* Hornem.), which are expected to slow down the dispersion of freshwater and control turbidity and nutrient peaks during river overflows [61]. The creation of a new freshwater input, scheduled for autumn 2019, is expected to restore many of the transitional attributes in the lagoon area. In particular, the project foresees the enhancement of habitat functionality for nekton fauna [61].

This paper aims at defining the structure of nekton assemblage to be expected after the restoration of the salinity gradient in an inner, euhaline area of the Venice lagoon, following the predictive approach proposed by Scapin et al. [36] for seagrass nekton communities. The work is structured into two phases: (i) an assemblage-level model was calibrated, in order to explain nekton variability with temporal, environmental and habitat factors characterising sites located along natural salinity gradients in the Venice lagoon; (ii) a set of target scenarios were defined in terms of salinity and other environmental factors that are expected at the end of the restoration process. The taxonomical and functional structure of nekton assemblage expected according to such scenarios was subsequently predicted, using the model developed in the first phase.

2. Materials and Methods

The Venice lagoon is the largest Mediterranean coastal lagoon, with a surface of approximately 550 km^2, and is characterised by a microtidal regime, experiencing a tidal range of ±0.50 m during spring tides [62]. Two main watersheds are present in the lagoon, identifying three large sub-basins (northern, central and southern) [63]. Each sub-basin exchanges water with the sea through an inlet. The lagoon is mostly composed of shallow water areas, with an average depth of 1.2 m [64], which are intersected by a network of deeper channels leading inwards from the inlets and branching inside each sub-basin [63,65]. The lagoon is also characterised by several freshwater inputs from the drainage basin, the most consistent of which are located in the northern sub-basin [66] (Figure 1).

Shallow water areas are therefore characterised by strong gradients in environmental conditions such as salinity, dissolved oxygen, turbidity, trophic status and sediment granulometry, these being driven not only by natural processes, but also by multiple anthropogenic pressures [49]. Overall, salinity levels range from polyhaline (18–30) to euhaline (>30) conditions, although values can reach 5 in areas closer to freshwater inputs, depending on the intensity of river discharges [66]. This heterogeneity contributes to creating a complex mosaic of islands, saltmarshes, mud and sand-flats, seagrass meadows and man-made structures. Saltmarshes and intertidal and subtidal flats dominate the mosaic of shallow waters in the inner lagoon areas. Seasonal beds of macroalgae often occur on flats, and sparse patches of *Zostera noltei* and *Ruppia cirrhosa* can be present along marsh edges and in marsh creeks. Reedbed is now rare, and limited to areas more directly influenced by freshwater [61]. The lagoon area that will be subjected to the restoration of freshwater inflow is a mostly shallow, inner portion of the northern sub-basin. It is characterised by high water residence times (>20 days; [63]) and euhaline conditions, with salinity levels often higher than 30 PSU, and only occasionally lowered by intense rainfall events [61,66].

In this study, nekton and environmental sampling was carried out in 39 shallow water sites in the three sub-basins. The sites were located along the major salinity gradients, either in confined saltmarsh creeks or at marsh edges exposed to shallow flats, and encompassed the range of environmental and habitat variability that characterises the inner areas of the lagoon (Figure 1).

Nekton and environmental data were gathered from 10 years of surveys in the Venice lagoon (between 2004 and 2018; [44,67–70]) and included observations performed either in spring, summer or autumn depending on the specific survey. A total of 179 observations were included in the dataset. Nekton sampling was carried out by seine netting, following the protocol described in Franco et al. [67]. The specimens were identified at the species level, and total biomass (g) per species was registered.

The data were standardised per area unit (g 100 m^{-2}), allowing comparison between samples. Together with nekton sampling, water temperature (°C), salinity (PSU), dissolved oxygen (DO, percentage of saturation) and turbidity (FNU) were measured with a multi-parameter probe, and the presence of macroalgae was recorded. A value of sediment grain size (percentage of sand in the 10 cm surface layer) was finally associated to each sampling site using data from previous studies [71–73].

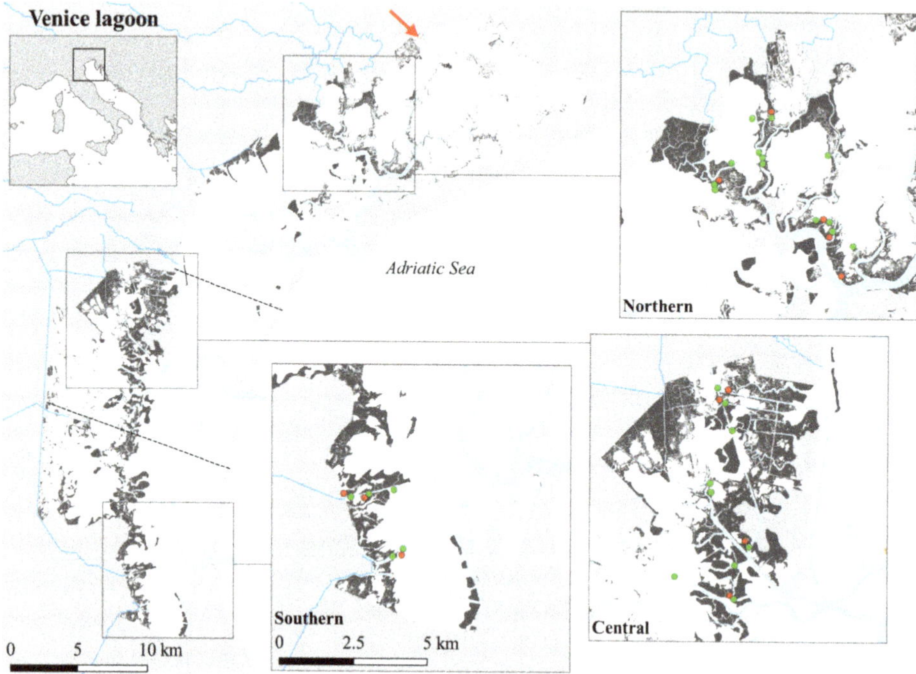

Figure 1. The Venice lagoon and location of sampling sites. The sites are located either in saltmarsh creeks (red dots) or at saltmarsh edges (green). The dashed lines indicate the approximate positions of the main watersheds. The distribution of the saltmarsh (dark grey), lagoon channels (light blue) and main freshwater courses flowing into the lagoon (blue) are also shown. The red arrow indicates the location of the planned river diversion into the lagoon planned under the LIFE Lagoon ReFresh project.

Field data were employed in a model framework to identify factors driving nekton variability and then predict assemblage characteristics under scenarios of salinity reduction, following the approach proposed for the restoration of seagrass meadows in the Venice lagoon by Scapin et al. [36]. Negative binomial generalised linear models (GLMs) were fitted to biomass of each species contributing to 98% of total assemblage biomass. A set of different model formulations was considered, taking into account different combinations of temporal, geographical, and environmental predictors. Five model categories were included, investigating the following hypotheses (Table 1): None of the predictors considered affect the response variable (null model: category m0); the response variable is affected by the temporal factor only (seasonal and inter-annual variability; category m1); the response variable responds to both temporal factors and the sub-basin (category m2); the response variable is affected by temporal factors, the sub-basin and environmental characteristics of water and bottom surface (category m3); the response variable responds to temporal factors, the sub-basin, environmental characteristics and location (the latter specified either as saltmarsh creek or edge) (category m4). The relative influence of each predictor on variability of nekton assemblage was hence evaluated by comparing the different model formulations (Table 1) in a hierarchical framework [36,11]. Following the approach of the *manyglm* function contained in the *mvabund* software package [74], the inference was carried out at the assemblage level by combining species-specific results in a global analysis. The significance of the contribution of additive variables to the simpler model was assessed by means of likelihood ratio tests with 1000 bootstrap iterations. This allowed to investigate a series of hypotheses on the different contribution of each predictor on the overall assemblage variability. Test t1 tested the hypothesis that the temporal factor would improve the null model; test t2 tested the hypothesis that sub-basin would improve the temporal model; tests belonging to category t3 tested the hypothesis that each water parameter and bottom characteristic would progressively improve the previous model; test t4 tested the hypothesis that location would improve a model already taking into account all the previous predictors (Table 1). The model formulation resulting from the outcomes of the tests (i.e., that including only relevant predictors) was then selected, and employed in the second phase of the work to predict the expected nekton assemblages.

Table 1. Generalised linear models (GLMs) formulations considered in this study, and a summary of likelihood ratio tests performed between pairs of models. For each model comparison, the predictors being tested are specified.

Model Category	Formula	Predictors Included
m0	$Y_i \sim \text{intercept} + \varepsilon_i$	None
m1	m0 + season + year	Temporal
m2	m1 + sub-basin	Temporal + geographical
m3.1	m2 + temperature	
m3.2	m2 + temperature + salinity	
m3.3	m2 + temperature + salinity + DO	Temporal + geographical +
m3.4	m2 + temperature + salinity + DO + turbidity	environmental
m3.5	m2 + temperature + salinity + DO + turbidity + grain	
m3.6	m2 + temperature + salinity + DO + turbidity + grain + algae	
m4	m3.x + location	Temporal + geographical + environmental + location (marsh creek or edge)
Test		**Testing the Effect of:**
t1	m0 vs. m1	Temporal factor
t2	m1 vs. m2	Geographical factor, when only temporal factor was considered before
t3.1	m2 vs. m3.1	
t3.2	m3.1 vs. m3.2	Each environmental factor, when
t3.3	m3.2 vs. m3.3	temporal, geographical and all
t3.4	m3.3 vs. m3.4	the significant previous
t3.5	m3.4 vs. m3.5	environmental factors were
t3.6	m3.5 vs. m3.6	considered before
t4	m3.x vs. m4	Location factor, when temporal, geographical and all the significant environmental factors were considered before

The predictive capability of the selected model was assessed by means of Spearman's r coefficients, calculated between observed and predicted biomass values by means of a k-fold cross-validation (k = 5, [75]). The coefficients were computed for the whole assemblage (i.e., using species, sampling sites and dates as replicates) and for each species separately.

Predictor coefficients estimated by the selected model were used to quantify the magnitude and the sign (either positive or negative) of species response to physico-chemical variables and location (coefficients for seasons, years and sub-basins not shown). In order to provide an assessment of the response at the assemblage level, the absolute (i.e., without sign) values of estimated coefficients for each predictor were averaged among species, using the mean species biomass as weight.

Following the approach of Scapin et al. [36], the values of predictor variables were defined in a set of scenarios, as expected at the end of the restoration scheme. This allowed the prediction of the assemblages expected after establishing the freshwater input and creating the salinity gradient in the project area.

Three restoration scenarios were included, accounting for three different target levels of salinity reduction: 10, 18 and 25 PSU, as expected at the end of the project (see also Figure A1 in Appendix A, [61]). A fourth scenario was taken into account, describing the current salinity conditions in the project area (ca. 30 PSU; [61]). All the scenarios were based on the northern lagoon sub-basin (i.e., where restoration will take place) and in all the scenarios, levels of environmental parameters were set as the inter-annual average values measured during each season in northern sub-basin sites employed in the calibration phase. The year was set as the average sampling year considered in calibration. Since saltmarsh creeks are not expected to be a prominent habitat feature in the restored area in the near future [61], in all the scenario explorations, predictions were carried out for the exposed marsh edge location. The scenarios were defined for spring, summer and autumn separately (Table 2).

Species predicted with low accuracy (i.e., species associated with a Spearman's cross-validation coefficient lower than 0.25) were excluded from the predicted assemblage. In order to provide a functional assessment of the expected assemblage, species were subsequently grouped into functional guilds. Both ecological guilds, summarising the different use of transitional water habitats by species, and feeding guilds, grouping species with similar food targets and foraging strategies, were considered. Predicted species densities were then aggregated accordingly. Eight ecological guilds (estuarine use functional guilds, EUFGs) were taken into account, adapting the classification approach of Potter et al. [76] to the Venice lagoon: Solely estuarine resident species (ESs), found exclusively within the lagoon ecosystem; estuarine resident species (ES), spending all or most of their life cycle within the lagoon, but represented also by marine populations; marine estuarine-dependents (ME-D), marine-spawning species that require transitional water habitats during the juvenile stages; marine estuarine-opportunists (ME-O), marine-spawning species that regularly enter the lagoon but can alternatively use other coastal habitats; marine stragglers (MS), stenohaline marine species irregularly found within the lagoon in areas most influenced by the sea; freshwater stragglers (FS), stenohaline freshwater species that occur occasionally in the lagoon near river mouths; catadromous (C), species entering the lagoon during periodic migrations from marine spawning areas to freshwaters; anadromous (A), species entering the lagoon during periodic migrations from freshwater spawning areas to the sea. Seven feeding guilds (feeding mode functional guilds, FMFGs) were taken into account, adapting the classification of Franco et al. [77] to the Venice lagoon: Detritivores (D), feeding on small organisms and organic matter associated to the substratum; microbenthivores (Bmi), feeding on benthic fauna smaller than 1 cm; macrobenthivores (Bma), feeding on benthic fauna larger than 1 cm; hyperbenthivores-zooplanctivores (HZ), feeding either on small (<1 cm) hyperbenthos or zooplankton; hyperbenthivores-piscivores (HP), feeding either on large (>1 cm) hyperbenthos or fish; omnivores (OV), which ingest both plant and animal material; planktivores (PL), feeding predominantly on zooplankton and occasionally on phytoplankton. Since a species could be allocated to multiple feeding guilds, the contribution of each species to a guild was expressed as a proportion (0 to 1), by identifying the importance of different food resources within the diet based on literature [78] and available data for the Venice lagoon [41,79].

Table 2. The scenarios defined for predicting the expected nekton assemblage.

Scenario	Salinity (PSU)	Season	Year	Sub-Basin	Other Environmental Variables	Location
Low	10	Either spring, summer or autumn	Average sampling year considered in calibration	Northern	Average values measured in the northern sub-basin during each season	Saltmarsh edge
Mid	18					
High	25					
Current	30					

3. Results

3.1. Environmental Variability

Physico-chemical parameters measured at natural sites employed for model calibration varied markedly according to the sampling season, sub-basin and location along the environmental gradients. Temperature, for instance, was characterised by seasonal fluctuations, showing average values of 19.5 (standard deviation 4.3), 25.1 (sd 2.8) and 18 °C (sd 5.6) in spring, summer and autumn respectively. Similarly, average dissolved oxygen values were 94.3 (sd 19.2), 73.4 (sd 19.1) and 85.6% (sd 15.2) during the three seasons. Other environmental parameters, in turn, varied spatially. Salinity ranged from 4 to 40, showing lower values at sites located in the inner portions of the lagoon, closer to freshwater inputs. Sediment grain size (range 1 to 72% sand) and water turbidity (range 1 to 69 FNU) varied both along the salinity gradient (with lower grain size and higher turbidity in inner lagoon areas) and between the two habitat locations considered (with lower grain size and lower turbidity inside saltmarsh creeks). Finally, the sampling sites in the northern, central and southern sub-basins showed some differences in terms of salinity, with average values of 24.2 (sd 8.8), 28.7 (sd 6.2) and 24.3 (sd 8.6) respectively, and sediment grain size, with average values of 16.9 (sd 11.9), 17.2 (sd 10.5) and 27.7% sand (sd 21.3) respectively.

3.2. Model Calibration

Twelve species accounted for 98% of total assemblage biomass at the sampling sites selected for the model calibration phase, including 10 fish and two decapod taxa (Table 3).

The likelihood ratio tests among pairs of model formulations highlighted that the temporal factor significantly explained assemblage biomass (*p*-value = 0.002; test t1), and that the geographical factor (sub-basin) significantly explained assemblage biomass (*p*-value = 0.002) when added to a model already including temporal factor (t2). Among physico-chemical predictors, temperature significantly (*p*-value = 0.041) improved a model already including both temporal and geographical factors (t3.1), and salinity significantly (*p*-value = 0.012) improved a model already including temporal and geographical factors and temperature (t3.2). Dissolved oxygen (t3.3), turbidity (t3.4), sediment grain size (t3.5) and presence of macroalgae (t3.6) did not improve the model significantly (*p*-values > 0.05). Finally, the inclusion of the location factor to the model already taking into account temporal and geographical factors, temperature and salinity significantly (*p*-value = 0.045) explained assemblage biomass (t4). As a result, the GLM formulation including temporal factor, geographical factor, temperature, salinity and location was selected as the best model explaining the variability of nekton biomass.

Most of the species included in the analysis were predicted accurately (cross-validated average Spearman's coefficients between 0.28 and 0.60), including big-scale sand smelt *Atherina boyeri* Risso, 1810, Mediterranean banded killifish *Aphanius fasciatus* Valenciennes, 1821, rockpool prawn *Palaemon elegans* Rathke, 1837, brown shrimp *Crangon crangon* Linnaeus, 1758, black-spotted goby *Pomatoschistus canestrinii* Ninni, 1883, marbled goby *P. marmoratus* Risso, 1810, Adriatic dwarf goby *Knipowitschia panizzae* Verga, 1841, thinlip grey mullet *Chelon ramada* Risso, 1827 and european flounder *Platichthys flesus* Linnaeus, 1758 (Table 3). Conversely, golden grey mullet *Chelon auratus* Risso, 1810, leaping mullet *C. saliens* Risso, 1810 and gilthead seabream *Sparus aurata* Linnaeus, 1758 were predicted with lower accuracy (cross-validated average coefficients <0.25), and therefore they were excluded from the prediction of the expected assemblage. Overall, the whole assemblage was predicted with good accuracy (cross-validated average coefficient = 0.46; Table 3).

Table 3. Average and standard error values of Spearman's coefficients calculated by cross-validation (k = 5) on species accounting for 98% of total biomass and on the whole assemblage. For each species the estuarine use functional guild (EUFG), the feeding mode functional guilds (FMFG) and the average biomass are reported. Guilds are abbreviated as follows. ES: estuarine resident species; ESs: solely estuarine resident species; MED: marine estuarine-dependent species; Bmi: microbenthivores; DV: detritivores; Bma: macrobenthivores; HZ: hyperbenthivores-zooplanctivores; HP: hyperbenthivores-piscivores; OV: omnivores.

Group	Species	EUFG	FMFG	Average Biomass (g 100 m^{-2})	Spearman's Coefficient	
					Average	Standard Error
Decapods	*Palaemon elegans*	ES	Bmi, OV	0.93	0.60	0.004
	Crangon crangon	MED	Bmi	3.63	0.60	0.22
Fish	*Atherina boyeri*	ES	HZ	25.59	0.49	0.11
	Pomatoschistus canestrinii	ESs	Bmi, HZ	1.38	0.48	0.09
	Chelon ramada	MED	HZ, DV	22.43	0.42	0.15
	Platichthys flesus	MED	Bmi, Bma, HP	0.65	0.39	0.04
	Pomatoschistus marmoratus	ES	Bmi, HZ	0.46	0.35	0.08
	Knipowitschia panizzae	ESs	Bmi, HZ	2.13	0.31	0.03
	Aphanius fasciatus	ESs	Bmi, OV	9.55	0.28	0.10
	Chelon saliens	MED	HZ, DV	2.04	0.18	0.08
	Sparus aurata	MED	Bmi, Bma, HZ	0.53	0.16	0.003
	Chelon auratus	MED	HZ, DV	4.68	0.14	0.22
	Whole assemblage				0.46	0.05

Among the significant environmental factors, both site location within saltmarsh creeks and salinity showed a strong average influence on the whole assemblage biomass according to the selected model. By comparison, temperature had a weaker average effect (Figure 2). Nekton response to environmental and location factors varied markedly at species level (Figure 3). Overall, biomass of most of the species showed a clear response to the variables investigated, with only *A. boyeri* and *K. panizzae* exhibiting weak relationships with water temperature, salinity and site location compared to the other species. The marginal effect of temperature (i.e., with seasonal factor already taken into account) was positive for biomass of most of the species, while both salinity and location showed variable effects. In particular, salinity was negatively associated to biomass of *C. ramada*, *C. saliens*, *P. canestrinii* and *P. elegans*, and positively to biomass of *A. fasciatus* and *P. marmoratus*. The effect of sites located within saltmarsh creeks, as opposed to exposed saltmarsh edges, was positive for most of the species, with only *C. crangon*, *P. flesus* and *P. marmoratus* showing a negative relationship with creeks.

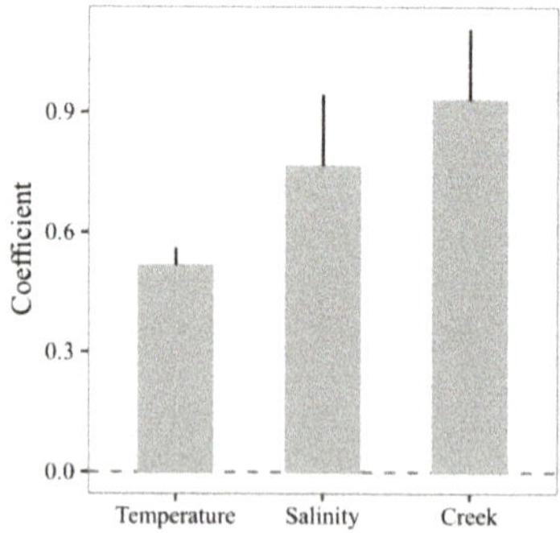

Figure 2. The average magnitude of the effect (and standard error) of relevant environmental and location factors estimated by the selected model. The values are calculated as the mean of the absolute (i.e., without sign) coefficients estimated for each species, weighted by average species biomass.

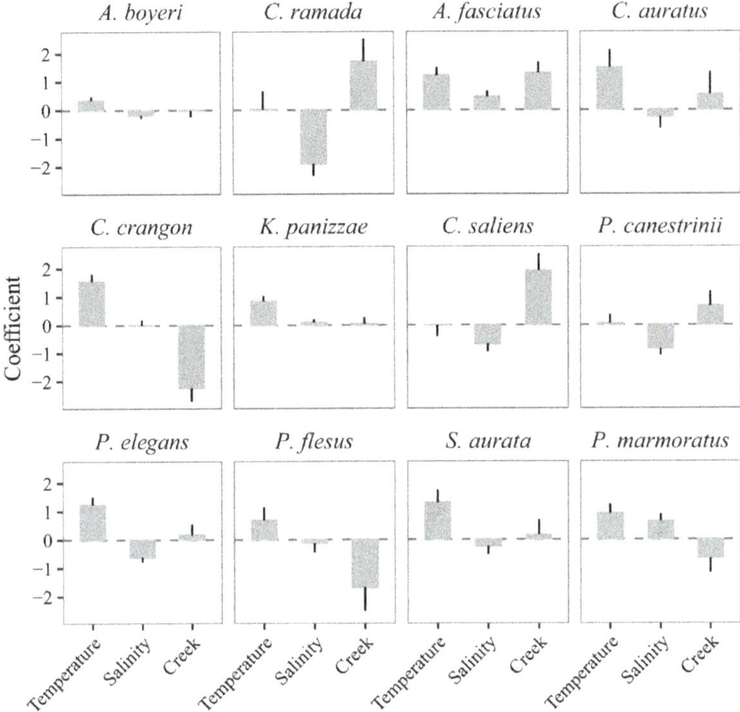

Figure 3. The estimated coefficients (and standard errors) of relevant environmental and location factors for species accounting for 98% of assemblage biomass.

3.3. Predicting the Expected Assemblages

Model predictions applied to the selected scenarios of salinity reduction showed that the whole nekton assemblage (calculated as the cumulative biomass of species included in the analysis) would markedly respond to salinity variations, showing a progressive increase in biomass from the current to the low-salinity scenario. The predicted increase in biomass would occur similarly in all the seasons investigated, although this appears to be stronger and associated to a lower uncertainty in spring (Figure 4). Despite the overall trend observed in whole assemblage, some major differences could be found in the response at the species level (Figure 5). Only minor variations of species response among seasons were detected (see Figure A2 in Appendix B), hence average values were shown (Figure 5). Most of the species investigated showed an increase in biomass with decreasing salinity, including *A. boyeri*, *C. ramada*, *P. canestrinii*, *P. elegans* and *P. flesus*. Among them, *P. canestrinii* and *P. elegans* were associated to the most marked increase in biomass. In turn, *A. fasciatus*, *K. panizzae* and *P. marmoratus* exhibited an opposite pattern, with higher biomass densities under the current conditions and progressively lower values under reduced salinity scenarios. *C. crangon* did not show any relevant response to salinity.

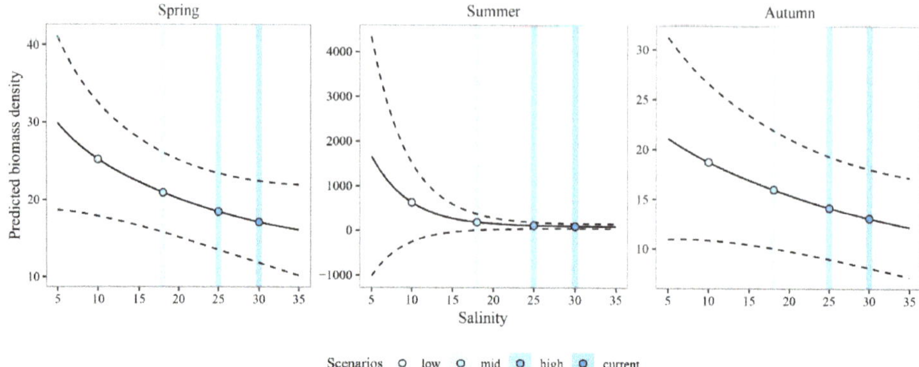

Figure 4. Whole nekton assemblage biomass (g 100 m^{-2}) predicted along the salinity gradient for each season investigated. The current salinity and the three scenarios of salinity reduction are highlighted. The dashed lines indicate the standard error of the prediction.

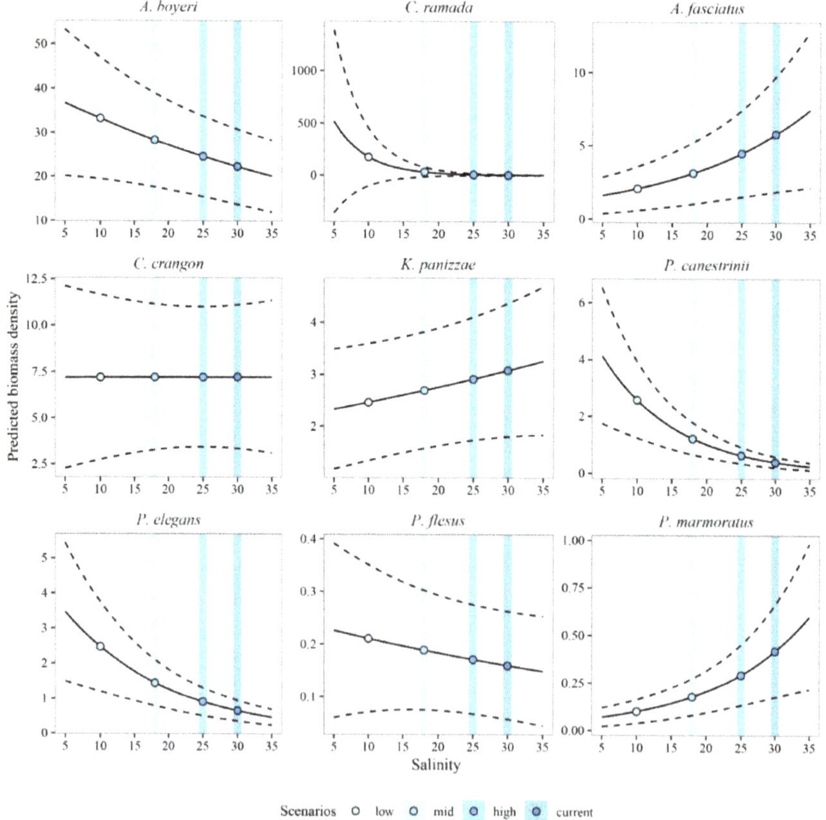

Figure 5. Species biomass (g 100 m^{-2}) predicted along the salinity gradient. The seasonal values were averaged. The current salinity and the three scenarios of salinity reduction are highlighted. The dashed lines indicate the standard error of the prediction.

Aggregating species biomasses in ecological and trophic guilds highlighted the potential changes in assemblage functional attributes after the restoration (Figure 6). On the whole, biomass of estuarine resident and marine estuarine-dependent species are expected to increase accordingly with the decrease in salinity, although the latter guild showed a marked increase only under the low-salinity scenario. In contrast, solely estuarine residents exhibited a pattern of decrease from the current conditions to mid and low-salinity scenarios. In terms of trophic structure of the nekton assemblage, salinity reduction would drive an increase in biomass of detritivorous, macrobenthivorous, hyperbenthivorous-zooplanctivorous and hyperbentivorous-piscivorous species (Figure 6). Guild response did not exhibit major variations among seasons (see Figure A3 in Appendix B), hence the average values were shown (Figure 6).

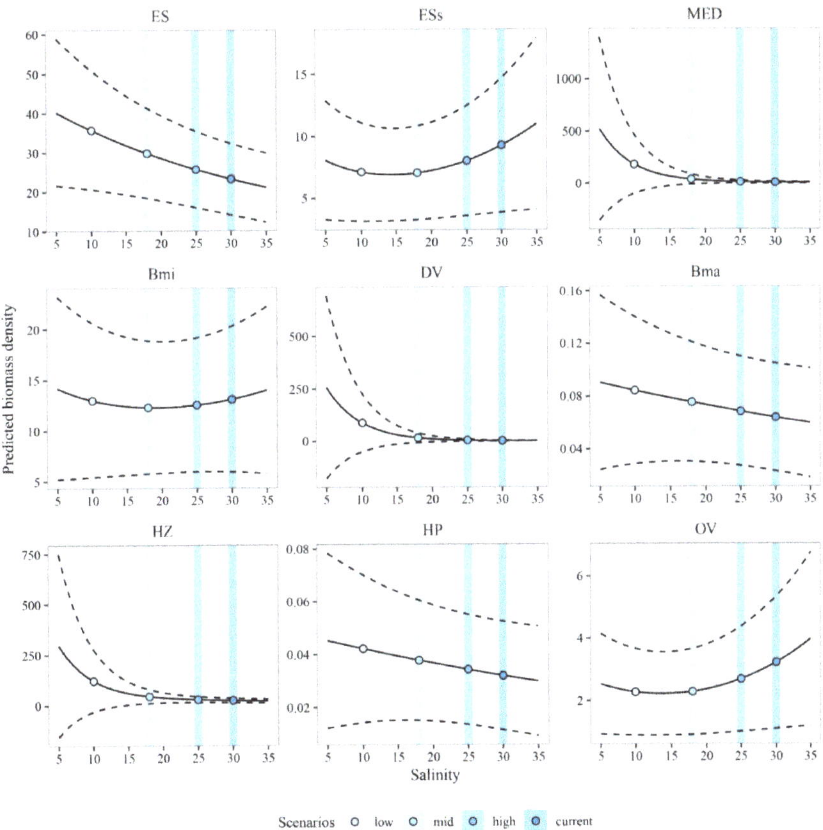

Figure 6. Biomass (g 100 m^{-2}) predicted along the salinity gradient for ecological and trophic guilds. The seasonal values were averaged. The current salinity and the three scenarios of salinity reduction are highlighted. The dashed lines indicate the standard error of the prediction. Guilds are abbreviated as follows. ES: estuarine resident species; ESs: solely estuarine resident species; MED: marine estuarine-dependent species; Bmi: microbenthivores; DV: detritivores; Bma: macrobenthivores; HZ: hyperbenthivores-zooplanctivores; HP: hyperbenthivores-piscivores; OV: omnivores.

4. Discussion

4.1. Predicting the Expected Nekton Changes after Salinity Reduction

This study developed a model to predict the expected shifts in nekton assemblage resulting from the restoration of a freshwater input in an inner portion of the Venice lagoon, taking into account temporal, geographic, environmental and habitat variability found along natural salinity gradients in the basin. The model calibration phase revealed a significant influence of the sampling season and year on overall assemblage and species biomass, confirming the high variability of temporal dynamics in nekton communities in the Venice lagoon [41,67,70,80]. Reproduction, recruitment and migrations are among the major factors causing seasonal variations in community structure in transitional water ecosystems [69,81]. In addition, seasonal patterns may show marked differences among years due to interannual changes in weather, abiotic conditions and trophic status [80,82,83]. Among the physico-chemical and habitat factors taken into account in the present analysis, only three variables significantly affected biomass of the nekton assemblage as a whole, namely water temperature, water salinity and habitat location within the saltmarsh. This confirms previous observations made on fish communities in the Venice lagoon [67,84,85] and in other Mediterranean transitional water ecosystems [8,82]. As many studies previously pointed out, nekton response to physico-chemical and habitat variables would probably be more complex, and show a significant influence of additional factors (e.g., type of substratum, trophic conditions, tidal cycles and weather) at the species level [80,83,86–89]. Exploring all the factors driving nekton changes at the species level was, however, beyond the scope of this work.

The primary effect on nekton fauna of reducing salinity in the lagoon area investigated would be an overall increase in total assemblage biomass compared to the current conditions, mostly due to the increase in biomass of five species. Among them, the solely resident *Pomatoschistus canestrinii* and the marine estuarine-dependent *Platichthys flesus* and *Chelon ramada* are known to prefer low-salinity, saltmarsh-dominated areas in the inner portions of the Venice lagoon, which they use as recruitment and feeding grounds and, for *P. canestrinii*, as elective reproductive habitats [80,87,90]. Also the estuarine residents *Atherina boyeri* and *Palaemon elegans*, widespread species that can tolerate a wide range of salinity levels in transitional water habitats [7,69], contributed to the expected increase in nekton biomass. The positive response to salinity decrease shown by these species could be related to broader changes in the restoration area, including an increase in trophic status. In the Venice lagoon, trophic levels indeed show marked changes along the salinity gradients, with higher nutrient and chlorophyll concentrations recorded in the more confined portions of the basin [91,92]. Common estuarine species, integrating complex ecological processes such as trophic interactions in transitional water ecosystems, could therefore play a key role as indicators of such changes in the inner areas of the lagoon [41,93–95].

This work emphasises the importance of adopting a functional perspective when employing predictive models in restoration ecology. Species traits, or a functional classification like the guild approach, could provide a better understanding of the processes involved in community colonisation of restored habitats and in overall ecosystem functioning [35,96]. The expected nekton response at the functional level further supports the hypothesis that the restoration of a salinity gradient in the Venice lagoon may cause relevant changes to the food web of the area, with biomass of hyperbenthivores-zooplanctivores, hyperbenthivores-piscivores and detritivores benefiting from the decrease in salinity. The newly-created freshwater input could have a major influence on distribution and community structure of hyperbenthic and planktonic prey due to the salinity decrease, as pointed out by Das et al. [28]. More generally, the nutrient loads from the drainage basin, although buffered by the restoration of vegetation, could enhance the productivity of the system, hence the overall availability of food items for nekton species in the area [8,67,97]. Future studies following the restoration of the salinity gradient should therefore take into account the potential changes in trophic status, which could explain a significant portion of variability in functional structure of nekton assemblages in the area.

The model's accuracy proved to be good in predicting the assemblage as a whole, as cross-validation highlighted. While predictive performances varied markedly among species, most of them were predicted with high accuracy, and could then be included in the set of species to be predicted in the second phase of the work. In particular, this approach identifies a pool of species that best explains the expected assemblage shifts under changing environmental conditions. Overall, the ecological and trophic structure of such assemblage largely corresponds to the typical nekton community composition found in the inner areas of the Venice lagoon and other Mediterranean transitional water ecosystems [11,69,90,98].

As Scapin et al. [36] highlighted, the methodology presented in this work may serve to develop concise tools guiding the management and conservation of biodiversity and ecosystem functions. In a perspective of reinstatement of a salinity gradient in the Venice lagoon, for instance, predicting the expected outcomes in terms of nekton structure and comparing them to the observed conditions after the restoration will provide a mean to quantitatively estimate the degree of restoration success. Applying the same protocol in a suitable time frame will also allow the tracking of restoration trajectories towards the endpoint conditions [36]. Similarly, the present approach may also be employed in the framework of WFD, to evaluate the efficacy of management responses aiming to enhance the ecological status of hydrologically impaired transitional water bodies.

4.2. Implications for Management and Conservation

The present analysis highlighted that restoring the salinity gradient could have major implications for management and conservation of the inner Venice lagoon. Most of the tools developed for the evaluation of ecological status of fish in transitional waters under the WFD, including the one employed in Italian transitional waters, incorporate metrics calculated on assemblage structure and composition, often based on functional guilds [79,99–102]. Given the expected increase in whole assemblage biomass and the shift in taxonomical and trophic structure, significant changes in the ecological status of fish may therefore occur after the restoration. As a result, nekton surveys in the area following the reinstatement of the freshwater input will be of critical importance, in order to measure the actual effect of creating a salinity gradient on the ecological status.

While restoration could benefit the overall ecological status of fish fauna, the consequences on species conservation may vary. The salinity decrease would contribute to creating suitable environmental conditions, particularly in areas characterised by lower salinity levels, for *Pomatoschistus canestrinii*, an endemic species of northern Adriatic coastal lagoons listed in the Annex II of the Habitats Directive. One of the major threats for this species is indeed the loss of estuarine and brackish habitats [80,103]. Other species protected at the European level, namely *Aphanius fasciatus* and *Knipowitschia panizzae*, could in turn be limited to areas that are less influenced by the salinity gradient, or gather in habitats different from saltmarsh edges, such as creeks. Small saltmarsh channels are often preferred by small estuarine resident species, by providing better shelter and higher food availability [80,90]. Overall, this emphasises the importance of preserving habitat diversity, even in a scenario of beneficial restoration, as already pointed out for the Venice lagoon by Cavraro et al. [90] and Scapin et al. [36,70].

The results of this work also indicate a positive influence of restoring a salinity gradient on the recruitment of *Chelon ramada* and *Platichthys flesus*, which exploit transitional water habitats during juvenile stages and represent important resources for local fisheries [41,87,104–109]. Transitional water ecosystems play a central role in supporting the populations of many marine migrant species of commercial value. Multiple environmental and geographical factors contribute to sustain this function, including water turbidity and sediment characteristics, trophic status and prey availability as well as the availability of structured habitats such as saltmarshes, and the degree of habitat connectivity [3,110–112]. In this light, salinity may not be the only factor regulating the entrance and growth of marine migrant juveniles in the Venice lagoon shallow waters and, while the restoration could contribute to support juveniles of *C. ramada* and *P. flesus*, the overall nursery role of the area should be evaluated by future ad hoc studies.

Transitional water habitats rely on the delicate balance between sediment accretion and erosion, as well as on quality and availability of freshwater inputs. The consequences of climate change could therefore pose additional threats to their survival and quality in the next decades. For instance, the expected sea level rise would result in the loss of major intertidal and subtidal shallow water areas in northern Adriatic coastal lagoons due to erosion and submergence in this century [49,58,113]. Moreover, relevant alterations in temperature and precipitation patterns all over the world are already affecting river flows and overall freshwater availability [114]. In the Venice lagoon, climate change could lead to longer periods of high salinity levels and more marine-like conditions due to more severe drought episodes, resulting in further loss of transitional water features in the inner portions of the basin [115]. Restoration schemes aiming at enhancing the riverine influence on the inner lagoon, such as the Lagoon ReFresh project, may therefore represent a viable way to mitigate the negative effects of climate change on transitional water ecosystems, in particular by preserving suitable environmental conditions for nekton assemblages.

Author Contributions: Conceptualization, L.S. and P.F.; methodology, L.S., M.Z. and A.F.; validation, L.S., M.Z. and P.F.; formal analysis, L.S. and M.Z.; investigation, L.S., P.F.; resources, P.F.; writing—original draft preparation, L.S.; writing—review and editing, All the authors; visualization, L.S. and M.Z.; supervision, P.F. and A.S.; project administration, A.B. and R.B.B.; funding acquisition, A.B., R.B.B. and A.S.

Funding: This research was funded by European Union's LIFE+ financial instrument (grant LIFE16 NAT/IT/000663—LIFE LAGOON REFRESH, which contributes to the environmental recovery of a Natura 2000 site, SIC IT3250031—Northern Venice Lagoon). Field data used in this study were collected under various projects funded by the Italian Ministry of Education, Universities and Research (PRIN grant 2009W2395), by Corila (Consorzio Ricerche Lagunari) and by European Union's LIFE+ financial instrument (grant LIFE12 NAT/IT/000331).

Conflicts of Interest: The authors declare no conflicts of interest.

Appendix A

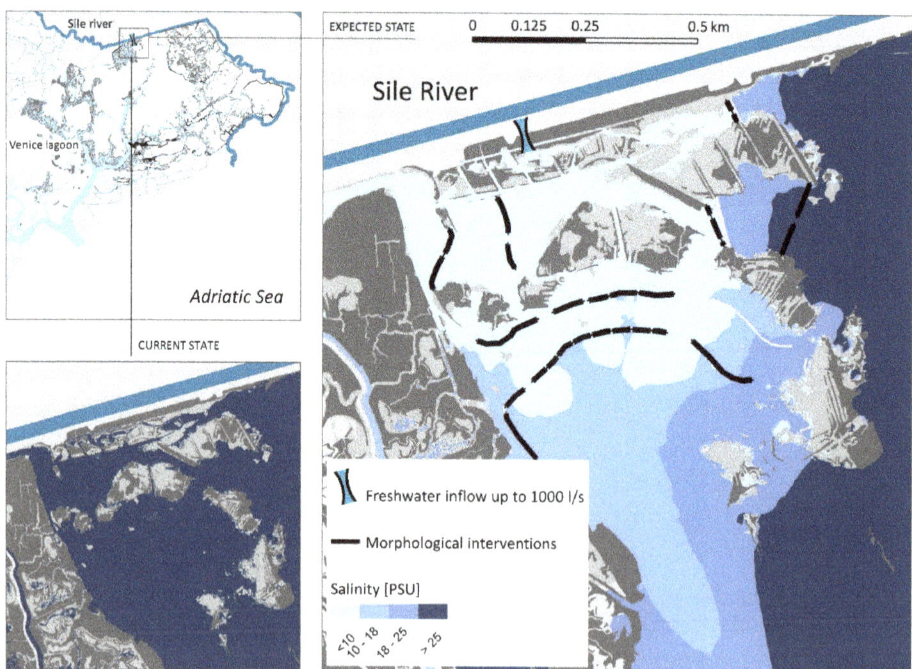

Figure A1. Project area of LIFE Lagoon ReFresh in the northern Venice Lagoon. Current state of salinity condition (small panel, bottom) and expected state after the creation of a freshwater input from the Sile river (low tide conditions). The freshwater inflow and morphological interventions are also represented. The diffusion of freshwater during low tide condition was obtained from numerical simulation [61].

Appendix B

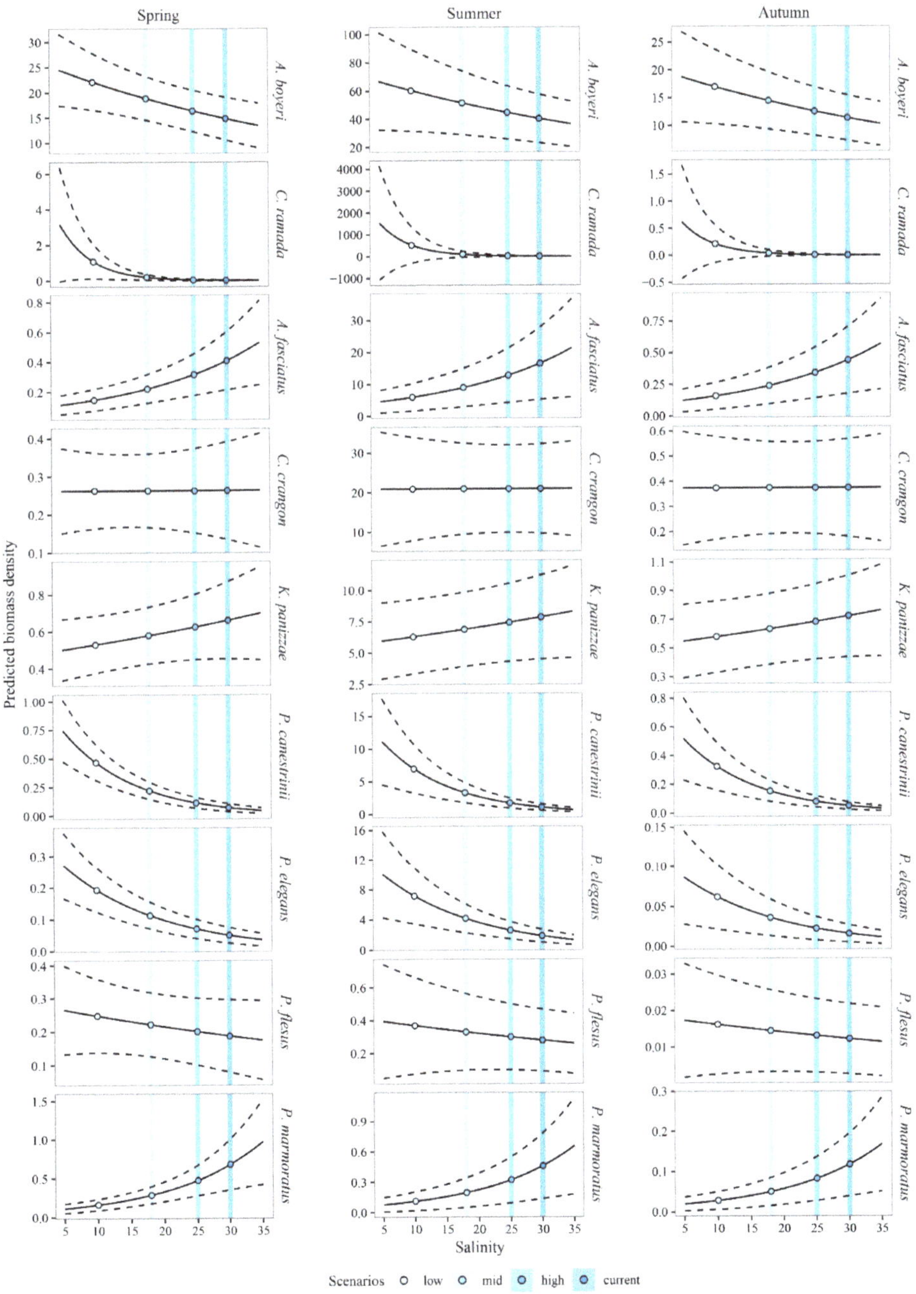

Figure A2. Species biomass (g 100 m^{-2}) predicted along the salinity gradient for each season investigated. The current salinity and the three scenarios of salinity reduction are highlighted. Dashed lines indicate the standard error of the prediction.

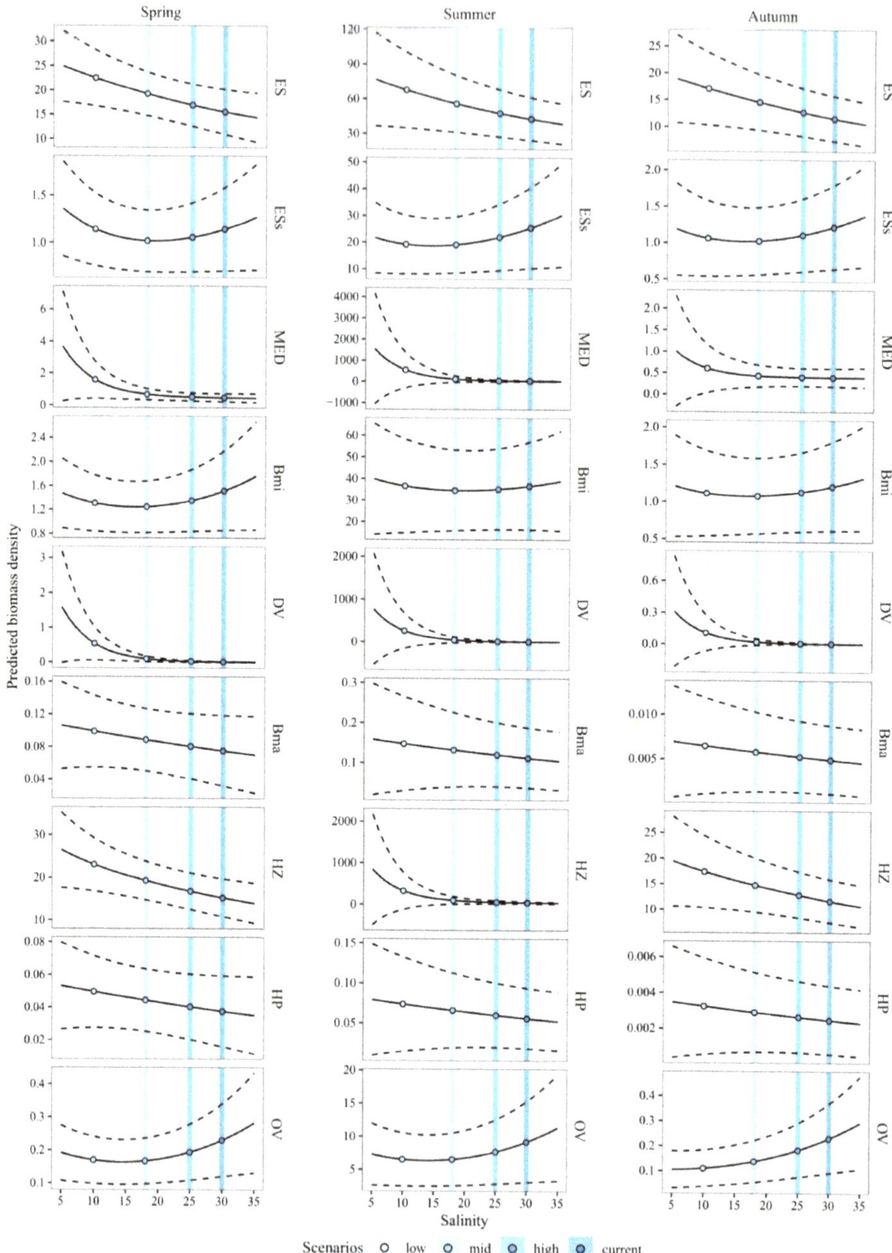

Figure A3. Guild biomass (g 100 m^{-2}) predicted along the salinity gradient for each season investigated. The current salinity and the three scenarios of salinity reduction are highlighted. Dashed lines indicate the standard error of the prediction. Guilds are abbreviated as follows. ES: estuarine resident species; ESs: solely estuarine resident species; MED: marine estuarine-dependent species; Bmi: microbenthivores; DV: detritivores; Bma: macrobenthivores; HZ: hyperbenthivores-zooplanctivores; HP: hyperbenthivores-piscivores; OV: omnivores.

References

1. McLusky, D.S.; Elliott, M. *The Estuarine Ecosystem: Ecology, Threats and Management*, 3rd ed.; Oxford University Press: Oxford, UK, 2004.

2. Sheaves, M. Consequences of ecological connectivity: The coastal ecosystem mosaic. *Mar. Ecol. Prog. Ser.* **2009**, *391*, 107–115. [CrossRef]

3. Elliott, M.; Hemingway, K.L. *Fishes in Estuaries*; Blackwell Science: Oxford, UK, 2002.

4. Kneib, R.T. Salt marsh ecoscapes and production transfers by estuarine nekton in the Southeastern United States. In *Concepts and Controversies in Tidal Marsh Ecology*; Weinstein, M.P., Kreeger, D.A., Eds.; Springer: Dordrecht, The Netherlands, 2000.

5. Pérez-Ruzafa, A.; Marcos, C.; Pérez-Ruzafa, I.M. Mediterranean coastal lagoons in an ecosystem and aquatic resources management context. *Phys. Chem. Earth* **2011**, *36*, 160–166. [CrossRef]

6. Smyth, K.; Elliott, M. Effects of changing salinity on the ecology of the marine environment. In *Stressors in the Marine Environment*; Solan, M., Witheley, N., Eds.; Oxford University Press: Oxford, UK, 2016; pp. 161–174.

7. Rodríguez-Climent, S.; Caiola, N.; Ibáñez, C. Salinity as the main factor structuring small-bodied fish assemblages in hydrologically altered Mediterranean coastal lagoons. *Sci. Mar.* **2013**, *77*, 37–45.

8. Prado, P.; Vergara, C.; Caiola, N.; Ibáñez, C. Influence of salinity regime on the food-web structure and feeding ecology of fish species from Mediterranean coastal lagoons. *Estuar. Coast. Shelf Sci.* **2014**, *139*, 1–10. [CrossRef]

9. Martino, E.J.; Able, K.W. Fish assemblages across the marine to low salinity transition zone of a temperate estuary. *Estuar. Coast. Shelf Sci.* **2003**, *56*, 969–987. [CrossRef]

10. Harrison, T.D.; Whitfield, A.K. Temperature and salinity as primary determinants influencing the biogeography of fishes in South African estuaries. *Estuar. Coast. Shelf Sci.* **2006**, *66*, 335–345. [CrossRef]

11. Franco, A.; Franzoi, P.; Torricelli, P. Structure and functioning of Mediterranean lagoon fish assemblages: A key for the identification of water body types. *Estuar. Coast. Shelf Sci.* **2008**, *79*, 549–558. [CrossRef]

12. Marchand, J.; Codling, I.; Drake, P.; Elliott, M.; Pihl, L.; Rebelo, J. Environmental quality of estuaries. In *Fishes in Estuaries*; Elliott, M., Hemingway, K.L., Eds.; Blackwell Publishing Ltd.: Hoboken, NJ, USA, 2002; pp. 322–409.

13. Vasconcelos, R.P.; Reis-Santos, P.; Fonseca, V.; Maia, A.; Ruano, M.; França, S.; Vinagre, C.; Costa, M.J.; Cabral, H. Assessing anthropogenic pressures on estuarine fish nurseries along the Portuguese coast: A multi-metric index and conceptual approach. *Sci. Total Environ.* **2007**, *374*, 199–215. [CrossRef]

14. Perring, M.P.; Standish, R.J.; Price, J.N.; Craig, M.D.; Erickson, T.E.; Ruthrof, K.X.; Whiteley, A.S.; Valentine, L.E.; Hobbs, R.J. Advances in restoration ecology: Rising to the challenges of the coming decades. *Ecosphere* **2015**, *6*, 1–25. [CrossRef]

15. Teichert, N.; Borja, A.; Chust, G.; Uriarte, A.; Lepage, M. Restoring fish ecological quality in estuaries: Implication of interactive and cumulative effects among anthropogenic stressors. *Sci. Total Environ.* **2016**, *542*, 383–393. [CrossRef]

16. Borja, A.; Dauer, D.M.; Elliott, M.; Simenstad, C.A. Medium-and long-term recovery of estuarine and coastal ecosystems: Patterns, rates and restoration effectiveness. *Estuar. Coasts* **2010**, *33*, 1249–1260. [CrossRef]

17. Van Katwijk, M.M.; Thorhaug, A.; Marbà, N.; Orth, R.J.; Duarte, C.M.; Kendrick, G.A.; Althuizen, I.H.J.; Balestri, E.; Bernard, G.; Cambridge, M.L.; et al. Global analysis of seagrass restoration: The importance of large-scale planting. *J. Appl. Ecol.* **2015**, *53*, 567–578. [CrossRef]

18. Society for Ecological Restoration. *Tidal Marsh Restoration. A Synthesis of Science and Management*; Roman, C.T., Burdick, D.M., Eds.; Island Press: Washington, DC, USA, 2012; ISBN 9781597265768.

19. Farrugia, T.J.; Espinoza, M.; Lowe, C.G. The fish community of a newly restored southern California estuary: Ecological perspective 3 years after restoration. *Environ. Biol. Fishes* **2014**, *97*, 1129–1147. [CrossRef]

20. Castro, N.; Félix, P.M.; Neto, J.; Cabral, H.; Marques, J.C.; Costa, M.J.; Costa, J. Fish communities' response to implementation of restoring measures in a highly artificialized estuary. *Ecol. Indic.* **2016**, *67*, 743–752. [CrossRef]

21. Reese, M.M.; Stunz, G.W.; Bushon, A.M. Recruitment of estuarine-dependent nekton through a new tidal inlet: The opening of Packery Channel in Corpus Christi, TX, USA. *Estuar. Coasts* **2008**, *31*, 1143–1157. [CrossRef]

22. Facca, C.; Bonometto, A.; Boscolo, R.; Buosi, A.; Parravicini, M.; Siega, A.; Volpe, V.; Sfriso, A. Coastal lagoon recovery by seagrass restoration. A new strategic approach to meet HD & WFD objectives. In Proceedings of the 9th European Conference on Ecological Restoration, Oulu, Finland, 3–8 August 2014.

23. Simenstad, C.; Reed, D.; Ford, M. When is restoration not? *Ecol. Eng.* **2006**, *26*, 27–39. [CrossRef]

24. Elliott, M.; Burdon, D.; Hemingway, K.L.; Apitz, S.E. Estuarine, coastal and marine ecosystem restoration: Confusing management and science—A revision of concepts. *Estuar. Coast. Shelf Sci.* **2007**, *74*, 349–366. [CrossRef]

25. Saunders, K.M.; Mcminn, A.; Roberts, D.; Hodgson, D.A.; Heijnis, H. Recent human-induced salinity changes in Ramsar-listed Orielton Lagoon, south-east Tasmania, Australia: A new approach for coastal lagoon conservation and management. *Aquat. Conserv. Mar. Freshw. Ecosyst.* **2007**, *17*, 51–70. [CrossRef]

26. Mohapatra, A.; Mohanty, R.K.; Mohanty, S.K.; Bhatta, K.S.; Das, N.R. Fisheries enhancement and biodiversity assessment of fish, prawn and mud crab in Chilika lagoon through hydrological intervention. *Wetl. Ecol. Manag.* **2007**, *15*, 229–251. [CrossRef]

27. Twilley, R.R.; Rivera-Monroy, V.H.; Chen, R.; Botero, L. Adapting an ecological mangrove model to simulate trajectories in restoration ecology. *Mar. Pollut. Bull.* **1998**, *37*, 404–419. [CrossRef]

28. Das, A.; Justic, D.; Inoue, M.; Hoda, A.; Huang, H.; Park, D. Impacts of Mississippi River diversions on salinity gradients in a deltaic Louisiana estuary: Ecological and management implications. *Estuar. Coast. Shelf Sci.* **2012**, *111*, 17–26. [CrossRef]

29. Koutrakis, E.T.; Sylaios, G.; Kamidis, N.; Markou, D.; Sapounidis, A. Fish fauna recovery in a newly re-flooded Mediterranean coastal lagoon. *Estuar. Coast. Shelf Sci.* **2009**, *83*, 505–515. [CrossRef]

30. Zedler, J.B.; Callaway, J.C. Evaluating the progress of engineered tidal wetlands. *Ecol. Eng.* **2000**, *15*, 211–225. [CrossRef]

31. McAlpine, C.; Catterall, C.P.; Mac Nally, R.; Lindenmayer, D.; Reid, J.L.; Holl, K.D.; Bennett, A.F.; Runting, R.K.; Wilson, K.; Hobbs, R.J.; et al. Integrating plant- and animal- based perspectives for more effective restoration of biodiversity. *Front. Ecol. Environ.* **2016**, *14*, 37–45. [CrossRef]

32. Bourque, A.S.; Fourqurean, J.W. Effects of common seagrass restoration methods on ecosystem structure in subtropical seagrass meadows. *Mar. Environ. Res.* **2014**, *97*, 67–78. [CrossRef] [PubMed]

33. Dolbeth, M.; Cardoso, P.; Grilo, T.; Raffaelli, D.; Pardal, M.A. Drivers of estuarine benthic species distribution patterns following a restoration of a seagrass bed: A functional trait analyses. *Mar. Pollut. Bull.* **2013**, *72*, 47–54. [CrossRef] [PubMed]

34. Fraser, L.H.; Harrower, W.L.; Garris, H.W.; Davidson, S.; Hebert, P.D.N.; Howie, R.; Moody, A.; Polster, D.; Schmitz, O.J.; Sinclair, A.R.E.; et al. A call for applying trophic structure in ecological restoration. *Restor. Ecol.* **2015**, *23*, 503–507. [CrossRef]

35. Brudvig, L.A. Toward prediction in the restoration of biodiversity. *J. Appl. Ecol.* **2017**, *54*, 1013–1017. [CrossRef]

36. Scapin, L.; Zucchetta, M.; Sfriso, A.; Franzoi, P. Predicting the response of nekton assemblages to seagrass transplantations in the Venice lagoon: An approach to assess ecological restoration. *Aquat. Conserv. Mar. Freshw. Ecosyst. Mar. Freshw. Ecosyst.* **2019**, *26*, 849–864. [CrossRef]

37. Brudvig, L.A.; Barak, R.S.; Bauer, J.T.; Caughlin, T.T.; Laughlin, D.C.; Larios, L.; Matthews, J.W.; Stuble, K.L.; Turley, N.E.; Zirbel, C.R. Interpreting variation to advance predictive restoration science. *J. Appl. Ecol.* **2017**, *54*, 1018–1027. [CrossRef]

38. Ebberts, B.D.; Zelinsky, B.D.; Karnezis, J.P.; Studebaker, C.A.; Lopez-Johnston, S.; Creason, A.M.; Krasnow, L.; Johnson, G.E.; Thom, R.M. Estuary ecosystem restoration: Implementing and institutionalizing adaptive management. *Restor. Ecol.* **2017**, *26*, 360–369. [CrossRef]

39. Duarte, C.M.; Borja, A.; Carstensen, J.; Elliott, M.; Krause-Jensen, D.; Marbà, N. Paradigms in the recovery of estuarine and coastal ecosystems. *Estuar. Coasts* **2015**, *38*, 1202–1212. [CrossRef]

40. Elliott, M.; Quintino, V. The Estuarine Quality Paradox, Environmental Homeostasis and the difficulty of detecting anthropogenic stress in naturally stressed areas. *Mar. Pollut. Bull.* **2007**, *54*, 640–645. [CrossRef] [PubMed]

41. Zucchetta, M.; Scapin, L.; Cavraro, F.; Pranovi, F.; Franco, A.; Franzoi, P. Can the effects of anthropogenic pressures and environmental variability on nekton fauna be detected in fishery data? Insights from the monitoring of the artisanal fishery within the Venice lagoon. *Estuar. Coasts* **2016**, *39*, 1164–1182. [CrossRef]

42. Pasquaud, S.; Courrat, A.; Fonseca, V.F.; Gamito, R.; Gonçalves, C.I.; Lobry, J.; Lepage, M.; Costa, M.J.; Cabral, H. Strength and time lag of relationships between human pressures and fish-based metrics used to assess ecological quality of estuarine systems. *Estuar. Coast. Shelf Sci.* **2013**, *134*, 119–127. [CrossRef]

43. Fonseca, V.F.; Vasconcelos, R.P.; Gamito, R.; Pasquaud, S.; Gonçalves, C.I.; Costa, J.L.; Costa, M.J.; Cabral, H. Fish community-based measures of estuarine ecological quality and pressure–impact relationships. *Estuar. Coast. Shelf Sci.* **2013**, *134*, 128–137. [CrossRef]

44. Franco, A.; Torricelli, P.; Franzoi, P. A habitat-specific fish-based approach to assess the ecological status of Mediterranean coastal lagoons. *Mar. Pollut. Bull.* **2009**, *58*, 1704–1717. [CrossRef] [PubMed]

45. Milbrandt, E.C.; Bartleson, R.D.; Coen, L.D.; Rybak, O.; Thompson, M.A.; DeAngelo, J.A.; Stevens, P.W. Local and regional effects of reopening a tidal inlet on estuarine water quality, seagrass habitat, and fish assemblages. *Cont. Shelf Res.* **2012**, *41*, 1–16. [CrossRef]

46. Able, K.W.; Grothues, T.M.; Hagan, S.M.; Kimball, M.E.; Nemerson, D.M.; Taghon, G.L. Long-term response of fishes and other fauna to restoration of former salt hay farms: Multiple measures of restoration success. *Rev. Fish Biol. Fish.* **2008**, *18*, 65–97. [CrossRef]

47. Scapin, L.; Zucchetta, M.; Facca, C.; Sfriso, A.; Franzoi, P. Using fish assemblage to identify success criteria for seagrass habitat restoration. *Web Ecol.* **2016**, *16*, 33–36. [CrossRef]

48. Boys, C.A.; Williams, R.J. Succession of fish and crustacean assemblages following reinstatement of tidal flow in a temperate coastal wetland. *Ecol. Eng.* **2012**, *49*, 221–232. [CrossRef]

49. Carniello, L.; Defina, A.; D'Alpaos, L. Morphological evolution of the Venice lagoon: Evidence from the past and trend for the future. *J. Geophys. Res. Earth Surf.* **2009**, *114*, 1–10. [CrossRef]

50. Sarretta, A.; Pillon, S.; Molinaroli, E.; Guerzoni, S.; Fontolan, G. Sediment budget in the Lagoon of Venice, Italy. *Cont. Shelf Res.* **2010**, *30*, 934–949. [CrossRef]

51. Ibanez, C.; Curco, A.; Day, J.W.; Prat, N. Structure and productivity of microtidal Mediterranean coastal marshes. In *Concepts and Controversies in Tidal Marsh Ecology*; Weinstein, M.P., Kreeger, D.A., Eds.; Kluwer Academic Publishers: Dordrecht, The Netherlands, 2000; pp. 107–136.

52. Bondesan, M.; Castiglioni, G.B.; Elmi, C.; Gabbbianelli, G.; Marocco, R.; Pirazzoli, P.A.; Tomasin, A. Coastal areas at risk from storm surges and sea-level rise in northeastern Italy. *J. Coast. Res.* **1995**, *11*, 1354–1379.

53. Cencini, C. Physical processes and human activities in the evolution of the Po delta, Italy. *J. Coast. Res.* **1998**, *14*, 774–793.

54. Fontolan, G.; Pillon, S.; Bezzi, A.; Villalta, R.; Lipizer, M.; Triches, A.; D'Aietti, A. Human impact and the historical transformation of saltmarshes in the Marano and Grado Lagoon, northern Adriatic Sea. *Estuar. Coast. Shelf Sci.* **2012**, *113*, 41–56. [CrossRef]

55. Marani, M.; Lanzoni, S.; Silvestri, S.; Rinaldo, A. Tidal landforms, patterns of halophytic vegetation and the fate of the lagoon of Venice. *J. Mar. Syst.* **2004**, *51*, 191–210. [CrossRef]

56. Tambroni, N.; Seminara, G. Are inlets responsible for the morphological degradation of Venice Lagoon? *J. Geophys. Res.* **2006**, *111*, F03013. [CrossRef]

57. Ferrarin, C.; Ghezzo, M.; Umgiesser, G.; Tagliapietra, D.; Camatti, E.; Zaggia, L.; Sarretta, A. Assessing hydrological effects of human interventions on coastal systems: Numerical applications to the Venice Lagoon. *Hydrol. Earth Syst. Sci.* **2013**, *17*, 1733–1748. [CrossRef]

58. Defina, A.; Carniello, L.; Fagherazzi, S.; D'Alpaos, L. Self-organization of shallow basins in tidal flats and salt marshes. *J. Geophys. Res. Earth Surf.* **2007**, *112*, 1–11. [CrossRef]

59. Molinaroli, E.; Guerzoni, S.; Sarretta, A.; Masiol, M.; Pistolato, M. Thirty-year changes (1970 to 2000) in bathymetry and sediment texture recorded in the Lagoon of Venice sub-basins, Italy. *Mar. Geol.* **2009**, *258*, 115–125. [CrossRef]

60. D'Alpaos, L. *Fatti e Misfatti di Idraulica Lagunare. La Laguna di Venezia dalla Diversione dei Fiumi alle Nuove Opere delle Bocche di Porto*; Istituto Veneto di Scienze Lettere ed Arti.: Venice, Italy, 2010.

61. Feola, A.; Bonometto, A.; Ponis, E.; Cacciatore, F.; Oselladore, F.; Matticchio, B.; Canesso, D.; Sponga, S.; Volpe, V.; Lizier, M.; et al. LIFE LAGOON REFRESH. Ecological restoration in Venice Lagoon (Italy): Concrete actions supported by numerical modeling and stakeholder involvement. In Proceedings of the Citizen Observatories for natural hazards and Water Management—2nd International Conference, Venice, Italy, 27–30 November 2018; Available online: http://www.lifelagoonrefresh.eu/file/pubblicazioni/COWM_2018_Extendedabstract.pdf (accessed on 26 April 2019).

62. Umgiesser, G.; Melaku canu, D.; Cucco, A.; Solidoro, C. A finite element model for the Venice Lagoon. Development, set up, calibration and validation. *J. Mar. Syst.* **2004**, *51*, 123–145. [CrossRef]

63. Solidoro, C.; Melaku Canu, D.; Cucco, A.; Umgiesser, G. A partition of the Venice Lagoon based on physical properties and analysis of general circulation. *J. Mar. Syst.* **2004**, *51*, 147–160. [CrossRef]

64. Molinaroli, E.; Guerzoni, S.; Sarretta, A.; Cucco, A.; Umgiesser, G. Links between hydrology and sedimentology in the Lagoon of Venice, Italy. *J. Mar. Syst.* **2007**, *68*, 303–317. [CrossRef]

65. Ghezzo, M.; Guerzoni, S.; Cucco, A.; Umgiesser, G. Changes in Venice Lagoon dynamics due to construction of mobile barriers. *Coast. Eng.* **2010**, *57*, 694–708. [CrossRef]

66. Ghezzo, M.; Sarretta, A.; Sigovini, M.; Guerzoni, S.; Tagliapietra, D.; Umgiesser, G. Modeling the inter-annual variability of salinity in the lagoon of Venice in relation to the water framework directive typologies. *Ocean Coast. Manag.* **2011**, *54*, 706–719. [CrossRef]

67. Franco, A.; Franzoi, P.; Malavasi, S.; Riccato, F.; Torricelli, P.; Mainardi, D. Use of shallow water habitats by fish assemblages in a Mediterranean coastal lagoon. *Estuar. Coast. Shelf Sci.* **2006**, *66*, 67–83. [CrossRef]

68. Malavasi, S.; Franco, A.; Fiorin, R.; Franzoi, P.; Torricelli, P.; Mainardi, D. The shallow water gobiid assemblage of the Venice Lagoon: Abundance, seasonal variation and habitat partitioning. *J. Fish Biol.* **2005**, *67*, 146–165. [CrossRef]

69. Franzoi, P.; Franco, A.; Torricelli, P. Fish assemblage diversity and dynamics in the Venice lagoon. *Rendiconti Lincei* **2010**, *21*, 269–281. [CrossRef]

70. Scapin, L.; Zucchetta, M.; Sfriso, A.; Franzoi, P. Local habitat and seascape structure influence seagrass fish assemblages in the Venice lagoon: The value of conservation at multiple spatial scales. *Estuar. Coasts* **2018**, *41*, 2410–2425. [CrossRef]

71. ARPAV. *Piano di Monitoraggio dei Corpi Idrici della Laguna di Venezia Finalizzato alla Definizione dello Stato Ecologico, ai Sensi della Direttiva 2000/60/CE*; Relazione Finale; Agenzia Regionale per la Prevenzione e Protezione Ambientale del Veneto: Venezia, Italy, 2012.

72. Magistrato alle Acque di Venezia (ora Provveditorato Interregionale alle OO. PP. del Veneto—Trentino Alto Adige—Friuli Venezia Giulia)—Selc Studio B.12.3/III. *La funzionalità dell'ambiente lagunare attraverso rilievi delle risorse alieutiche, dell'avifauna e dell'ittiofauna. Erodibilità del fondale e fattori di disturbo: Rilievi dell'erodibilità del fondale*; Rapporto Intermedio; 2005.

73. Magistrato alle Acque di Venezia (ora Provveditorato Interregionale alle OO. PP. del Veneto—Trentino Alto Adige—Friuli Venezia Giulia)—Thetis. *Programma Generale delle Attività di Approfondimento del Quadro Conoscitivo di Riferimento per Gli Interventi Ambientali. 2° Stralcio Triennale (2003–2006) "Progetto ICSEL"*; Attività A; Prodotto dal Concessionario, Consorzio Venezia Nuova: Venice, Italy, 2005.

74. Wang, Y.; Naumann, U.; Wright, S.T.; Warton, D.I. Mvabund—An R package for model-based analysis of multivariate abundance data. *Methods Ecol. Evol.* **2012**, *3*, 471–474. [CrossRef]

75. Hastie, T.; Tibshirani, R.; Friedman, J. *The Elements of Statistical Learning*, 2nd ed.; Springer: New York, NY, USA, 2009; ISBN 978-0-387-84857-0.

76. Potter, I.C.; Tweedley, J.R.; Elliott, M.; Whitfield, A.K. The ways in which fish use estuaries: A refinement and expansion of the guild approach. *Fish Fish.* **2013**, *16*, 230–239. [CrossRef]

77. Franco, A.; Elliott, M.; Franzoi, P.; Torricelli, P. Life strategies of fishes in European estuaries: The functional guild approach. *Mar. Ecol. Progr. Ser.* **2008**, *354*, 219–228. [CrossRef]

78. Froese, R.; Pauly, D. FishBase. Available online: http://www.fishbase.org (accessed on 29 June 2019).

79. Catalano, B.; Penna, M.; Riccato, F.; Fiorin, R.; Franceschini, G.; Antonini, C.; Zucchetta, M.; Cicero, A.M.; Franzoi, P. *Manuale per la classificazione dell'Elemento di Qualità Biologica "Fauna Ittica" nelle lagune costiere italiane. Applicazione dell'indice nazionale HFBI (Habitat Fish Bio-Indicator) ai sensi del D.Lgs 152/2006*; ISPRA: Roma, Italy, 2017; ISBN 9788844808716.

80. Franco, A.; Franzoi, P.; Malavasi, S.; Zucchetta, M.; Torricelli, P. Population and habitat status of two endemic sand gobies in lagoon marshes—Implications for conservation. *Estuar. Coast. Shelf Sci.* **2012**, *114*, 31–40. [CrossRef]

81. Rountree, R.A.; Able, K.W. Spatial and temporal habitat use patterns for salt marsh nekton: Implications for ecological functions. *Aquat. Ecol.* **2007**, *41*, 25–45. [CrossRef]

82. Poizat, G.; Rosecchi, E.; Chauvelon, P.; Contournet, P.; Crivelli, A.J. Long-term fish and macro-crustacean community variation in a Mediterranean lagoon. *Estuar. Coast. Shelf Sci.* **2004**, *59*, 615–624. [CrossRef]

83. Milardi, M.; Gavioli, A.; Lanzoni, M.; Fano, E.A.; Castaldelli, G. Meteorological factors influence marine and resident fish movements in a brackish lagoon. *Aquat. Ecol.* **2019**, *53*, 251–263. [CrossRef]

84. Franco, A.; Riccato, F.; Torricelli, P.; Franzoi, P. Fish assemblage response to environmental pressures in the Venice lagoon. *Trans. Waters Bull.* **2009**, *3*, 29–44.

85. Franco, A.; Malavasi, S.; Zucchetta, M.; Torricelli, P.; Franzoi, P. Environmental influences on fish assemblage in the Venice Lagoon, Italy. *Chem. Ecol.* **2006**, *22*, S105–S118. [CrossRef]

86. Cavraro, F.; Varin, C.; Malavasi, S. Lunar-induced reproductive patterns in transitional habitats: Insights from a Mediterranean killifish inhabiting northern Adriatic Saltmarshes. *Estuar. Coast. Shelf Sci.* **2014**, *139*, 60–66. [CrossRef]

87. Zucchetta, M.; Franco, A.; Torricelli, P.; Franzoi, P. Habitat distribution model for European flounder juveniles in the Venice lagoon. *J. Sea Res.* **2010**, *64*, 133–144. [CrossRef]

88. Verdiell-Cubedo, D.; Oliva-Paterna, F.J.; Ruiz-Navarro, A.; Torralva, M. Assessing the nursery role for marine fish species in a hypersaline coastal lagoon (Mar Menor, Mediterranean Sea). *Mar. Biol. Res.* **2013**, *9*, 739–748. [CrossRef]

89. Tournois, J.; Darnaude, A.M.; Ferraton, F.; Aliaume, C.; Mercier, L.; McKenzie, D.J. Lagoon nurseries make a major contribution to adult populations of a highly prized coastal fish. *Limnol. Oceanogr.* **2017**, *62*, 1219–1233. [CrossRef]

90. Cavraro, F.; Zucchetta, M.; Malavasi, S.; Franzoi, P. Small creeks in a big lagoon: The importance of marginal habitats for fish populations. *Ecol. Eng.* **2017**, *99*, 228–237. [CrossRef]

91. Sfriso, A.; Facca, C.; Ceoldo, S.; Marcomini, A. Recording the occurrence of trophic level changes in the lagoon of Venice over the '90s. *Environ. Int.* **2005**, *31*, 993–1001. [PubMed]

92. Sfriso, A.; Facca, C. Distribution and production of macrophytes and phytoplankton in the lagoon of Venice: Comparison of actual and past situation. *Hydrobiologia* **2007**, *577*, 71–85. [CrossRef]

93. Whitfield, A.K.; Elliott, M. Fishes as indicators of environmental and ecological changes within estuaries: A review of progress and some suggestions. *J. Fish Biol.* **2002**, *61*, 229–250. [CrossRef]

94. Cabral, H.; Fonseca, V.F.; Gamito, R.; Gonçalves, C.I.; Costa, J.L.; Erzini, K.; Gonçalves, J.; Martins, J.; Leite, L.; Andrade, J.P.; et al. Ecological quality assessment of transitional waters based on fish assemblages in Portuguese estuaries: The Estuarine Fish Assessment Index (EFAI). *Ecol. Indic.* **2012**, *19*, 144–153. [CrossRef]

95. Cavraro, F.; Bettoso, N.; Zucchetta, M.; D'Aietti, A.; Faresi, L.; Franzoi, P. Body condition in fish as a tool to detect the effects of anthropogenic pressures in transitional waters. *Aquat. Ecol.* **2019**, *53*, 21–35. [CrossRef]

96. Lavorel, S.; Garnier, E. Predicting changes in community composition and ecosystem functioning from plant traits: Revisting the Holy Grail. *Funct. Ecol.* **2002**, *16*, 545–556. [CrossRef]

97. Franco, T.P.; Neves, L.M.; Araújo, F.G. Better with more or less salt? The association of fish assemblages in coastal lagoons with different salinity ranges. *Hydrobiologia* **2019**, *828*, 83–100. [CrossRef]

98. Manzo, C.; Fabbrocini, A.; Roselli, L.; D'Adamo, R. Characterization of the fish assemblage in a Mediterranean coastal lagoon: Lesina Lagoon (central Adriatic Sea). *Reg. Stud. Mar. Sci.* **2016**, *8*, 192–200. [CrossRef]

99. Pérez-Domínguez, R.; Maci, S.; Courrat, A.; Lepage, M.; Borja, A.; Uriarte, A.; Neto, J.M.; Cabral, H.; St. Raykov, V.; Franco, A.; et al. Current developments on fish-based indices to assess ecological-quality status of estuaries and lagoons. *Ecol. Indic.* **2012**, *23*, 34–45.

100. Alvarez, M.C.; Franco, A.; Pérez-Domínguez, R.; Elliott, M. Sensitivity analysis to explore responsiveness and dynamic range of multi-metric fish-based indices for assessing the ecological status of estuaries and lagoons. *Hydrobiologia* **2013**, *704*, 347–362. [CrossRef]

101. Reyjol, Y.; Argillier, C.; Bonne, W.; Borja, A.; Buijse, A.D.; Cardoso, A.C.; Daufresne, M.; Kernan, M.; Ferreira, M.T.; Poikane, S.; et al. Assessing the ecological status in the context of the European Water Framework Directive: Where do we go now? *Sci. Total Environ.* **2014**, *497–498*, 332–344. [CrossRef] [PubMed]

102. Lepage, M.; Harrison, T.; Breine, J.; Cabral, H.; Coates, S.; Galván, C.; García, P.; Jager, Z.; Kelly, F.; Mosch, E.C.; et al. An approach to intercalibrate ecological classification tools using fish in transitional water of the North East Atlantic. *Ecol. Indic.* **2016**, *67*, 318–327. [CrossRef]

103. Franco, A.; Fiorin, R.; Franzoi, P.; Torricelli, P. Threatened fishes of the world: Pomatoschistus canestrinii Ninni, 1883 (Gobiidae). *Environ. Biol. Fishes* **2005**, *72*, 32. [CrossRef]

104. Franco, A.; Fiorin, R.; Zucchetta, M.; Torricelli, P.; Franzoi, P. Flounder growth and production as indicators of the nursery value of marsh habitats in a Mediterranean lagoon. *J. Sea Res.* **2010**, *64*, 457–464. [CrossRef]

105. Pranovi, F.; Caccin, A.; Franzoi, P.; Malavasi, S.; Zucchetta, M.; Torricelli, P. Vulnerability of artisanal fisheries to climate change in the Venice Lagoon. *J. Fish Biol.* **2013**, *83*, 847–864.

106. Pérez-Ruzafa, A.; Mompeán, M.C.; Marcos, C. Hydrographic, geomorphologic and fish assemblage relationships in coastal lagoons. *Hydrobiologia* **2007**, *577*, 107–125. [CrossRef]

107. Pérez-Ruzafa, A.; Marcos, C. Fisheries in coastal lagoons: An assumed but poorly researched aspect of the ecology and functioning of coastal lagoons. *Estuar. Coast. Shelf Sci.* **2012**, *110*, 15–31. [CrossRef]

108. *Management of Coastal Lagoon Fisheries*; Kapetsky, J.M., Lasserre, G., Eds.; FAO Studies and Reviews, GFCM 61; FAO: Rome, Italy, 1984.

109. Ardizzone, G.D.; Cataudella, S.; Rossi, R. Management of coastal lagoon fisheries and aquaculture in Italy. *AO Fish. Tech. Paper* **1988**, *293*, 1–103.

110. Nagelkerken, I.; Sheaves, M.; Baker, R.; Connolly, R.M. The seascape nursery: A novel spatial approach to identify and manage nurseries for coastal marine fauna. *Fish Fish.* **2015**, *16*, 362–371. [CrossRef]

111. Brown, C.J.; Harborne, A.R.; Paris, C.B.; Mumby, P.J. Uniting paradigms of connectivity in marine ecology. *Ecology* **2016**, *97*, 2447–2457. [CrossRef] [PubMed]

112. Whitfield, A.K.; Pattrick, P. Habitat type and nursery function for coastal marine fish species, withemphasis on the Eastern Cape region, South Africa. *Estuar. Coast. Shelf Sci.* **2015**, *160*, 49–59. [CrossRef]

113. Day, J.W.; Rybczyk, J.; Scarton, F.; Rismondo, A.; Are, D.; Cecconi, G. Soil accretionary dynamics, sea-level rise and the survival of wetlands in Venice lagoon: A field and modelling approach. *Estuar. Coast. Shelf Sci.* **1999**, *49*, 607–628. [CrossRef]

114. IPCC. *Climate Change 2007: The Physical Science Basis. Contribution of Working Group I to the Fourth Assessment Report of the IPCC*; Cambridge University Press: Cambridge, UK, 2007.

115. Bellafiore, D.; Ghezzo, M.; Tagliapietra, D.; Umgiesser, G. Climate change and artificial barrier effects on the Venice Lagoon: Inundation dynamics of salt marshes and implications for halophytes distribution. *Ocean Coast. Manag.* **2014**, *100*, 101–115. [CrossRef]

Article

Contribution of Biological Effects to the Carbon Sources/Sinks and the Trophic Status of the Ecosystem in the Changjiang (Yangtze) River Estuary Plume in Summer as Indicated by Net Ecosystem Production Variations

Yifan Zhang, Dewang Li, Kui Wang * and Bin Xue

Key Laboratory of Marine Ecosystem and Biogeochemistry, State Oceanic Administration & Second Institute of Oceanography, Ministry of Natural Resources, P.R. China, Hangzhou 310012, China; angela06z@163.com (Y.Z.); dwli@sio.org.cn (D.L.); xuebin119@sio.org.cn (B.X.)

* Correspondence: wangkui@sio.org.cn

Received: 6 May 2019; Accepted: 13 June 2019; Published: 17 June 2019

Abstract: We conducted 24-h real-time monitoring of temperature, salinity, dissolved oxygen, and nutrients in the near-shore (M4-1), front (M4-8), and offshore (M4-13) regions of the 31° N section of the Changjiang (Yangtze) River estuary plume in summer. Carbon dioxide partial pressure changes caused by biological processes (pCO_2bio) and net ecosystem production (NEP) were calculated using a mass balance model and used to determine the relative contribution of biological processes (including the release of CO_2 from organic matter degradation by microbes and CO_2 uptake by phytoplankton) to the CO_2 flux in the Changjiang River estuary plume. Results show that seawater in the near-shore region is a source of atmospheric CO_2, and the front and offshore regions generally serve as atmospheric CO_2 sinks. In the mixed layer of the three regions, pCO_2bio has an overall positive feedback effect on the air–sea CO_2 exchange flux. The contribution of biological processes to the air–sea CO_2 exchange flux (Cont) in the three regions changes to varying extents. From west to east, the daily means (±standard deviation) of the Cont are 32% (±40%), 34% (±216%), and 9% (±13%), respectively. In the front region, the Cont reaches values as high as 360%. Under the mixed layer, the daily means of potential Conts in the near-shore, front, and offshore regions are 34% (±43%), 8% (±13%), and 19% (±24%), respectively. The daily 24-hour means of NEP show that the near-shore region is a heterotrophic system, the front and offshore regions are autotrophic systems in the mixed layer, and all three regions are heterotrophic under the mixed layer.

Keywords: biological processes; air–sea CO_2 exchange flux; net ecosystem production; potential CO_2 emissions; trophic status; Changjiang River estuary plume

1. Introduction

The Changjiang estuary plume is a typical marginal sea with a coastal continental shelf that has large spatial and temporal variations in carbon sinks/sources. In summer, the East China Sea generally acts as a carbon sink for atmospheric CO_2 (-4.6 ± 1.3 mmol m^{-2} day^{-1}) [1–3]. The influence of physical processes, such as strong winds, and the large amount of dissolved inorganic carbon produced by respiration under the mixed layer turns the region into an atmospheric carbon source [4]. The water mass compositions in the mixed layer of the Changjiang River estuary plume are determined primarily by the Changjiang Diluted Water and the Kuroshio Surface Water. However, the originally deep (50 m) subsurface water of the Kuroshio [5] will rise and form an upwelling around 123° E, where there is a trough [6]. On shorter time scales (e.g., 24-h), the complicated physical (upwelling, wind, tidal mixing, etc.) and biogeochemical (including the release of CO_2 from organic matter degradation by microbes and CO_2 uptake by phytoplankton) effects on the coastal and shelf ecosystems lead to complex transitions between carbon sinks and sources [7]. Thus, observations at high temporal resolutions are urgently needed to study the effects of biological processes on carbon sinks and sources.

The difference between gross primary production (GPP) and respiration (R) in an ecosystem is defined as the net ecosystem production (NEP) [8]. Negative NEP indicates that the ecosystem is heterotrophic, and positive NEP indicates that it is autotrophic; therefore, NEP can be used as an indicator of the trophic status, which is an important factor in the assessment of a specific ecosystem [9,10]. For example, Li [11] estimated the nutrient flux, primary production, and NEP in the Changjiang River estuary in the four seasons using the budget box model. Xu [12] used in situ sampling data and the "muddy" LOICZ (land–ocean interaction in the coastal zone) model to evaluate the tropical status of the Changjiang River estuary plume in summer and winter. NEP is also used to distinguish biogeochemical controls from other controls of carbon sinks and sources in marginal environments [13,14]. For instance, Borges established the relationship between mixed-layer NEP and air–sea CO_2 flux in order to detail the function of biogeochemical processes in European coastal seas [15]. Studies of NEP in the Changjiang River estuary plume have mostly applied the biogeochemical budget model on a seasonal scale. However, the contributions of biological processes to the impact of air–sea CO_2 flux using continuous monitoring data have rarely been reported. In addition, the quantification of potential CO_2 flux under the mixed layer using NEP remains to be studied in depth.

In this study, data from 24 hours of continuous monitoring in the Changjiang River estuary plume in summer were used to further explore these processes. The diel variations in parameters such as carbon dioxide partial pressure (pCO_2) and NEP in the near-shore, front, and offshore regions were calculated by a mass balance model to separate the controlling processes of pCO_2. We differentiated the air–sea CO_2 exchange flux associated with physical and biological processes and then quantified the contribution of biological processes to the total air–sea CO_2 exchange flux in the mixed layer. We also attempted to calculate the quantitative potential CO_2 emission under the mixed layer. Moreover, the NEP vales of the three regions were compared to assess the trophic statuses of the different ecosystems. The results demonstrate the importance of biological processes in the regulation of estuarine carbon sources and sinks, and they also show the gradients of trophic statuses that are influenced by Changjiang-diluted water in the Changjiang River estuary plume.

Our research on the carbon sinks and sources and assessment of the trophic statuses is based on a 24-h dataset. Although this maybe a shorter period than the timescale at which pCO_2 variation occurs in a carbonate system because of the buffer capacity of seawater, this study is meaningful from the perspective of the steady state over several months in summer in the Changjiang River estuary plume [16].

2. Materials and Methods

2.1. Study Area

Changjiang-diluted water has a strong influence on the Changjiang River estuary plume by virtue of a water discharge of about 944×10^9 m^{-3} year^{-1} [17] that carries a large amount of nutrients and sediments [18]. The eutrophic Changjiang-diluted water enters the upper estuary area, resulting in phytoplankton blooms that can absorb a substantial quantity of atmospheric CO_2 [19]. At the same time, fluvial carbon input [20], as well as the decomposition and regeneration of organic matter in primary production, causes the estuary to release CO_2 into the atmosphere [21]. In addition, the Changjiang River estuary plume has a regular semidiurnal tide [22], which results in periodic changes in sea surface temperature and salinity. The largest monthly water discharge at Datong Station, which is 624 km from the river mouth, occurred in July, with the second largest occurring in August [23].

2.2. Sampling Collection

Samples from M4-1 (122.13° E, 31.04° N), M4-8 (122.97° E, 31° N), and M4-13 (124.01° E, 31° N) were collected on 26–27 July, 13–14 August, and 14–15 August in 2006 during cruises on the Changjiang River estuary plume (Figure 1). No tropical cyclones, typhoons, or rainstorms occurred during the sampling period [24]. Although the sampling period spanned nearly 17 days because the three stations are regulated by regular semidiel tides, we considered the water properties of each station to be quasisynchronized within almost one month, so each station is representative of a typical summer in a specific location. This is consistent with approaches used by other studies in summer [25–27]. The water depths at each station were 6, 52, and 37 m, respectively. Samples from the surface layer (2 m), depths of 5, 10, 30, and the bottom layer (with a height of 2 m above the seabed) were collected every three hours for 24 hours. In particular, we collected 5 m when the high slack tide impacted the M4-1 station and 4 m otherwise. In this study, M4-1, M4-8, and M4-13 denote the near-shore, front, and offshore regions, respectively.

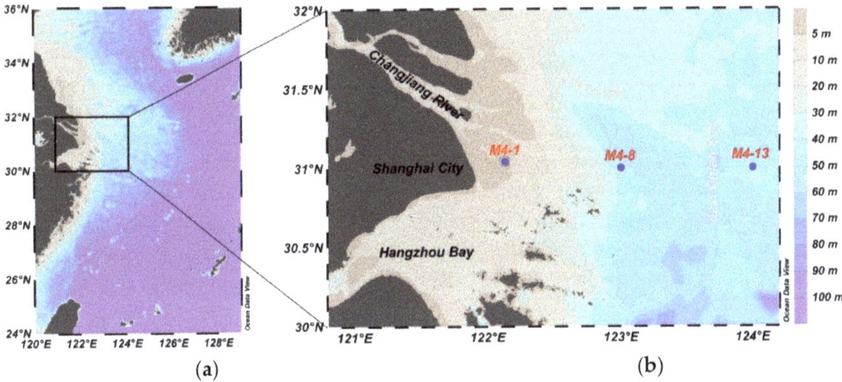

Figure 1. Map of Changjiang River estuary plume (**a**) and sampling stations (**b**): stations M4-1, M4-8, and M4-13 denote the near-shore, front, and offshore regions, respectively.

2.3. Hydrographic Measurements

Seawater samples were collected using a rosette water collector. Temperature, salinity, and depth data were measured in situ with a Hydro-bios®MWS6 conductivity-temperature-depth (CTD) recorder. The data were recorded every three hours and monitored continuously for 24 hours. pH was measured with an ORION Ross-type combination electrode, which was calibrated on the NBS scale. The measurement precision was ±0.01 pH units. Total alkalinity (TA) was calculated using the TA–salinity relationship (equation 1), which was acquired by averaging the slopes and intercepts of the TA–salinity relationships in Table 1. The partial pressures of CO_2 (pCO_2) and dissolved inorganic carbon (DIC) were calculated from pH and TA using the program CO2SYS [28].

$$\text{TA } (\mu\text{mol kg}^{-1}) = (13.38 \pm 0.15)S + (1788.40 \pm 32.63) \tag{1}$$

Table 1. Summary of correlation between total alkalinity (TA, μmol kg^{-1}) and salinity.

Sampling Date	Sampling Area	Correlation	Reference
27 August 2013	31–31.5° N, 121.5–124° E (with a salinity of 5.17–34.26)	TA = 13.3507S + 1797.39	[7]
August 2009	31° N, 122.5–125° E (Transect C)	TA = 13.2S + 1744.7	[29]
8–27 April and 2–7 May 2007	30.0–31.8° N, 122.5–123.5° E (with a salinity of 13.00–34.49)	TA = 13.5875S + 1823.1	[30]

2.4. Mass Balance Model Based on Separating pCO₂ Controlling Processes

The volumetric flow equation [31] was used to calculate the air–sea CO_2 exchange flux:

$$F_{CO2} = k \times K_0 \times (pCO_{2water} - pCO_{2air}) \tag{2}$$

where pCO_{2air} and pCO_{2water} are the partial pressures of CO_2 in the atmosphere and surface water (µatm), respectively; pCO_{2air} was 380 and 377 µatm in July and August 2006, respectively (ftp://aftp. cmdl.noaa.gov/data/trace_gases/co2/flask/surface/co2_tap_surface-flask_1_ccgg_month.txt). F_{CO2} is the air–sea CO_2 exchange flux (mmol m^{-2} day^{-1}), where $F_{CO2} > 0$ indicates that seawater releases CO_2 into the atmosphere, and $F_{CO2} < 0$ means that seawater absorbs atmospheric CO_2. K_0 is the solubility coefficient of CO_2 in seawater [32], and k is the gas transfer velocity. For short-term wind, k was calculated using the empirical formula proposed by Wanninkhof [33] and revised by Sweeney [34]:

$$k = 0.27 \times U_{10}^2 \times (Sc/660)^{-0.5} \tag{3}$$

$$Sc = Sc_0 \times (1 + 3.14S/1000) \tag{4}$$

$$Sc_0 = 0.0476T^3 + 3.7818T^2 - 1.201T + 1800.6 \tag{5}$$

where U_{10} is the wind speed (m s^{-1}) at a height of 10 m above the sea surface (Remote Sensing Systems, CCMP Wind Vector Analysis Product, http://www.remss.com/measurements/ccmp/); Schmidt number (Sc) is expressed as a function of temperature (T, Celsius) and salinity (S, psu) [33,35].

We chose to use the mass balance method [36,37] that was modified for the calculation of NEP. At the initial time (t_1), the sea surface temperature (SST), sea surface salinity (SSS), and carbonate system parameters, including dissolved inorganic carbon (DIC), total alkalinity (TA), and pCO_2, are T_1, S_1, TA_1, DIC_1, and $(pCO_2)_1$, respectively. At time t_2, the above parameters are change to T_2, S_2, TA_2, DIC_2, and $(pCO_2)_2$.

$$\Delta pCO_2 = (pCO_2)_2 - (pCO_2)_1 = \Delta pCO_{2tem} + \Delta pCO_{2a\text{-}s} + \Delta pCO_{2mix} + \Delta pCO_{2bio} + \Delta pCO_{2non} \tag{6}$$

$$\Delta DIC = \Delta DIC_{a\text{-}s} + \Delta DIC_{mix} + \Delta DIC_{bio} \tag{7}$$

The subscripts "tem", "a-s", "mix", and "bio" of the specific parameter denote temperature, air–sea exchange, mixing, and in situ biological processes (including the release of CO_2 from organic matter degradation by microbes and CO_2 uptake by phytoplankton), respectively. "Δ" refers to the change in a particular parameter within a certain period of time (from t_1 to t_2). On a short timescale (three hours or each day), the nonlinear term (ΔpCO_{2non}) is essentially zero. The four different factors in Equation (6) were calculated as described below.

First, the thermal effect on ΔpCO_2 was calculated by Equation (8).

$$\Delta pCO_{2tem} = (pCO_2)_1 \times \exp(0.0423 \times (T_2 - T_1)) - (pCO_2)_1 \tag{8}$$

where 0.0423 is the temperature dependence coefficient of pCO_2 presented by Takahashi [38].

Second, air–sea CO_2 exchanges only change DIC and pCO_2 but have no effect on TA.

$$\Delta DIC_{a\text{-}s} = -F_{CO2} \times \Delta t/(\rho \times MLD) \tag{9}$$

$$(DIC_2)_{a\text{-}s} = DIC_1 + \Delta DIC_{a\text{-}s} \tag{10}$$

$$\Delta pCO_{2a\text{-}s} = f((DIC_2)_{a\text{-}s}, TA_1, S_1, T_1) - (pCO_2)_1 \tag{11}$$

where ρ is seawater density (kg m^{-3}), MLD is the mixed-layer depth, and $(DIC_2)_{a\text{-}s}$ is the DIC concentration at time t_2 and is affected only by the air–sea exchange from t_1 to t_2. The functions $f((DIC_2)_{a\text{-}s}, TA_1, S_1, T_1)$ were calculated using the CO2SYS program [28], and the dissociation constants were taken from Dickson et al. [39].The evaluation of the mixed-layer depth (MLD) was based on the sigma-*t* criterion proposed by Sprintall [40], and it was calculated as follows:

$$\sigma_{t,MLD} = \sigma_{t,0} + \Delta T \times (\partial t / \partial T) \tag{12}$$

$$\sigma_t = \rho - 1000 \tag{13}$$

where $\sigma_{t,0}$ is the σ_t value in the surface layer. ΔT is the desired temperature difference, and $\Delta T = 0.5\,°C$ in this study. The coefficient of thermal expansion $(\partial t / \partial T)$ was calculated from the surface temperature and salinity.

Third, using the interaction with the above-mentioned Kuroshio current, the original sources of the three end-member water masses were determined to be Changjiang diluted water (CDW), Kuroshio surface water (KSW), and Kuroshio subsurface water (KSSW) (Figure 2). The equations and characteristics of the three end-member mixing model are as follows (Table 2).

$$m_{CDW} + m_{KSW} + m_{KSSW} = 1 \tag{14}$$

$$m_{CDW} \times S_{CDW} + m_{KSW} \times S_{KSW} + m_{KSSW} \times S_{KSSW} = S \tag{15}$$

$$m_{CDW} \times \theta_{CDW} + m_{KSW} \times \theta_{KSW} + m_{KSSW} \times \theta_{KSSW} = \theta \tag{16}$$

where the subscripts CDW, KSW, and KSSW denote the three end-member water masses CDW, KSW, and KSSW, respectively; m_{CDW}, m_{KSW}, m_{KSSW} respectively denote the proportion of three end-members water masses; S_{CDW}, S_{KSW}, S_{KSSW} and θ_{CDW}, θ_{KSW}, θ_{KSSW} denote the salinity and bit temperature of the three-terminal element, respectively; S and θ denote the measured salinity and potential temperature, respectively. From this calculation, the theoretical values of total alkalinity $(TA_2)_{mix}$ and dissolved inorganic carbon $(DIC_2)_{mix}$ due to mixing during a given time period (from t_1 to t_2) can be determined. Further, ΔpCO_{2mix} can be calculated. The equations are

$$m_{CDW} \times (TA_2)_{CDW} + m_{KSW} \times (TA_2)_{KSW} + m_{KSSW} \times (TA_2)_{KSSW} = (TA_2)_{mix} \tag{17}$$

$$m_{CDW} \times (DIC_2)_{CDW} + m_{KSW} \times (DIC_2)_{KSW} + m_{KSSW} \times (DIC_2)_{KSSW} = (DIC_2)_{mix} \tag{18}$$

$$\Delta pCO_{2mix} = f((DIC_2)_{mix}, (TA_2)_{mix}, S_2, T_1) - (pCO_2)_1 \tag{19}$$

where $(TA_2)_{CDW}$, $(TA_2)_{KSW}$, $(TA_2)_{KSSW}$ and $(DIC_2)_{CDW}$, $(DIC_2)_{KSW}$, $(DIC_2)_{KSSW}$ denote the TA and DIC concentrations of the three end-member at time t_2, respectively.

Table 2. Three end-member characteristics of water mass from measurements obtained during cruises in July and August 2006.

Sampling Date	θ (°C)	S	TA (μmol kg^{-1})	DIC (μmol kg^{-1})
CDW	27.76 ± 0.20	7.88 ± 0.28	1898 ± 3.6	1863 ± 3.6
KSW	29.49 ± 0.10	33.22 ± 0.33	2232 ± 4.4	1808 ± 0.6
KSSW	19.48 ± 0.09	34.11 ± 0.05	2244 ± 0.6	2105 ± 13

Finally, the pCO_2 changes caused by biological processes (ΔpCO_{2bio}) were calculated from the other DIC changes. Thus,

$$\Delta DIC_{bio} = \Delta DIC - (\Delta DIC_{a\text{-}s} + \Delta DIC_{mix}) \tag{20}$$

$$(DIC_2)_{bio} = DIC_1 + \Delta DIC_{bio} \tag{21}$$

$$\Delta pCO_{2bio} = f((DIC_2)_{bio}, TA_1, S_1, T_1) - (pCO_2)_1 \tag{22}$$

where $(DIC_2)_{bio}$ is the theoretical value of DIC at time t_2 due to biological processes that occurred during a given time period (from t_1 to t_2).

According to the definition, the NEP calculation formula is

$$NEP = -\Delta DIC_{bio}/\Delta t \tag{23}$$

The NEP values in or under the mixed layer (mmol C m^{-2} day^{-1}) were calculated using the integral of the NEP over different water layers (mmol C m^{-3} day^{-1}).

Finally, we calculated the CO_2 flux caused by biological processes and its contribution to the air–sea CO_2 exchange flux as

$$F_{CO2bio} = k \times K_0 \times \Delta pCO_{2bio} \tag{24}$$

$$F_{CO2non\text{-}bio} = F_{CO2} - F_{CO2bio} \tag{25}$$

$$Cont = (F_{CO2bio}/F_{CO2}) \times 100\% \tag{26}$$

where F_{CO2bio} is the change in CO_2 flux caused by biological processes (mmol m^{-2} day^{-1}) and $F_{CO2non\text{-}bio}$ is the change in CO_2 flux caused by other processes. $F_{CO2bio} > 0$ indicates that biological processes, such as the degradation of organic matter by microorganisms, cause seawater to release CO_2. $F_{CO2bio} < 0$ indicates that biological processes, such as absorption of CO_2 by phytoplankton photosynthesis, cause seawater to absorb CO_2 from the atmosphere. Cont is the contribution of CO_2 flux changes caused by biological processes to the air–sea CO_2 exchange flux. Cont > 0 means that the variation in CO_2 caused by biological processes has the same direction as the variation in air–sea CO_2 exchange, which indicates a positive feedback progress; Cont < 0 indicates a negative feedback progress.

Under the mixed layer, potential pCO_2 and pCO_{2bio} were evaluated with the CO2SYS program using DIC_2, TA_2, S_2, T_2 and $(DIC_2)_{bio}$, TA_1, S_1, and T_1, respectively. *F_{CO2} and *F_{CO2bio} for each depth were calculated using Equations (1) and (23), and then the potential carbon flux (*F_{CO2}) and the potential carbon flux caused by biological processes (*F_{CO2bio}) in the three regions at each time point were integrated for the water layers beneath the MLD.

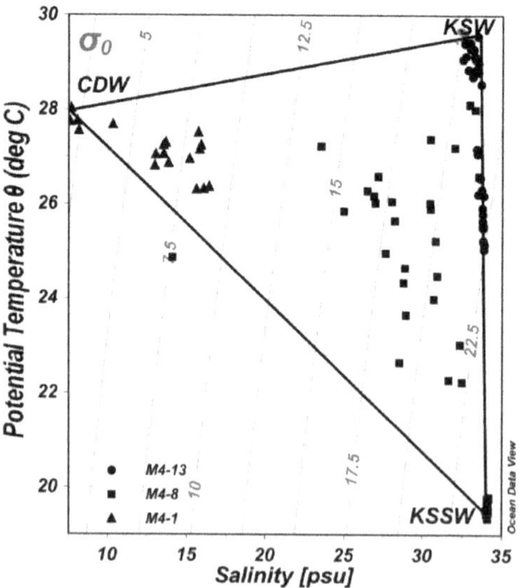

Figure 2. Scatter plots of potential temperature and salinity: M4-1 (triangles), M4-8 (squares), and M4-13 (circles). The labeled vertices denote the three end-members from the three water masses: Changjiang diluted water (CDW), Kuroshio surface water (KSW), and Kuroshio subsurface water (KSSW). Isoclines of potential density are shown in this figure.

2.5. Error Analysis

The uncertainty in pH arose from the pH measurement process. The uncertainty in TA is from the measured salinity and the TA–S Equation (1). The uncertainty in $(TA_2)_{mix}$ and $(DIC_2)_{mix}$ is introduced during the determination of the three endmembers. The uncertainty in DIC, pCO_{2water}, pCO_{2bio}, potential pCO_2, and pCO_{2bio} originates from CO2SYS with the equilibrium constants established by Mehrbach et al. [41] and refit by Dickson and Millero [39] (i.e., with carbonic acid dissociation constants omitted from calculations). The uncertainty in F_{CO2}, F_{CO2bio}, $*F_{CO2}$, and $*F_{CO2bio}$ arises from the calculation using the daily gas transfer velocity (k) and deviations in pCO_{2water} and pCO_{2bio}. In this study, we used error propagation formulas to estimate the uncertainties [42].

Assuming that the errors of the variables X, Y, and Z are δX, δY, and δZ, respectively, for linear sum functions, the error of R is

$$R = X + Y + Z \tag{27}$$

$$\delta R = \delta X + \delta Y + \delta Z \tag{28}$$

For multiplication and division, the error of R is

$$R = (X \times Y)/Z \tag{29}$$

$$(\delta R/R)^2 = (\delta X/X)^2 + (\delta Y/Y)^2 + (\delta Z/Z)^2 \tag{30}$$

Overall, the uncertainty in the salinity-based TA calculation is less than 3%; the uncertainties in $(TA_2)_{mix}$ and $(DIC_2)_{mix}$ are ~0.4% and ~0.8%, respectively; the uncertainty in k is ~13%; the uncertainty of F_{CO2}, F_{CO2bio}, $*F_{CO2}$, and $*F_{CO2bio}$ is ± 1.61, ± 2.10, ± 2.61, and ± 0.86 mmol m^{-2} day^{-1}, respectively.

3. Results

3.1. 24 Hourly Variations in Temperature and Salinity

The trend of the surface temperature in the three regions was offshore > near-shore > front, and the bottom temperature showed a trend of near-shore > offshore > front (Figure 3a–c). The trend of salinity in the surface and bottom layer showed a distribution trend of offshore > front > near-shore (Figure 3d–f). The difference between the surface and bottom temperature in the front region was the largest, followed by the offshore region, and the temperature difference in the near-shore region was the smallest. The temperature and salinity changes in the near-shore region fluctuated with a semidiurnal frequency. The temperature and salinity at 06:00 and 18:00 both had extreme values (Figure 3a,d). The relative standard deviation of surface salinity changes was as high as 25.82% in 24 hours. In the front region, the relative standard deviation of the temperature variation at 10 m reached 8.90%, and the salinity variation at 5 m was as high as 21.01%. In the offshore region, the temperature and salinity changes were small in 24 hours: the relative standard deviation of the temperature at 10 m was 5.15%, and the relative standard deviation of the changes in salinity at the surface in 24 hours was 1.01%; the others were less than 1%.

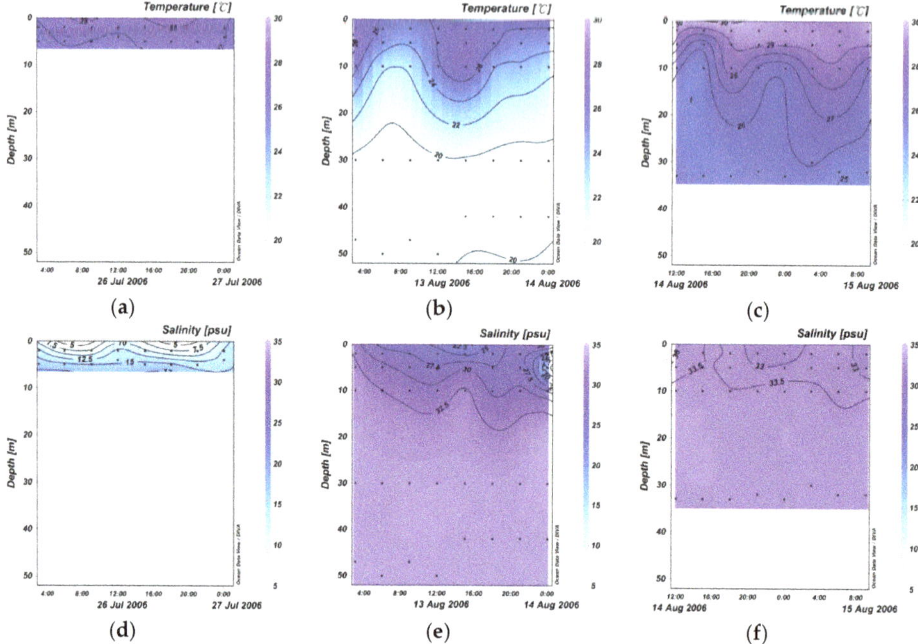

Figure 3. Twenty-four-hour variations in temperature in the near-shore (**a**), front (**b**), and offshore (**c**) regions and salinity in the near-shore (**d**), front (**e**), and offshore (**f**) regions in summer.

3.2. Variation in pH, TA, DIC, and Sea Surface pCO$_2$ within 24 Hours

In the near-shore region, the surface daily averages (standard deviations in brackets) of pH increased from 7.92 ($\pm$0.02) to 7.95 ($\pm$0.02) at the bottom (Figure 4a), TA increased from 1936.06 ($\pm$40.21) to 1993.43 ($\pm$15.01) μmol kg^{-1} at the bottom (Figure 4d), DIC increased from 1889.75 ($\pm$28.04) to 1919.82 ($\pm$9.35) μmol kg^{-1} at the bottom (Figure 4g), and pCO$_2$ decreased from 996 ($\pm$71) to 868 ($\pm$53) μatm at the bottom (Figure 4j).

In the front region, the surface daily averages (standard deviations in brackets) of pH decreased from 8.33 ($\pm$0.11) to 7.94 ($\pm$0.03) at the bottom (Figure 4b). TA increased from 2157.80 ($\pm$34.17) to 2244.12 ($\pm$0.68) μmol kg^{-1} at 30 m and then decreased to 2244.06 ($\pm$0.50) μmol kg^{-1} at the bottom (Figure 4e). DIC increased from 1833.97 ($\pm$68.63) to 2102.68 ($\pm$12.88) μmol kg^{-1} at the bottom (Figure 4h), and pCO$_2$ increased from 283 ($\pm$87) to 735 ($\pm$59) μatm at the bottom (Figure 4k).

In the offshore region, the surface daily averages (standard deviations in brackets) of pH decreased from 8.38 ($\pm$0.03) to 8.05 ($\pm$0.02) at the bottom (Figure 4c), TA increased from 2229.48 ($\pm$4.72) to 2240.15 ($\pm$0.53) μmol kg^{-1} at the bottom (Figure 4f), and DIC increased from 1791.73 ($\pm$24.87) to 2018.65 ($\pm$13.21) μmol kg^{-1} at the bottom (Figure 4i). Daily average pCO$_2$ was 227 ($\pm$23) μatm at the surface, and it decreased to 226 ($\pm$32) μatm at 5 m and then increased to 566 ($\pm$38) μatm at the bottom (Figure 4l).

Overall, from the vertical distribution of the water column, pH was generally highest at the surface and lowest at the bottom. On the contrary, TA, DIC, and pCO$_2$ were generally lowest at the surface and highest at the bottom. Spatially, pH and TA generally increased from the near-shore to the offshore region. On the contrary, DIC and pCO$_2$ generally decreased from the near-shore to the offshore region.

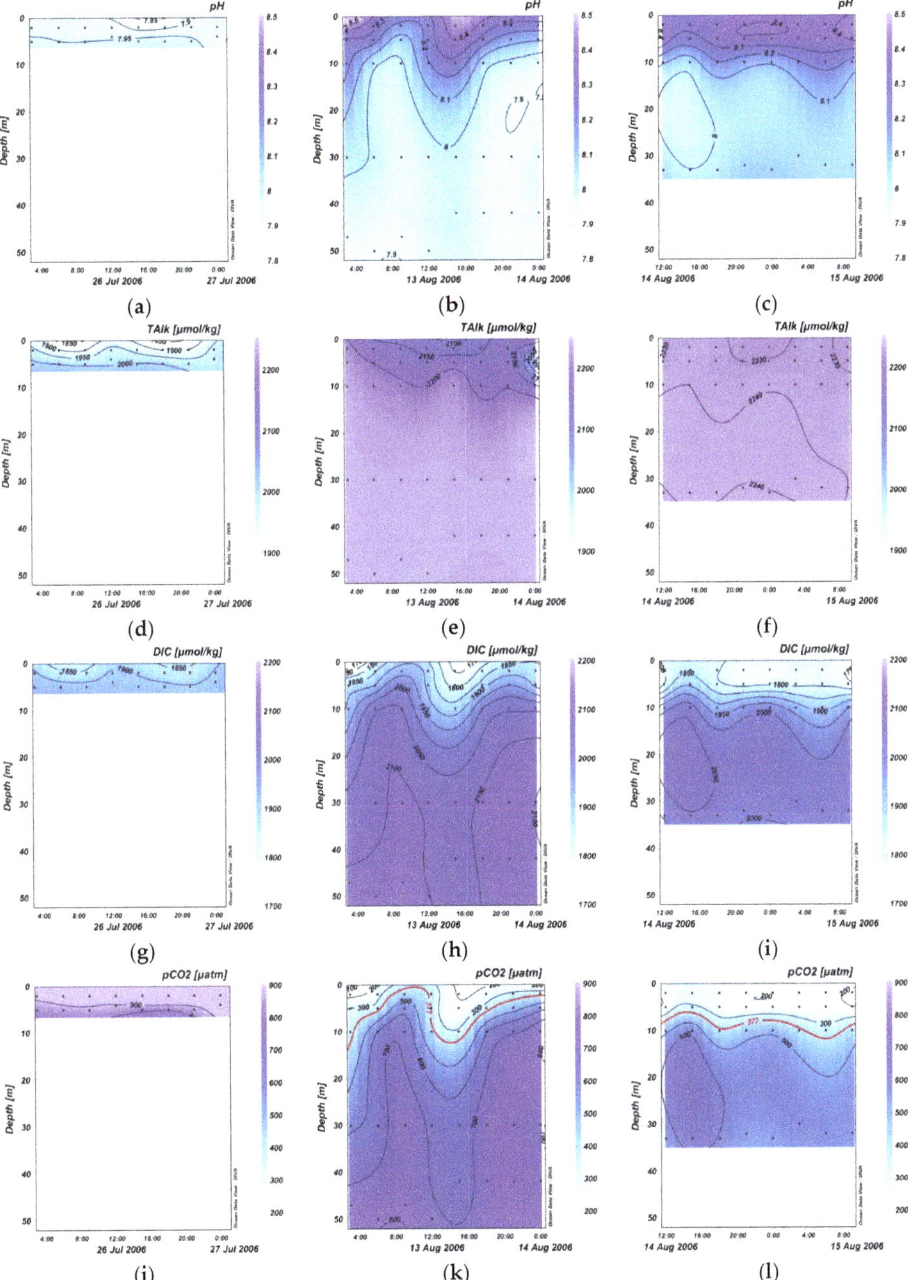

Figure 4. Twenty-four hour variations in pH in the near-shore (**a**), front (**b**), and offshore (**c**) regions; TA in the near-shore (**d**), front (**e**), and offshore (**f**) regions; DIC in the near-shore (**g**), front (**h**), and offshore (**i**) regions; and sea surface pCO$_2$ in the near-shore (**j**), front (**k**), and offshore (**l**) regions in summer.

3.3. Variation in NEP within 24 Hours

In the near-shore region, there were negative NEP values, and the NEP at the bottom was slightly larger than that at the surface (Table 3). The minimum NEP value was −0.36 mmol C m^{-3} day^{-1} at 12:00 at the surface, and the maximum value was 0.13 mmol C m^{-3} day^{-1} at 15:00 at the bottom (Figure 5a).

Table 3. Minimum, maximum, mean, and standard deviation of NEP in the three regions in summer (mmol C m^{-3} day^{-1}).

Regions	Depth	Minimum	Maximum	Mean	Standard Deviation
Near-shore	Surface	−0.36	0.09	−0.12	0.16
	Bottom	−0.34	0.13	−0.17	0.18
Front	Surface	−0.04	1.89	1.07	0.62
	5 m	0.19	1.26	0.65	0.38
	10 m	−0.32	0.28	−0.05	0.20
	30 m	−0.15	0.21	−0.01	0.12
	Bottom	−0.16	0.11	−0.08	0.09
Offshore	Surface	−0.14	0.43	0.16	0.19
	5 m	−0.08	0.52	0.22	0.17
	10 m	−0.54	0.28	−0.08	0.31
	Bottom	−0.31	0.03	−0.09	0.12

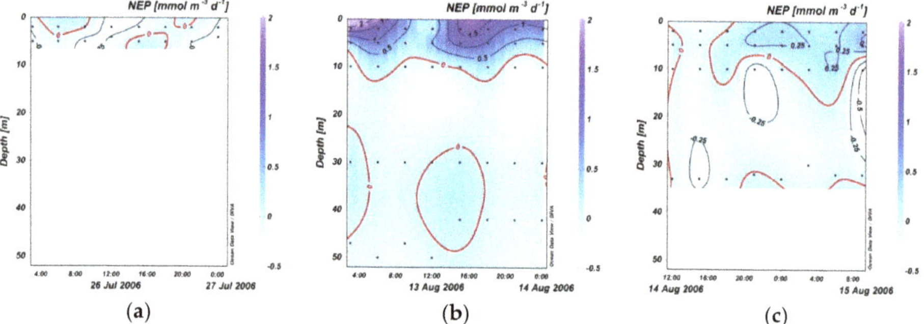

Figure 5. Twenty-four-hour variation in NEP in the near-shore (**a**), front (**b**), and offshore (**c**) regions in summer.

In the front region, the maximum NEP value (1.89 mmol C m^{-3} day^{-1}) was observed at 03:00 at the surface, and the minimum value (−0.32 mmol C m^{-3} day^{-1}) was observed at 21:00 at 10 m (Figure 5b). In the vertical direction, the daily variation in the surface NEP was slightly larger than that at the bottom. In the front region, the daily mean NEP from the surface to the bottom generally decreased, and the 24-h variation in NEP in the mixed layer (Table 3) was larger than that under the mixed layer (Table 4).

Table 4. Mixed layer depth (m) at each measurement time within 24-h in three regions.

Regions	00:00	03:00	06:00	09:00	12:00	15:00	18:00	21:00
Near-shore	2.73	2.14	2.06	2.06	2.72	2.09	2.06	2.09
Front	8.49	2.94	2.18	2.36	2.83	2.07	5.11	2.35
Offshore	3.29	6.03	8.84	5.03	3.05	2.45	6.67	3.44

In the offshore region, the maximum NEP (0.52 mmol C m^{-3} day^{-1}) was observed at 09:00 at 5 m, while the minimum NRP (−0.54 mmol C m^{-3} day^{-1}) was observed at 10 m (Figure 5c). The largest variation in NEP within 24-h was at 10 m, and the smallest variation was at the bottom (Table 3).

4. Discussion

4.1. Variations in F_{CO2bio} and F_{CO2} in the Mixed Layer

F_{CO2} is strongly positive in each timeslot (Figure 6a), indicating that this region acts as a source of atmospheric CO_2 [43,44] because the near-shore region is affected by the CDW [45] which has abundant pCO_2 [25,26]. F_{CO2bio} is strongly positive most of the time (Figure 6a), meaning that heterotrophic respiration releases CO_2 to the atmosphere for most of the day in the near-shore region. Because this study region is located in the largest turbid zone of the Changjiang River estuary plume [46,47], we infer that the mixing effect and extremely limited light may reduce the primary production by phytoplankton photosynthesis and that planktonic community respiration may dominate the biological processes, which maintain a high pCO_2 value in the near-shore region.

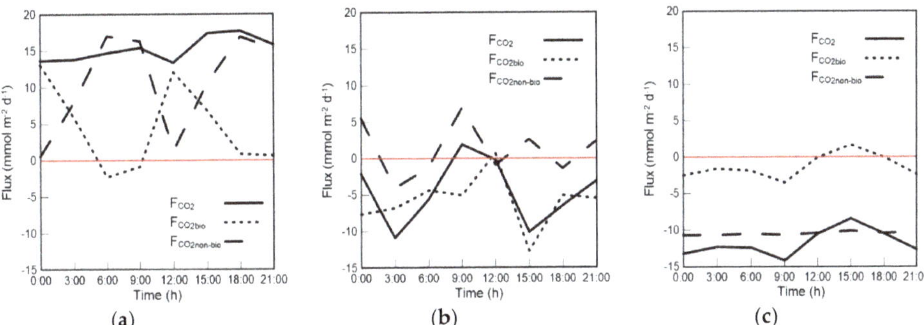

Figure 6. Twenty-four hour variations in F_{CO2}, F_{CO2bio}, and $F_{CO2non-bio}$ in the near-shore (**a**), front (**b**), and offshore (**c**) regions in summer.

The 24-h F_{CO2} in the front region is almost negative (Figure 6b), indicating that the front region acts as a sink for atmospheric CO_2. The 24-h F_{CO2bio} in the front region is also almost negative which indicates the biological processes absorb CO_2 (Figure 6b). Because of the high values of NEP (Figure 5b, Table 3) and Chl *a* [6] in this region, we infer that the front region has a great capacity for biological productivity and that a large amount of CO_2 is fixed in the surface water by phytoplankton.

Most F_{CO2} values in the offshore region are mostly slightly less than 0 (Figure 6c), indicating that the offshore region acts as a sink for atmospheric CO_2. This finding is in agreement with the study by Song et al. [2]. F_{CO2bio} in the offshore region was mostly slightly less than 0 (Figure 6c), indicating that the photosynthesis rate of fixed CO_2 by phytoplankton is higher than degradation rates of organic matter releasing CO_2 by microbial action in the offshore region.

4.2. The Contribution of Biological Processes to the Air–Sea CO_2 Exchange Flux in the Mixed Layer

F_{CO2} in the mixed layer of the three regions shows that the near-shore region acts as a strong source of atmospheric CO_2 (Figure 6a) and that the front and offshore regions act as sinks for atmospheric CO_2 (Figure 6b,c), similar to the results of other studies [1,48,49]. The daily average Cont in the mixed layer shows that the biological processes have a positive feedback effect on air–sea CO_2 exchange in the near-shore, front, and offshore regions (Table 5). This agrees with the conclusion of Borges et al. [15]. The air–sea CO_2 flux is inversely proportional to the NEP in the mixed layer, indicating that the contribution to the variation in air–sea CO_2 flux in these coastal waters is dominated by biological processes during a diel cycle. However, the average Cont in the offshore region is lower than that in the near-shore and front regions. This could be related to the fact that primary production in the offshore region is very low, even in summer, and other effects such as wind, temperature, and water mixing may play more important roles in controlling air–sea CO_2 flux.

Table 5. The contribution of CO_2 flux variation caused by biological processes to F_{CO2} (Cont) in the mixed layer.

Regions	00:00	03:00	06:00	09:00	12:00	15:00	18:00	21:00	Mean
Near-shore	96%	42%	−16%	−6%	90%	39%	5%	4%	32%
Front	360%	63%	79%	−269%	−341%	126%	78%	175%	34%
Offshore	19%	13%	15%	25%	1%	−20%	1%	19%	9%

4.3. Potential Carbon Sources under the Mixed Layer

Under the mixed layer (Table 4), the water column is determined to be a potential carbon source of atmospheric CO_2 in the three regions (Figure 7a–c). The variations in *F_{CO2} and *F_{CO2bio} show that the near-shore, front, and offshore regions could be potential atmospheric carbon sources, and a large amount of CO_2 produced by biological processes (e.g., respiration) is stored under the mixed layer (Figure 7). Although the surface water in the front and offshore regions acts as a sink for atmospheric CO_2, respiration under the mixed layer will result in the degradation of organic matter with substantial CO_2 release, which could be observed when vertical mixing occurred [27]. Hence, the CO_2 sink region in the Changjiang River estuary plume will become a source region when there is a tropical storm or an upwelling process. In a relevant study of the East China Sea, Chen et al. [4] also proposed that phytoplankton and planktonic bacteria could store dissolved inorganic carbon in the subsurface and might affect the surface air–sea CO_2 flux. Further, the daily means (standard deviations in brackets) of the potential contribution of biological processes to air–sea CO_2 exchange flux in the near-shore, front, and offshore regions are 34% (±43%), 8% (±13%), and 19% (±24%) in 24 hours, respectively, indicating that local respiration accounts for a large part of the total potential CO_2 release under the mixed layer. Other factors probably include KSSW intrusion, temperature elevation, and so on, which need further exploration.

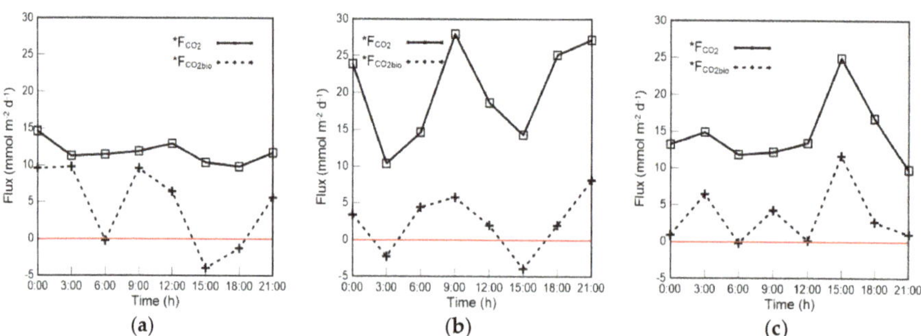

Figure 7. Twenty-four-hour variations in potential carbon flux (*F_{CO2}) and the biological contribution to carbon flux (*F_{CO2bio}) in the near-shore (**a**), front (**b**), and offshore (**c**) regions in summer.

4.4. Trophic Status Assessments and the Relationship between Cont and NEP

The mixed layer in the front and offshore regions is an autotrophic system (Table 3), but that in the near-shore region is a heterotrophic system. On the whole, we consider the Changjiang River estuary plume to be an autotrophic ecosystem in summer, similar to the conclusion of Li et al. [11], in August 2006. The daily mean NEP values of the study region are negative under the mixed layer, indicating that they are heterotrophic systems, which is in agreement with Chou et al. [27]. However, the positivity or negativity of the NEP values changes throughout a 24-hour period, and the trophic status of the same region varies as well. The Changjiang River estuary plume has a complex current structure featuring multiple eddies [50] or low salinity water detachment (LSW) [16]; however, eddies and LSW are on the mesoscale in terms of time and space (e.g., a couple of weeks and hundreds of kilometers). In 24 hours, eddies and LSW have little effect on the variation in water properties. Therefore, we suggest that trophic statuses in a day are regulated by the tide.

In order to explore the influences of trophic status on Cont, we compared the Cont and NEP in the mixed layer in the region (Figure 8). The significant correlations between Cont and NEP in the mixed layer in the near-shore and offshore regions show that trophic status can be used as an index of the contribution of biological process to the air–sea CO_2 flux. Cont in the near-shore region has a significantly negative correlation ($r^2 = 0.94$, $p < 0.05$) with NEP, indicating that the more heterotrophic the system, the greater the influence on the contribution of biological processes (e.g., organic matter degradation by microorganisms) to F_{CO2}. When there is no biological contribution to F_{CO2} (Cont = 0), the NEP background value is -0.003 mmol C m^{-3} day^{-1}. In the front region, the correlation between Cont and NEP is not significant (Figure 8c). This could be because there are opposing processes causing the trophic status on the east and west sides of the front region, the west side of the front region is dominated by the degradation of organic matter, while the east side is dominated by the absorption of dissolved inorganic carbon. When the tide has a continuous impact on the front region, the NEP in the front region would present a large fluctuation. The NEP of offshore region was significantly and positively correlated with Cont ($r^2 = 0.94$, $p < 0.05$), indicating that the more autotrophic the system, the greater the contribution of the biological processes (e.g., primary production) to F_{CO2}. Assuming that there are no biological processes in the offshore region (Cont = 0), the background value of NEP was also 0.03 mmol C m^{-3} day^{-1} (Figure 8c). In addition, the slopes of NEP and Cont show that the biological processes have a stronger influence on the variation in the air–sea CO_2 exchange flux in the near-shore region than that in the offshore region when the two systems have an equal trophic status level.

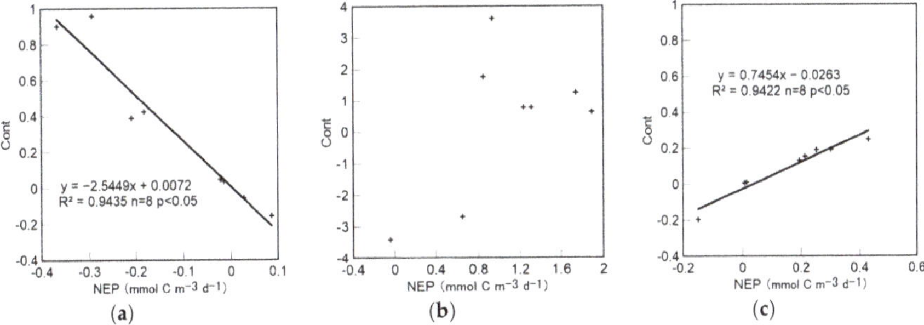

Figure 8. Correlations between Cont and NEP in the near-shore (**a**), front (**b**), and offshore (**c**) regions in the mixed layer in summer.

5. Conclusions

Using a mass balance model, we calculated the NEP, F_{CO2bio}, and F_{CO2} at eight time-points per day in the near-shore, front, and offshore regions of the Changjiang River estuary plume in summer. Then, we calculated the contribution of biological processes to F_{CO2} in the three regions. In the mixed layer, both F_{CO2} and F_{CO2bio} significantly varied at different times within the 24-h period. The near-shore region was found to be a source of atmospheric CO_2, and the offshore region is a sink for atmospheric CO_2. The front region is a sink for atmospheric CO_2 on the whole, but it transforms between a source and a sink from time to time. The biological processes in the mixed layer in the three regions were shown to have an overall positive feedback effect on the variation in the air–sea CO_2 exchange flux. Within the 24 hour period, the mean values of F_{CO2} and F_{CO2bio} were both positive in the near-shore region, indicating that CO_2 was being released into the atmosphere, and microbial degradation of organic matter accounted for a large part of this. In the front and offshore regions, the daily mean values of F_{CO2} and F_{CO2bio} were both negative, indicating that these areas absorb CO_2 from the atmosphere and that phytoplankton also fixes CO_2 from the atmosphere into the ocean. The daily averages of Cont of stations from west to east were 32% (±40%), 34% (±216%), and 9% (±13%), respectively. Cont reached 360% in the front region. Under the mixed layer, the near-shore, front, and offshore regions could be potential carbon sources for the atmosphere. Therefore, the CO_2 sink region might become a source when there is a tropical storm or upwelling process that overturns the water from the deep. Under the mixed layer, the daily means of the potential contribution of biological processes to air–sea CO_2 exchange flux were 34% (±43%), 8% (±13%), and 19% (±24%) within the 24-h period, respectively. In addition, in the mixed layer, the near-shore region was shown to be a typical heterotrophic system, while the front and offshore regions are both autotrophic systems. Conversely, in all three regions, under the mixed layer is heterotrophic. However, at different time points, the trophic statuses change, even in the same region.

At a short timescale or in a steady-state environment, these conclusions can accurately represent the influence of biological processes on the variation in air–sea CO_2 exchange flux and can be used to assess the trophic statuses in the Changjiang River estuary plume in summer. Nevertheless, the biochemical and hydrological conditions in coastal regions constantly change at high frequency; thus, the use of data with high spatial and temporal resolutions is necessary to study the contribution of biological processes to the air–sea CO_2 exchange flux and to more accurately quantify the potential carbon stock of deep water bodies. Further, variations in long-term trophic statuses require additional exploration, especially in coastal waters, given the intensity of human activities and quickly progressing climate change.

Author Contributions: Conceptualization, Y.Z. and K.W.; Formal analysis, Y.Z. and D.L.; Investigation, K.W. and B.X.; Writing–original draft, Y.Z.; Writing–review & editing, D.L. and K.W.

Funding: This study was jointly supported by the National Natural Science Foundation of China (U1609201, U1709201, 91128212, 41203085, and 41206085), the Public Science and Technology Research Funds Projects of Ocean (201105014 and 201205015), the Scientific Research Fund of the Second Institute of Oceanography, State Oceanic Administration, China (JT1603), the Natural Science Project of Zhejiang Province (Y5110171,LQ17D060006), and the Project of State Key Laboratory of Satellite Ocean Environment Dynamics, Second Institute of Oceanography (SOEDZZ1402 and SOEDZZ1521).

Acknowledgments: The authors would like to thank the crew of R/V "China Marine Surveillance 49" for their supports in sampling and logistics. We also thank Wei-jun Cai (University of Delaware), Zhaohui Aleck Wang (Woods Hole Oceanographic Institution), Quanzhen Chen and his group, Daji Huang and his group, Jianfang Chen, Bin Wang and Tianzhen Zhang (Second Institute of Oceanography, Ministry of Natural Resources) for their technical support and helpful comments.

References

1. Guo, X.; Zhai, W.; Dai, M.; Zhang, C.; Bai, Y.; Xu, Y.; Li, Q.; Wang, G. Air–Sea CO_2 fluxes in the East China Sea based on multiple-year underway observations. *Biogeosciences* **2015**, *12*, 5123–5167. [CrossRef]
2. Song, J.; Qu, B.; Li, X.; Yuan, H.; Li, N.; Duan, L. Carbon sinks/sources in the Yellow and East China Seas—Air-sea interface exchange, dissolution in seawater, and burial in sediments. *Sci. China Earth Sci.* **2018**, *61*, 1583. [CrossRef]
3. Jiao, N.; Liang, Y.; Zhang, Y.; Liu, J.; Zhang, Y.; Zhang, R.; Zhao, M.; Dai, M.; Gao, K.; Song, J.; et al. Carbon pools and fluxes in the China Seas and adjacent oceans. *Sci. China Earth Sci.* **2018**, *61*, 1535. [CrossRef]
4. Chen, C.; Chiang, K.; Gong, G.; Shiah, F.; Tseng, C.; Liu, K. Importance of planktonic community respiration on the carbon balance of the East China Sea in summer. *Glob. Biogeochem. Cycles* **2006**, *20*. [CrossRef]
5. Chen, C.A. The Kuroshio intermediate water is the major source of nutrients on the East China Sea continental shelf. *Oceanol. Acta* **1996**, *19*, 523–527.
6. Wang, K.; Chen, J.; Jin, H.; Gao, S.; Xu, J.; Lu, Y.; Haung, D.; Hao, Q.; Weng, H. Summer nutrient dynamics and biological carbon uptake rate in the Changjiang River plume inferred using a three end-member mixing model. *Cont. Shelf Res.* **2014**, *91*, 192–200. [CrossRef]
7. Li, D.; Chen, J.; Ni, X.; Wang, K.; Zeng, D.; Wang, B.; Jin, H.; Haung, D.; Cai, W. Effects of biological production and vertical mixing on sea surface pCO_2 variations in the Changjiang River plume during early autumn: A buoy-based time series study. *J. Geophys. Res. Ocean.* **2018**, *123*, 6156–6173. [CrossRef]

8.   Odum, H.T. Primary production in flowing waters. *Limnol. Oceanogr.* **1956**, *1*, 102–117. [CrossRef]

9.   Oviatt, C.A.; Doering, P.H.; Nowicki, B.L.; Zoppini, A. Net system production in coastal waters as a function of eutrophication, seasonality and benthic macrofaunal abundance. *Estuaries* **1993**, *16*, 247–254. [CrossRef]

10.   Carvalho, M.C.; Schulz, K.G.; Eyre, B.D. Respiration of new and old carbon in the surface ocean: Implications for estimates of global oceanic gross primary productivity. *Glob. Biogeochem. Cycles* **2017**, *31*, 975–984. [CrossRef]

11.   Li, X.; Yu, Z.; Song, X.; Cao, X.; Yuan, Y. Nitrogen and phosphorus budgets of the Changjiang River estuary. *Chin. J. Oceanol. Limnol.* **2011**, *29*, 762–774. [CrossRef]

12.   Xu, H.; Wolanski, E.; Chen, Z. Suspended particulate matter affects the nutrient budget of turbid estuaries: Modification of the LOICZ model and application to the Yangtze Estuary. *Estuar. Coast. Shelf Sci.* **2013**, *127*, 59–62. [CrossRef]

13.   Swaney, D.P.; Howarth, R.W.; Butler, T.J. A novel approach for estimating ecosystem production and respiration in estuaries: Application to the oligohaline and mesohaline Hudson River. *Limnol. Oceanogr.* **1999**, *44*, 1509–1521. [CrossRef]

14.   Palevsky, H.I.; Quay, P.D. Influence of biological carbon export on ocean carbon uptake over the annual cycle across the North Pacific Ocean. *Glob. Biogeochem. Cycles* **2017**, *31*, 81–95. [CrossRef]

15.   Borges, A.; Schiettecatte, L.-S.; Abril, G.; Delille, B.; Gazeau, F. Carbon dioxide in European coastal waters. *Estuar. Coast. Shelf Sci.* **2006**, *70*, 375–387. [CrossRef]

16.   Xuan, J.; Huang, D.; Zhou, F.; Zhu, X.; Fan, X.; Ni, X.; Xing, C. Application of data assimilation to synoptic temperature mapping of the coastal ocean survey. *Oceanol. Limnol. Sin.* **2012**, *43*, 17–26.

17.   Dai, A.; Trenberth, K.E. Estimates of freshwater discharge from continents: Latitudinal and seasonal variations. *J. Hydrometeorol.* **2002**, *3*, 660–687. [CrossRef]

18.   Liu, X.C. Concentration variation and flux estimation of dissolved inorganic nutrient from the Changjianag River into its estuary. *Oceanol. Limnol. Sin.* **2002**, *32*, 332–340.

19.   Wang, K.; Chen, J.; Ni, X.; Zeng, D.; Li, D.; Jin, H.; Glibert, P.; Qiu, W.; Haung, D. Real-time monitoring of nutrients in the Changjiang Estuary reveals short-term nutrient-algal bloom dynamics. *J. Geophys. Res. Ocean.* **2017**, *122*, 5390–5403. [CrossRef]

20.   Liu, Q.; Guo, X.; Yin, Z.; Zhou, K.; Roberts, E.; Dai, M. Carbon fluxes in the China Seas: An overview and perspective. *Sci. China* **2018**, *61*, 1564–1582. [CrossRef]

21.   Chen, C.; Shiah, F.; Chiang, K.; Gong, G.; Kemp, W. Effects of the Changjiang (Yangtze) River discharge on planktonic community respiration in the East China Sea. *J. Geophys. Res. Ocean.* **2009**, *114*. [CrossRef]

22.   Huang, H.; Wang, Y.; Zhang, W. Characteristics of tidal waves in the eastern Jiangsu coast and Changjiang Estuary. In Proceedings of the 2015 International Forum on Energy, Environment Science and Materials, Shenzhen, China, 25–26 September 2015.

23.   Chen, C.A.; Zhai, W.; Dai, M. Riverine input and air-sea CO_2 exchanges near the Changjiang (Yangtze River) Estuary: Status quo and implication on possible future changes in metabolic status. *Cont. Shelf Res.* **2008**, *28*, 1476–1482. [CrossRef]

24.   Zhejiang Meteorological Bureau. *Zhejiang Provincial Climate Bulletin*; Zhejiang Meteorological Bureau: Zhejiang, China, 2006.

25.   Zhai, W.; Dai, M. On the seasonal variation of air-sea CO_2 fluxes in the outer Changjiang (Yangtze River) Estuary, East China Sea. *Mar. Chem.* **2009**, *117*, 2–10. [CrossRef]

26.   Gao, X.; Song, J.; Li, X.; Li, N.; Yuan, H. pCO_2 and carbon fluxes across sea-air interface in the Changjiang Estuary and Hangzhou Bay. *Chin. J. Oceanol. Limnol.* **2008**, *26*, 289–295. [CrossRef]

27.   Chou, W.; Gong, G.; Sheu, D.D.; Jan, S.; Huang, C.; Chen, C. Reconciling the paradox that the heterotrophic waters of the East China Sea shelf act as a significant CO_2 sink during the summertime: Evidence and implications. *Geophys. Res. Lett.* **2009**, *36*, 139–156. [CrossRef]

28.   Lewis, E.; Wallace, D.; Allison, L.J. *Program Developed for CO_2 System Calculations*; Carbon Dioxide Information Analysis Center, managed by Lockheed Martin Energy Research Corporation for the US Department of Energy Tennessee: Nashville, TN, USA, 1998.

29.   Wang, B.; Chen, J.; Jin, H.; Li, H.; Liu, X.; Zhuang, Y.; Xu, Y.; Zhang, H. Preliminary study on the dissolved inorganic carbon system and its response mechanism in Changjiang River Estuary and its adjacent sea areas in summer. *J. Mar. Sci.* **2011**, *29*, 63–70.

30. Zhai, W.; Chen, J.; Jin, H.; Li, H.; Liu, J.; He, X.; Bai, Y. Spring carbonate chemistry dynamics of surface waters in the northern East China Sea: Water mixing, biological uptake of CO_2, and chemical buffering capacity. *J. Geophys. Res. Ocean.* **2015**, *119*, 5638–5653. [CrossRef]

31. Liss, P.S.; Slater, P.G. Flux of gases across the air-sea interface. *Nature* **1974**, *247*, 181–184. [CrossRef]

32. Weiss, R.F. Carbon dioxide in water and seawater: The solubility of a non-ideal gas. *Mar. Chem.* **1974**, *2*, 203–215. [CrossRef]

33. Wanninkhof, R. Relationship between wind speed and gas exchange over the ocean. *J. Geophys. Res.* **1992**, *97*, 7373–7382. [CrossRef]

34. Sweeney, C.; Gloor, E.; Jacobson, A.R.; Key, R.; Mckinley, G.; Sarmiento, J.; Wanninkhof, R. Constraining global air-sea gas exchange for CO_2 with recent bomb ^{14}C measurements. *Glob. Biogeochem. Cycles* **2007**, *21*. [CrossRef]

35. Chaturvedi, M.K.M.; Langote, S.D.; Kumar, D.; Asolekar, S. Significance and estimation of oxygen mass transfer coefficient in simulated waste stabilization pond. *Ecol. Eng.* **2014**, *73*, 331–334. [CrossRef]

36. Nixon, S.W. Physical energy inputs and the comparative ecology of lake and marine ecosystems: Physical energy inputs. *Limnol. Oceanogr.* **1988**, *33*, 1005–1025. [CrossRef]

37. Xue, L.; Cai, W.; Hu, X.; Sabine, C.; Jones, S.; Sutton, A.; Jiang, L.; Reimer, J. Sea surface carbon dioxide at the Georgia time series site (2006–2007): Air–sea flux and controlling processes. *Prog. Oceanogr.* **2016**, *140*, 14–26. [CrossRef]

38. Takahashi, T.; Olafsson, J.; Goddard, J.G.; Chipman, D.; Sutherland, S. Seasonal variation of CO_2 and nutrients in the high-latitude surface oceans: A comparative study. *Glob. Biogeochem. Cycles* **1993**, *7*, 843–878. [CrossRef]

39. Dickson, A.G.; Millero, F.J. A comparison of the equilibrium constants for the dissociation of carbonic acid in seawater media. *Deep Sea Res. Part A Oceanogr. Res. Pap.* **1987**, *34*, 1733–1743. [CrossRef]

40. Sprintall, J.; Tomczak, M. Evidence of the barrier layer in the surface layer of the tropics. *J. Geophys. Res. Ocean.* **1992**, *97*, 7305–7316. [CrossRef]

41. Mehrbach, C.; Culberson, C.H.; Hawley, J.E.; Pytkowicx, R.M. Measurement of the apparent dissociation constants of carbonic acid in seawater at atmospheric pressure. *Limnol. Oceanogr.* **1973**, *18*, 897–907. [CrossRef]

42. Taylor, J.R. *An Introduction to Error Analysis*; University Science Books: Sausalito, CA, USA, 1997.

43. Qu, B.; Song, J.; Li, X.; Yuan, H.; Li, N.; Ma, Q. pCO_2 distribution and CO_2 flux on the inner continental shelf of the East China Sea during summer 2011. *Chin. J. Oceanol. Limnol.* **2013**, *31*, 1088–1097. [CrossRef]

44. Chen, X.; Song, J.; Yuan, H.; Li, X.; Li, N.; Duan, L.; Qu, B. Preliminary study on the change of carbon exchange at sea-air interface and its regional carbon sink intensity in the summer of 2012 in the East China Sea. *Haiyang Xuebao* **2014**, *36*, 18–31.

45. Zhou, F.; Xuan, J.; Ni, X.; Huang, D. Preliminary Analysis of Dynamic Factors of Summer Yangtze River Freshwater Change between 1999 and 2006. *Haiyang Xuebao* **2009**, *31*, 1–12.

46. Milliman, J.D.; Shen, H.; Yang, Z.; Mead, R. Transport and deposition of river sediment in the Changjiang estuary and adjacent continental shelf. *Cont. Shelf Res.* **1985**, *4*, 37–45. [CrossRef]

47. Zhang, J.; Wu, Y.; Jennerjahn, T.C.; Ittekkot, V.; He, Q. Distribution of organic matter in the Changjiang (Yangtze River) Estuary and their stable carbon and nitrogen isotopic ratios: Implications for source discrimination and sedimentary dynamics. *Mar. Chem.* **2007**, *106*, 111–126. [CrossRef]

48. Borges, A.V.; Delille, B.; Frankignoulle, M. Budgeting sinks and sources of CO_2 in the coastal ocean: Diversity of ecosystems counts. *Geophys. Res. Lett.* **2005**, *32*, 301–320. [CrossRef]

49. Cai, W.; Dai, M.; Wang, Y. Air-sea exchange of carbon dioxide in ocean margins: A province-based synthesis. *Geophys. Res. Lett.* **2006**, *33*. [CrossRef]

50. Li, M.; Rong, Z. Effects of tides on freshwater and volume transports in the Changjiang River plume. *J. Geophys. Res. Ocean.* **2012**, *117*. [CrossRef]

Article

Modeling the Influence of Outflow and Community Structure on an Endangered Fish Population in the Upper San Francisco Estuary

Gonzalo C. Castillo

Lodi Fish and Wildlife Office, U.S. Fish and Wildlife Service, Suite 105, 850 S. Guild Avenue, Lodi, CA 95240, USA; gonzalo_castillo@fws.gov; Tel.: +1-209-334-2968 (ext. 418)

Received: 26 March 2019; Accepted: 24 May 2019; Published: 3 June 2019

Abstract: The aim of this community modeling study was to evaluate potential mechanisms by which freshwater outflow in the upper San Francisco Estuary, CA, controls the fall habitat and abundance of subadult delta smelt *Hypomesus transpacificus* and its community. Through analyses of the community matrix, community stability and the direction of change of community variables were qualitatively and quantitatively modeled under four outflow–input scenarios. Three subsystems were modeled in the low salinity zone (1–6 psu), each overlapping the location corresponding to the distance from the mouth of the estuary to upstream positions where the near-bottom 2 psu isohaline (X2) is at 74, 81, and 85 km (corresponding to high-, mid-, and low-outflows). Results suggested communities were qualitatively stable at each X2 position, but simulations showed the percent of stable models decreased from low- to high-X2 positions. Under all outflow–input scenarios, the predicted qualitative population responses of delta smelt were: (1) consistently positive for the low X2 position, and (2) uncertain under both mid- and high-X2 positions. Qualitative predictions were generally consistent with quantitative simulations and with the relations between relative abundance of delta smelt and X2. Thus, high outflow seems beneficial to subadult delta smelt when X2 reaches 74 km during fall.

Keywords: biological communities; transitional waters; community matrix; qualitative model; species interactions; freshwater flow; low salinity zone; simulation; ecological assessment; adaptive management

---

1. Introduction

The complexity of estuarine ecosystems is influenced by the broad range and extent of anthropogenic effects in these transitional waters [1,2], including morphological and hydrological alteration [3,4], entrainment of plankton and nekton to water pumps [5,6], contaminants [3,7], introduced species [8,9], and climate change [10,11]. Changes in freshwater flow, natural or otherwise, can perturb the estuarine salinity gradient [12,13], and influence the abundance and distribution of estuarine organisms through a variety of flow-mediated processes (e.g., [9,14,15]). Moreover, estuarine food webs can be sensitive to small variations in the freshwater input [16,17]. Thus, understanding how freshwater flow controls species abundance, their distribution and interactions requires explicit consideration of spatial and temporal scales and abiotic and biotic habitat components. Impaired freshwater flows can diminish the amount of habitat and the habitat quality for estuarine-dependent species [18,19]. Effects of impaired flows on fish populations include reductions in growth [20,21], survival [21], abundance [10,21], and biomass [22]. This highlights the variety of ecological roles freshwater flows may have on the population dynamics and sustainability of estuarine-dependent species.

The identification of specific ecological mechanisms linking species responses to freshwater flow in estuaries remains challenging due to the highly complex nature of such ecosystems [17,23]. In particular, there is a pressing need for analytical tools to evaluate ecological hypotheses in such ecosystems, and to inform adaptive management of freshwater flows to conserve fish species at risk of extinction (e.g., [9,24]). Because species interactions are notoriously difficult to quantify [25], qualitative analysis of the community matrix is particularly suited to model ecological interactions since essential community properties can be determined by ecosystem topology and sign structure of the interactions, rather than by the values of variables and parameters themselves [26–28]. Analysis of qualitative interactions provides a mathematically rigorous foundation for understanding ecosystem behavior [29–31] and has been used in a wide range of aquatic ecosystems to evaluate how community stability and community variables respond to abiotic and biotic perturbations, including, nutrient input in freshwater food webs (e.g., [26,32]); introduced species in freshwater communities (e.g., [30,33]) and environmental disturbance in kelp–urchin communities [34]. Qualitatively models in transitional waters have included evaluations of community responses to dumping of mine wastes in shallow-coastal communities (e.g., [35,36]); and increased ocean acidification in an estuarine community [37]. However, there is a need for further metrics to interpret community responses and to determine whether the predicted influence of a perturbation on a given community variable or species of interest is consistent with field observations.

The upper San Francisco Estuary (hereafter upper SF Estuary; California; Figure 1) is an imperiled ecosystem and the central hub of California's water supply infrastructure [38]. Outflow is the most ecologically important source of freshwater to the SF Estuary and represents the net flow at Chipps Island in the Suisun Region (river km 74; Figure 1) after subtracting water diversions and depletions in the upper SF Estuary and its watershed [6,9]. An ecologically meaningful metric of outflow forcing on the upstream distribution of the salinity field in the SF Estuary is X2, which is defined as the distance (km) from the mouth of the estuary (Golden Gate Bridge), to upstream locations where the near-bottom tidally averaged salinity is 2 psu [39]. The upper SF Estuary is home to several threatened and endangered species including the endemic delta smelt *Hypomesus transpacificus*, an osmerid particularly sensitive to human impacts due to its low fecundity, predominant annual life cycle, and limited spatial distribution [38,40]. Spawning of delta smelt occurs mostly in freshwater during spring and their planktonic larvae tend to gradually move downstream into the low salinity zone (hereafter LSZ, 1–6 psu), [24], where juveniles and subadults generally rear during summer and fall. Despite major long-term changes in pelagic food webs in the upper SF Estuary, calanoid copepods remain one of their most important prey [41]. Delta smelt was listed as threatened (federal and state) in 1993 and up-listed to endangered (state) in 2010 [42].

Understanding the ecological processes underlying the response of delta smelt to fall outflows is relevant to adaptive management in the upper SF Estuary [24,43], and such management actions are intended to improve fall habitat of delta smelt when preceding precipitation and runoff conditions result in high outflows [43]. Outflow controls the location and range of the LSZ (Figure 1), the extent of the abiotic habitat available for subadult delta smelt [10,44], and the position of X2 which controls the upstream distribution of subadult delta smelt during the fall [45]. Nevertheless, because X2 has not been found to be a simple predictor of delta smelt abundance, the need to investigate more complex mechanisms or additional variables has been suggested [39,44]. The high outflows observed in the upper SF Estuary during 2011 prompted monitoring of abiotic and biotic factors, which along with long-term monitoring in the SF Estuary, enabled the performance of interannual comparisons of abiotic and community responses to changes in the position of fall X2 [24]. Although several of the predicted abiotic and biotic responses in that study could not be confirmed from available field data, the abundance of delta smelt in 2011 and early 2012 was one the highest reported since the early 2000s. Yet, delta smelt became very rare during the 2012–2016 drought and failed to increase its abundance during the wet year 2017, indicating the need to evaluate new management options (e.g., [7,9,46]).

Given the complex abiotic and biotic responses to fall outflow in the upper SF estuary [24,38], there is a pressing need for further analytical approaches to evaluate how outflow controls the essential community structure and functional group interactions influencing delta smelt in the LSZ under particular outflow scenarios (hereafter delta smelt subsystems), and to investigate the likely mechanisms controlling the population response of delta smelt in different subsystems (Figure 1).

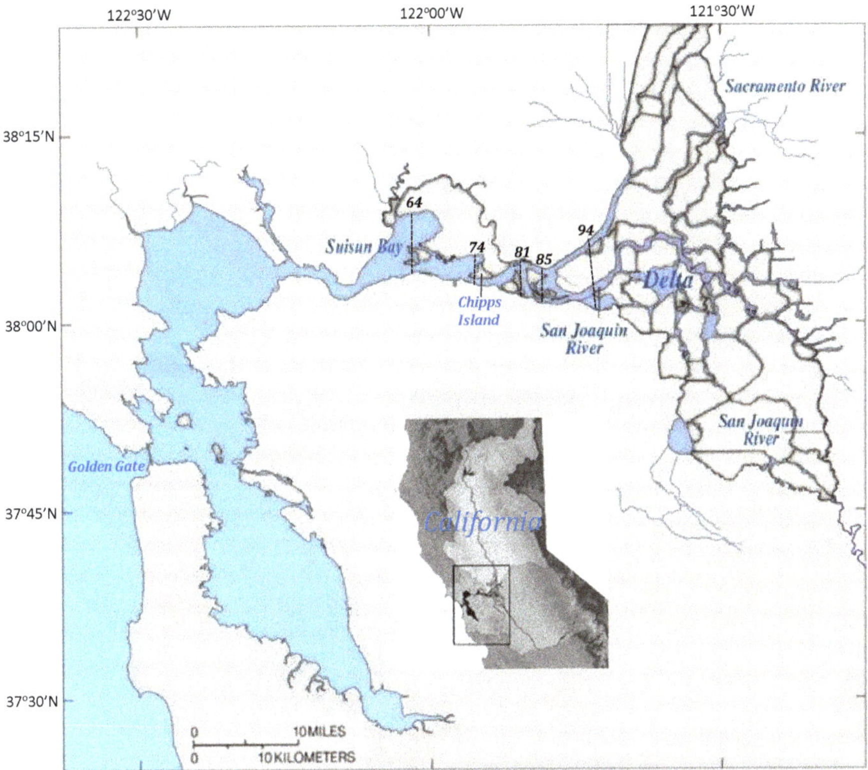

Figure 1. The San Francisco Estuary showing the Suisun Bay and Delta regions in the upper estuary. Dashed lines denote the positions of the 2 psu isohalines (X2) considered in community models (74, 81, and 85 km) and the range of the average X2 position in September–October (64–94 km) from 1967 to 2017. Map adapted from [24].

The objectives of this study were to evaluate: (1) the stability of the delta smelt subsystems in the LSZ when fall outflows are alternatively maintained at low-, mid-, and high-X2 positions corresponding to 74, 81, and 85 km, (2) the predicted direction of change and its uncertainty for the abundance of delta smelt and other species and trophic levels in each subsystem under four outflow input scenarios, and (3) whether field patterns of relative abundance for delta smelt and X2 are consistent with the predictions of community models. Three community metrics were adapted to model the response of delta smelt and other community variables to outflows, first, the weighted prediction matrix [47] was modified to denote both the direction of change and its uncertainty in a signed weighted prediction matrix (Ws), second, a modified community matrix specifying press inputs (A^P) was used to estimate the combined effect of outflow press perturbations on two or more community variables, and third, the proportional net change of adjoints in the negative community matrix ($\Delta\hat{p}$) was used as a measure of the determinacy in the direction of change of variables in quantitative simulations. Finally, statistical relations between the field relative abundance of subadult delta smelt and X2 enabled to evaluate the predicted population response of delta smelt in each of the modeled subsystems.

2. Material and Methods

2.1. Study Area

The SF Estuary is the largest estuary in the west coast of the United States (Figure 1), and has been profoundly altered physically, chemically, and biologically [6,8]. Multiple stressors in the upper SF Estuary have been implicated in the pelagic organism decline occurring in the early 2000s [48,49]. Reservoirs upstream of the SF Estuary, and substantial water diversions within an extensive network of interconnected channels surrounding leveed agricultural islands (the Delta) have greatly altered the hydrodynamics of this estuary [13,50]. Two dataset are central to the evaluation of hydrologic conditions in the SF Estuary: (1) Dayflow, a 2-D mass balance model of freshwater outflows from the tributaries of the Upper SF Estuary [51], and (2) the California Data Exchange Center, a query-based portal which maintains, and operates an extensive hydrologic data collection network [52]. In addition, the Interagency Ecological Program (IEP) provides long-term monitoring data of pelagic, planktonic, and benthic organisms in the SF Estuary [53]. The field data generated by the IEP have been collected consistently over time at multiple locations representing the study areas in the upper SF Estuary (e.g., [24,38,54]). The temporal and spatial coverage of these field surveys has enabled to evaluate the role of hydrodynamic forcing on the distribution and relative abundance of planktonic, pelagic, and benthic species (e.g., [24,38,55–57]). Additional monitoring of aquatic macrophytes by the University of California at Davis, has been conducted extensively using a combination of field sampling and remote sensing in the upper SF Estuary (e.g., [58,59]).

The LSZ area varies greatly with outflow, and is about 90 km^2 when X2 is positioned at 74 km, 50 km^2 when X2 is positioned at 81 km, and 40 km^2 when X2 is positioned at 85 km [24]. Altered fresh water outflow in the upper SF Estuary has greatly changed the abiotic habitat for pelagic fishes [9,54], and facilitated the introduction of species [60,61]. A high proportion of pelagic and planktonic organisms in the upper SF Estuary are also lost due to prevailing upstream flows in the south Delta caused by massive water pumps [5,6]. As is the case of the lower SF Estuary, most communities in the upper SF Estuary are dominated by introduced species [61,62]. The introduced macrophyte *Egeria densa* was first reported in the Delta in 1946 [63] and became the most abundant submersed plant in the Delta, where it has greatly altered community structure and functions [58,64,65]. The introduction of the clam *Potamocorbula amurensis* in the late 1980s caused major trophic changes in the benthic and pelagic communities of the SF Estuary [60,66], greatly adding to the filter feeding activity of the clam *Corbicula fluminea*, an earlier introduction in the upper SF Estuary [67]. In addition, the cyanobacterium *Microcystis aeruginosa* was first observed in the Delta in 1999 [68] and produces toxic algal blooms that can be harmful to upper trophic levels [69,70].

2.2. Community Matrix Models

The community structure and interactions in this study were based on qualitative analysis of the community matrix [29,71]. Given a classical dynamical ecosystem of *n* interacting variables (N_i), the change in the equilibrium level of species n_i^* is

$$\frac{dn_i^*}{dt} = fi(n_1, n_2, \ldots, n_n; c_1, c_2, \ldots, c_n) \quad (i = 1,\ n). \tag{1}$$

where the n_i represent species abundances or abiotic factors. The c_i are parameters that govern biological rates (e.g., birth, mortality, growth), and depend on the species and environment [72,73]. Modeled subsystems were represented by a signed digraphs, where element a_{ij} of a qualitatively defined community matrix (A^0) of order *n* denotes the net effect to variable *i* from variable *j* (for *i* and *j* from 1 to *n*). For example, a subsystem composed of three trophic levels can be represented by three variables and their interactions (Figure 2). Three possible qualitative effects to variable *i* from variable *j* were assigned to each matrix element: $a_{i,j} = 1$, $a_{i,j} = -1$ and $a_{i,j} = 0$, which in the neighborhood of any equilibrium point respectively denote positive, negative, and no effect on the instantaneous growth of variable i due to increase in the level of variable j. The stability of the community matrix was evaluated using the characteristic polynomial:

$$p(\lambda) = |A^o - \lambda I| = a_0\lambda^n + a_1\lambda^{n-1} + a_2\lambda^{n-2} + \ldots a_n = 0 \tag{2}$$

where, A^o is as defined earlier, I is the n-dimensional identity matrix, and λ are the roots or eigenvalues [71]. Following a perturbation, negative real roots means the community can return to equilibrium. Conversely, positive real roots indicate the ecosystem moves away from equilibrium, as in unstable ecosystems. Multiple zero roots can result in unstable equilibrium [73], and zero overall feedback indicates neutral stability [74].

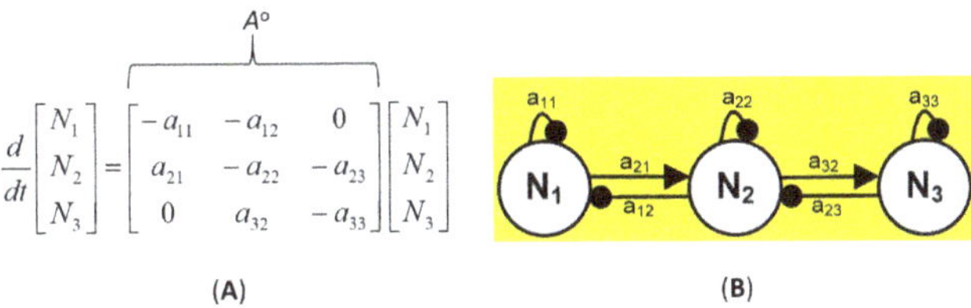

Figure 2. Example of (**A**) a 3×3 community matrix (A^o) showing qualitative interaction terms for 3 variables such as species or trophic groups, with diagonal matrix elements $-a_{11}$, $-a_{22}$, and $-a_{33}$ representing negative self-effects of each variable and zero denoting no effect, and (**B**) its corresponding signed-digraph. Open circles denote variables (N_i), with lines between variables ending in arrows and bubbles respectively denoting positive and negative effects, and curved lines originating and ending in the same variable denoting self-effects, after [29].

2.3. Community Structure and Outflow Scenarios

Qualitative analysis was used to model three flow-dependent fall community structures in the upper SF Estuary under high-, mid-, and low-net Delta outflows corresponding respectively to X2 positions of 74, 81, and 85 km [24]. The three essential components of qualitative models (community variables, interactions or links, and driving force or perturbation) were based on the following five steps:

(1) Compilation of long-term data and studies in the upper SF Estuary, including: monitoring data for delta smelt and other species [53], outflow and X2 data [51], interpretation of the spatio-temporal distribution of the salinity field in the LSZ based on the three-dimensional UnTRIM hydrodynamic model for the upper SF Estuary [24,75], and baseline knowledge of the modeled ecosystem (e.g., [6,10,39,40,54,55,63,76,77]).

(2) Consideration of dynamic and stationary factors for developing conceptual models of estuarine communities [78], where dynamic factors include physico-chemical and biological characteristics of the low salinity habitat corresponding to the LSZ each X2 position, and stationary factors include geographically fixed habitat features at each X2 position such as substrate, erodible sediment, and bathymetry in the low salinity habitat [24,38].

(3) Refinement of conceptual models into different subsystems describing the essential community variables influencing subadult delta smelt based on ecological syntheses of long-term field data and studies [24,38]. Community variables selected for each of the three modeled subsystems were based on functional groups (e.g., [74,79]), and their predominant spatio-temporal overlap with dynamic and stationary abiotic and biotic factors. Except for delta smelt and species with significant ecological impacts (the cyanobacterium *M. aeruginosa* [80], the clams *P. amurensis* and *C. fluminea* [56], and the macrophyte *E. densa* [58], Table 1), other functional groups included trophic levels to minimize redundant species interactions (e.g., [74,81]), (Table 1).

Table 1. Variables considered in qualitative community models for the upper San Francisco Estuary. Black circles denote community variables included in the low salinity zone at three positions of the 2 psu isohaline (X2).

Variable Type	Variable Code (Description)	Low X2 74 km	Mid X2 81 km	High X2 85 km	Functional Role	Community Response to Outflow
Explanatory	FL (Delta outflow, m^3 s^{-1})	323 [24]	227 [24]	142 [24]	Major abiotic forcing factor controlling the overlap between dynamic and stationary factors in the upper San Francisco Estuary, the position and size of the low salinity zone and the low salinity habitat of delta smelt and community interactions [24,38].	
Response (Community variables)	CF (*Corbicula fluminea*, Asian clam) [1]		● [24]	● [24]	Filter feeder exerting grazing pressure on phytoplankton [82,83].	(+) [56]
	DS (*Hypomesus transpacificus*, delta smelt)	● [38,45]	● [38,45]	● [38,45]	Zooplanktivorous preying primarily on copepods, mysids, and cladocerans [40,41].	(+) [10,54]
	ED (*Egeria densa*, Brazilian waterweed) [1]			● [58,59]	Primary producer of dense submersed vegetation beds which reduce open water habitat, alter the habitat for other aquatic plants [58,65] and provide habitat for centrarchids [84].	(−) [59,85]
	MA (*Microcystis aeruginosa*, cyanobacterium) [2]		● [24,86]	● [24,86]	Biological contaminant due to its poor nutritional value and transfer of toxic microcystins into the aquatic food web [24,69,70,80].	(−) [86]
	PA (*Potamocorbula amurensis*, overbite clam) [1]	● [24]	● [24]		Filter feeder exerting grazing pressure on phytoplankton and zooplankton [87,88].	(−) [56]
	PH (Phytoplankton)	● [38,89,90]	● [38,89,90]	● [38,89,90]	Primary producer fueling higher trophic levels of the food web in the open waters of the Delta and Suisun Bay [89,90].	(+) [38,91]
	PR (Predators of delta smelt)	● [24,38]	● [24,38]	● [24,38]	Piscivores deemed to exert predation pressure on delta smelt throughout the Delta and Suisun Bay [24], including two introduced species, striped bass *Morone saxatilis* [62,92] and largemouth bass *Micropterus salmoides* [64,84].	(0) [3] [38]
	ZO (Zooplankton)	● [24,38]	● [24,38]	● [24,38]	Primary and secondary consumer supporting pelagic and benthic food webs [41,88].	(+) [55]

Notes: [1] Introduced species [58,60,67]; [2] species likely introduced [68]; [3] assumed zero since vulnerability of delta smelt to predators is affected by turbidity and bioenergetics [38].

(4)   Reformulation of conceptual models into signed digraphs based on qualitative model guidelines (e.g., [71,79,93]). Negative self-effect (self-damping) was assumed to arise for each variable from density-dependent growth rate or a limited source, as in the case of nutrients [71]. Each community variable was then implicitly connected to other variables or abiotic factors through negative feedback [71,74]. Reported community interactions considered for the modeled subsystem included: predation (+, −); interference competition (−, −); and amensalism (0, −), (Appendix A).

(5)   Estimation of the direction of change of community variables (+, 0, −) in response to increased outflow (Table 1). Four outflow input scenarios were modeled, with the first scenario accounting for the effect of outflow on the previously referred species having significant ecological impacts. The outflow inputs for the three subsequent scenarios were used to evaluate whether cumulative outflow inputs in each subsystem could reinforce or reverse potential responses on delta smelt and other community variables. These scenarios included: scenario 1 + phytoplankton (scenario 2), scenario 2 + zooplankton (scenario 3), and scenario 3 + delta smelt (scenario 4).

2.4. Qualitative Analyses

The modeled subsystem at each X2 position was considered stable if the two Routh–Hurwitz (R–H) criteria [94] were met. The first R–H criterion requires all roots of the characteristic polynomial be different from zero and have the same sign. The second R–H criterion requires all Hurwitz determinants be positive, resulting in stronger feedback at lower feedback levels than at higher levels. The effect of sustained input (C_h), on the rate of change of a hth variable corresponding to the jth variable in the community matrix was predicted assuming an ecosystem in moving equilibrium [29,71]. Such input could result in positive, negative, or no effect on the equilibrium magnitude or carrying capacity (N_i^*) of a species. The inverse of the negative community matrix ($-A_{ij}^{-1}$) provides an estimate of change in the equilibrium level of variable N_i^* carrying capacity due to change in parameter C_h [95].

$$\frac{dN_i^*}{dC_h} = \frac{-C_{ji}^T}{\left|-A_{ij}\right|} \tag{3}$$

where $-C_{ji}^T$ is the adjoint of the negative community matrix ($Adj - A_{ij}$), and $\left|-A_{ij}\right|$ is the determinant of the negative community matrix. Since $\left|-A_{ij}\right|$ is constant and positive for each element of $-A^{-1}$ in stable ecosystems, the predicted response under local equilibrium was inferred from the $Adj - A$:

$$Adj - A_{ij} = \left(-A_{ij}^{-1}\right)^T \tag{4}$$

which corresponds to the transpose of the negative cofactor matrix [33]. The absolute number of positive and negative complementary feedback cycles (Tc) which contributes to each element of the $Adj - A$ was used to derive model uncertainty. Because Tc differed from zero in all present models, it was deemed practical to modify the weighted prediction matrix [47] to express both the direction of change in response to input and its uncertainty by defining a signed weighted prediction matrix (Ws):

$$Ws = \frac{\overrightarrow{Adj - A}}{Tc} \tag{5}$$

where $Adj - A$ is as defined in Equation (4), and "→" is a vectorized matrix operator denoting element-by-element division for each element of the Ws matrix [47], and Tc is as defined earlier and $\neq 0$. Hence, both the lowest Ws value (−1, negative response) and highest Ws value (+1, positive response) indicate unconditional sign determinacy and $Ws = 0$ denote total uncertainty in the direction of change, while absolute values of $Ws \geq 0.5$ and < 1 are indicative of high level of sign determinacy, after [27].

To estimate the overall influence of a press input (P_k) due to a disturbance k acting directly upon two or more variables, the combined effect of P_k on the direction of change for community variable i

and their corresponding uncertainty level was computed from the corresponding matrix element Ws_{ik} (Equation (5)) and these were derived from a modified community matrix specifying press inputs (A^P):

$$
A^P = \begin{bmatrix}
a_{11} & . & a_{1n} & P_{1k} \\
. & . & . & . \\
a_{n1} & . & a_{nn} & P_{nk} \\
0 & 0 & 0 & -1
\end{bmatrix}
\tag{6}
$$

where a_{11} to a_{nn} represent the interaction elements a_{ij} of community matrix A^o, and P_{ik} denotes the press input k on community variable i, which in a qualitatively defined ecosystem is specified as having positive (1), negative (−1) or no effect (0). The last row of matrix A^P includes zero elements to denote no effect of community variables on P_k, and a negative element (−1) to preserve the stability of matrix A^P provided the community matrix A^o is stable. Qualitative analyses were conducted in Maple 18 [47].

2.5. Quantitative Simulations

Simulations using A^P further enabled to quantitatively assess the stability of the community and the direction of change of its variables relative to the qualitatively specified matrix with same structure and sign of interactions. Simulations were performed by assigning interaction strengths between 0.01 and 1.0 to non-zero elements using a pseudorandom number generator from a uniform distribution [31]. The overall direction of change and determinacy in a quantitatively-specified ecosystem was derived from the proportional net change for each community variable ($\Delta \hat{p}$):

$$
\Delta \hat{p}_{ik} = \frac{f(+Adj - A_{ik})^* - f(-Adj - A_{ik})^*}{N}
\tag{7}
$$

where $f(+Adj - A_{ik})^*$ and $f(-Adj - A_{ik})^*$ respectively denote the absolute frequencies of positive and negative $Adj - A_{ik}$ in stable matrices, and N is the total number of simulations for each perturbation scenario, including stable and non-stable matrices ($N = 10{,}000$). As in the case of Ws, the value of $\Delta \hat{p}$ provided both the overall direction of change and its uncertainty for each community variable, with extreme potential values ($\Delta \hat{p} = -1$, negative response and $\Delta \hat{p} = 1$, positive response) both indicating no uncertainty in the direction of change among simulations, $\Delta \hat{p} \geq |0.5|$ and $< |1|$ indicating low uncertainty, and $\Delta \hat{p} = 0$ denoting total uncertainty. Quantitative simulations were performed in a program written in C# version 3.0 [96].

2.6. Statistical Analyses

To determine the extent to which the predicted response to outflow input by each variable in qualitative models represented the overall direction of change in quantitative simulations, $\Delta \hat{p}$ was plotted vs. $Adj - A$ for all scenarios at each X2 position. Values of $\Delta \hat{p}$ were also plotted vs. Ws to compare the direction of change and uncertainty between qualitative and quantitative predictions. These variables were fitted at each X2 position using Lowess smoothing. The predicted population response for subadult delta smelt by community models at each X2 position was compared with the relative abundance for subadult delta smelt as a function of X2 to evaluate model predictions. Relative abundance data for subadult delta smelt, monitored since 1967, was obtained from the California Department of Fish and Wildlife's Fall midwater trawl index for the period 1967–2017 [97]. The average position of X2 during September–October was derived from the California Department of Water Resources Dayflow program [51]. Linear regression analysis was used to evaluate the slope of the relative abundance for delta smelt during fall (Y) versus the average X2 position in September–October (X). Regressions were conducted for three ranges of X2, which included the three X2 positions in community models (X2 $\geq$ 74, $\geq$81, and $\geq$85 km). The entire historical record of subadult delta smelt abundance as a function of the average September–October X2, was further considered as a baseline relative to the three X2 positions evaluated in the present community models and a previous fall outflow evaluation [24]. Data were analyzed and plotted using Curve Expert Professional (version 2.6; D.G. Hyams).

3. Results

3.1. Community Stability

Compared to the models at the low X2 position, the modeled communities at the mid- and high-X2 positions included a higher number of introduced species and more non-trophic interactions (Figure 3, Appendix A). Although all qualitative community models were stable, the percent of stable simulations decreased greatly from downstream (low X2) to upstream (high X2) positions in the LSZ, and such decrease in stability seemed primarily due to the increasing percent of models with stronger high-level feedbacks than low-level feedbacks at the mid- and high-X2 positions (Table 2).

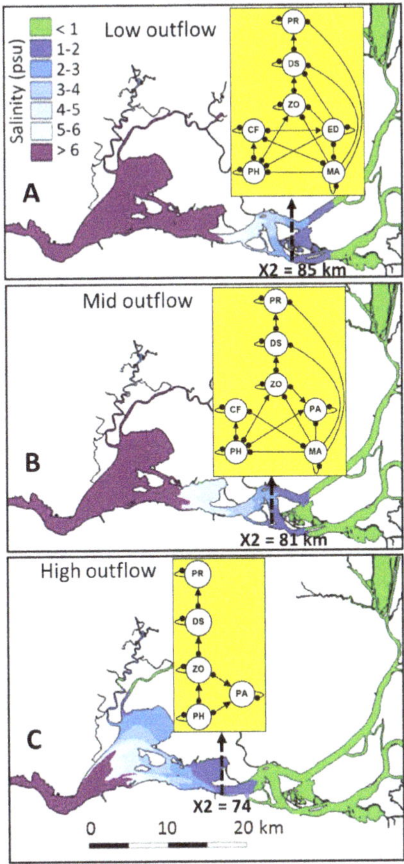

Figure 3. Alternative fall community models for the Delta smelt subsystem under three scenarios of the salinity field in upper San Francisco Estuary. Dashed arrows show where the low salinity zone overlaps the position of the 2 psu isohaline (X2) at (**A**) high X2 (85 km), (**B**) mid X2 (81 km), and (**C**) low X2 (74 km) at low-, mid- and high-outflow, respectively. Community variables included in signed digraphs: CF = *Corbicula fluminea*; DS = Delta smelt; ED = *Egeria densa*; MA = *Microcystis aeruginosa*; PA = *Potamocorbula amurensis*; PH = phytoplankton; PR = predators of delta smelt; ZO = zooplankton. Salinity field after [75].

Table 2. Percent of quantitative community models meeting zero, one and both Routh–Hurwitz (R–H) stability criteria at three X2 positions in the low salinity zone of the upper San Francisco Estuary (based on 10,000 simulations at each X2 position).

R–H Criteria	Low-X2 (74 km)	Mid-X2 (81 km)	High-X2 (85 km)
None	2.7	39.9	48.2
Only I	0.3	3.5	10.6
Only II	0.0	0.0	0.0
Both	97	56.6	41.2

3.2. Community Response to Outflow

The predicted change in abundance for delta smelt and other community variables to outflow input varied greatly among the three X2 positions (Figure 4), both in the case of qualitative models (Figure 5) and simulations (Figure 6). In response to outflow input for the delta smelt subsystem at the low X2 position (Figure 4), the predicted direction of change in the abundance of all community variables was positive, except for the clam *P. amurensis*, with predicted responses of species and trophic levels generally showing high sign determinacy, particularly for delta smelt, its predators, and zooplankton across most outflow input scenarios (Figures 5 and 6). Outflow input at the mid X2 position (Figure 4), revealed the cyanobacterium *M. aeruginosa* was the only community variable with a consistently negative response and high sign determinacy in both qualitative models and simulations. Yet, the clam *C. fluminea* and predators of delta smelt showed high sign determinacy under most quantitative simulation scenarios at the mid X2 position (Figures 5 and 6).

The cyanobacterium *M. aeruginosa* suggested a negative response to outflow at the high X2 position (Figure 4), but to a lower extent compared to the mid X2 position (Figures 5 and 6). The Brazilian waterweed, *E. densa* was the only other variable at the high X2 position indicating high degree of qualitative sign determinacy but only at one outflow scenario (Figure 5). In contrast, simulations at the high X2 position indicated all the predicted responses of community variables had low sign determinacy (Figure 6).

The relation between $\Delta\hat{p}$ and $Adj - A$ indicated the predicted direction of change for species and trophic groups in response to outflow input was generally consistent between qualitatively and quantitatively specified subsystems at each X2 position; this despite a plateau in $\Delta\hat{p}$ was towards its highest values (Figure 7A). As inferred from $\Delta\hat{p}$ and Ws, the direction of change and degree of certainty in the direction of change in quantitatively specified simulations were associated with the corresponding qualitative model, with such relations appearing progressively more sigmoidal toward higher-X2 positions (Figure 7B).

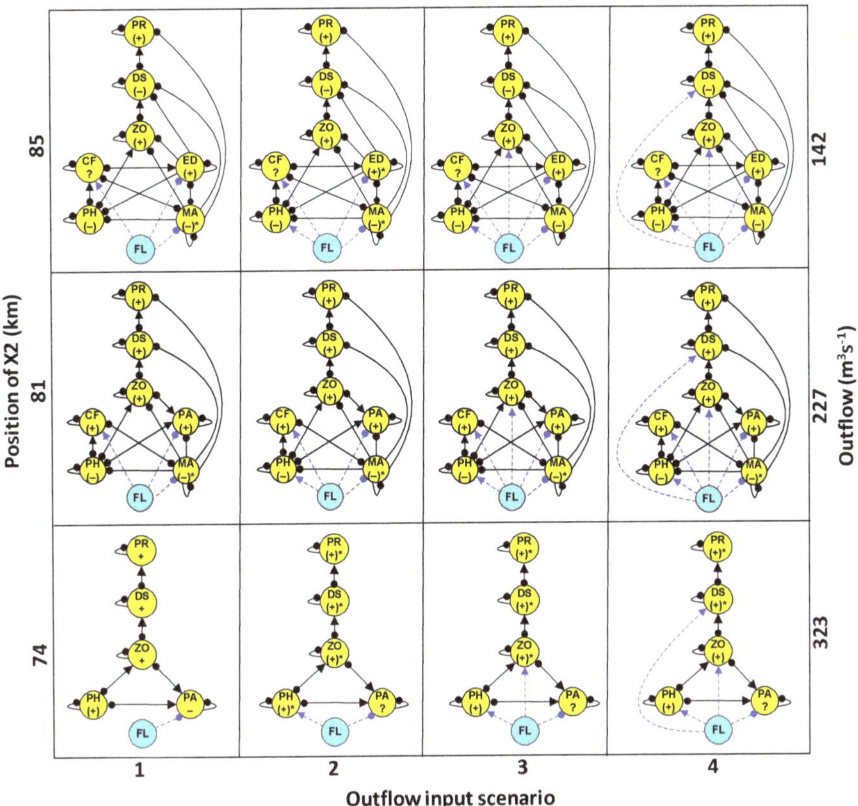

Figure 4. Signed digraphs of the modeled delta smelt subsystems at three positions of the 2 psu isohaline (X2) in the upper San Francisco Estuary. Dashed lines indicate the four alternative outflow (FL) input scenarios for community variables. Dashed lines in scenario 1 represent outflow perturbations to *Corbicula fluminea* (CF), *Egeria densa* (ED), *Microcystis aeruginosa* (MA), and *Potamocorbula amurensis* (PA). Dashed lines in outflow scenarios 2, 3, and 4 cumulatively add outflow perturbations to phytoplankton (PH, scenario 2), zooplankton (ZO, scenario 3), and delta smelt (DS, scenario 4). PR denotes predators of delta smelt. Community variables show the predicted direction of change in abundance (+, −) in response to sustained outflow input. Direction of change without parentheses is unconditional. Direction of change in parentheses denote high-certainty (asterisk) and low-conditional certainly (no asterisk), and ? denotes complete ambiguity.

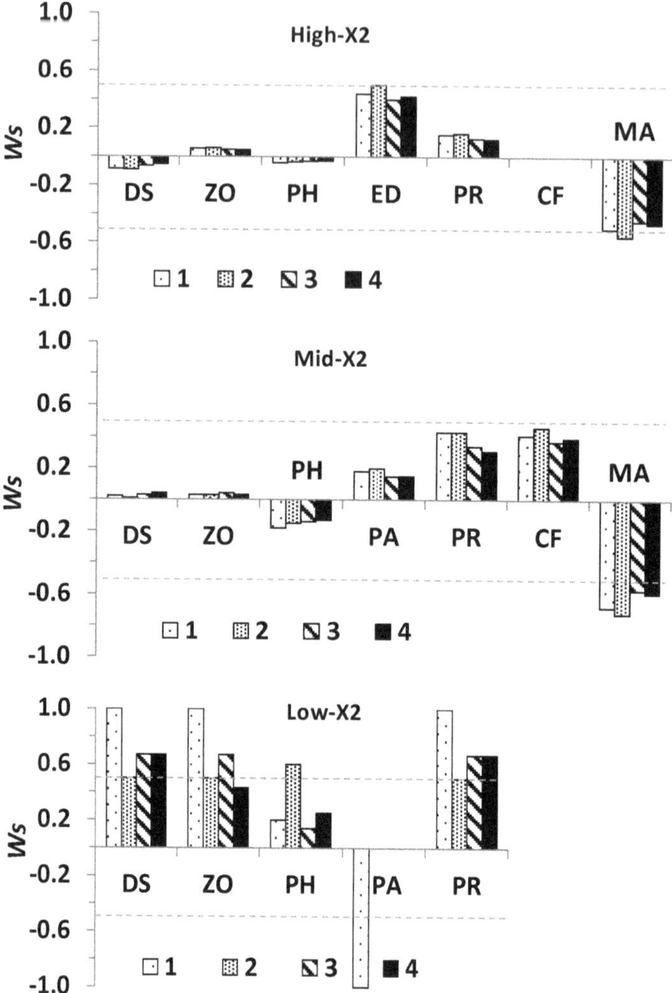

Figure 5. Predicted direction of change and its uncertainty for community variables based on signed weighted predictions (Ws) at low-, mid-, and high-X2 positions under outflow–input scenarios (1–4, Figure 4). Values of $|Ws|$ for each community variable range between unconditional sign determinacy (1) and complete uncertainty (0), with $|Ws|$ between 0.50 (dotted line) and ≤ 1 indicating high sign determinacy.

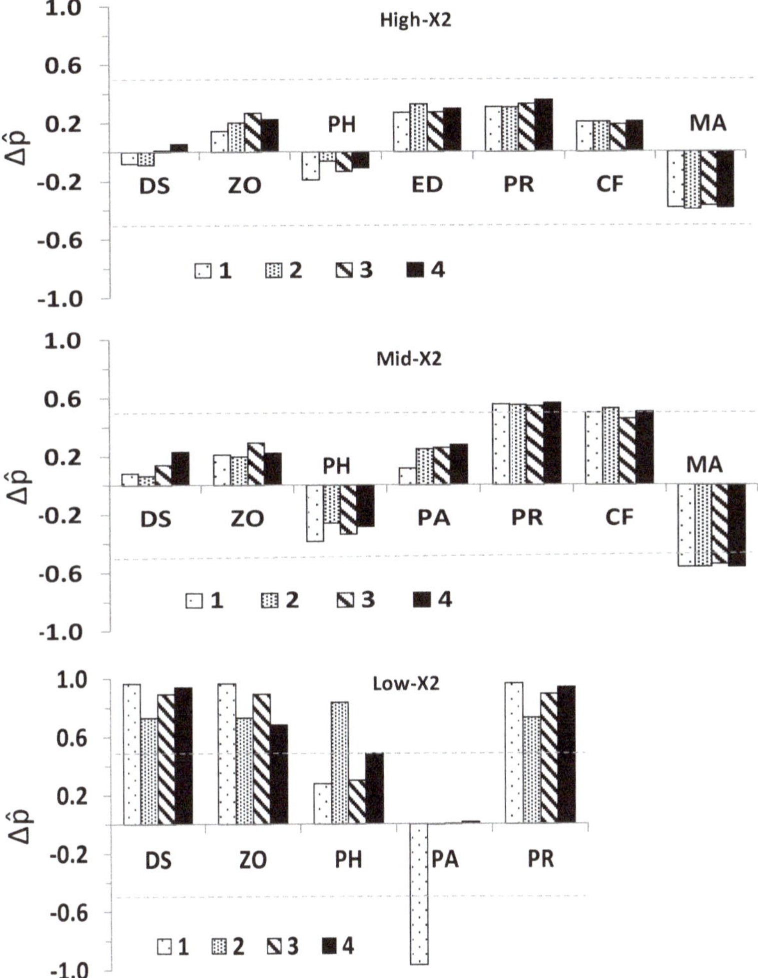

Figure 6. Predicted direction of change of community variables for low-, mid-, and high-X2 positions based on the difference between the proportions of quantitative simulations with positive and negative $Adj - A$ ($\Delta\hat{p}$). Acronyms represent species or trophic groups (Table 1, Figure 3) and legend denotes the four outflow input scenarios (Figure 4). Values of $|\Delta\hat{p}|$ can range between 1 (unconditional sign determinacy) and 0 (complete uncertainty), with $|\Delta\hat{p}|$ between 0.5 (dotted lines) and <1 implying low uncertainty in the predicted direction of change.

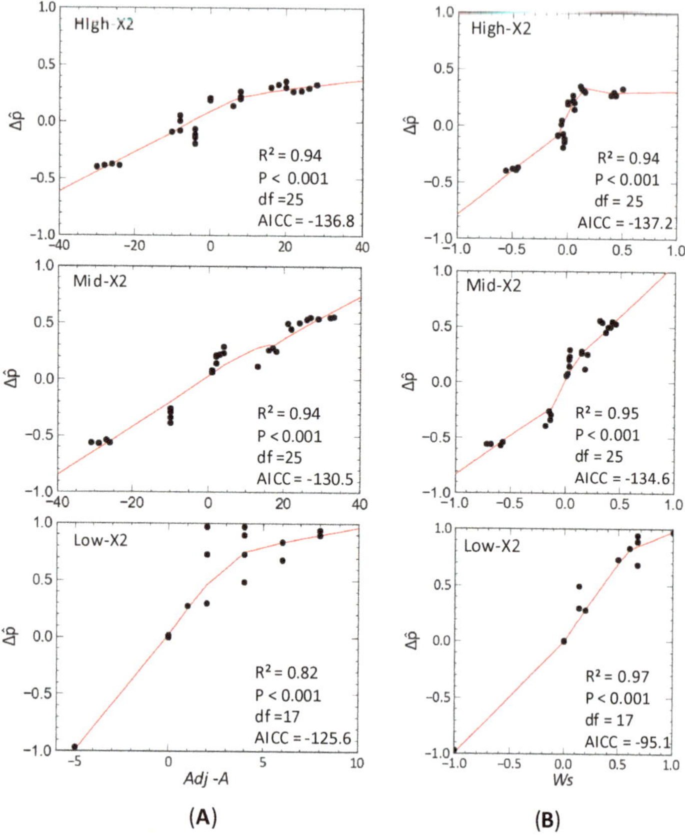

Figure 7. Relations between the net proportion of positive and negative *Adj − A* in stable quantitative models ($\Delta\hat{p}$) at low-, mid-, and high-X2 positions and: (**A**) the estimated direction of change of corresponding variables in qualitative models (*Adj − A*); (**B**) the signed weighted feedback in qualitative models (*Ws*), as indicated for modeled subsystems in the low salinity zone at low-, mid-, and high-X2 positions.

3.3. Delta Smelt Abundance and X2

When only considering years in which the average position of X2 in September–October was located at 81 km and further upstream, the relative abundance of subadult delta smelt was not associated to the position of X2 (Figure 8). However, when all years in which X2 was located at 74 km or further upstream were included, the relative abundance of subadult delta smelt was inversely associated to the average position of X2 in September and October (Figure 8). Moreover, when considering all the years with available relative abundance for subadult delta smelt over the period 1967–2017, an inverse association between the relative abundance of subadult delta smelt and the average position of X2 during September–October was also detected, and such relation included X2 positions as low as 65 km (Figure 8).

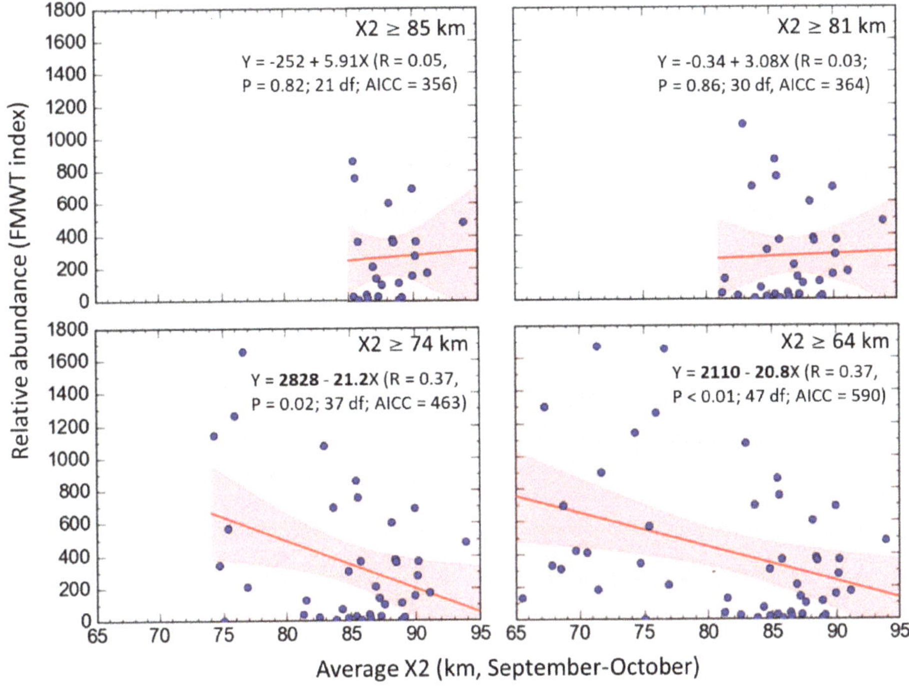

Figure 8. Relative abundance for subadult delta smelt based on the fall midwater trawl (FMWT) index (Y) versus the average position of the 2 psu isohaline (X2) during September and October (X) under four ranges of X2 over the period 1967–2017. Dark and light shaded areas denote 95% confidence intervals for regression lines and predicted values, respectively. Regression coefficients in bold type are significant (P < 0.05).

4. Discussion

4.1. Community Stability

While the modeled subsystems in the upper SF Estuary at the low-, mid-, and high-X2 positions were qualitative stable based on the R–H criteria, quantitative simulations using such stability criteria revealed a significantly higher percent of stable models at the low X2 position, which could promote enhanced ecosystem resilience when the LSZ is predominantly located in the Suisun region (Figure 1). In contrast, a decreasing number of simulations met both R–H stability criteria at the mid- and high-X2 positions, respectively (Table 2). Such spatial stability pattern supports the hypothesis that fall outflows maintaining the X2 position at 74 km are more conducive to sustaining a community structure supporting the delta smelt population compared to upstream X2 positions. The stability patterns inferred from simulations are consistent with marked reductions in upstream fall habitat quality for subadult delta smelt and with the low abundance for this species since the early 2000s [24,54]. These changes have implicated long-term stressors in the upper SF Estuary, including water diversions, introduced species and contaminants [6,13,50]. As in the case of other ecosystems where native estuarine fish species evolved in turbid waters [98], the decreased turbidity in the Delta is partly due to the spread of *E. densa* and other introduced macrophytes, which have favored the population expansion of introduced predators [38,99]. Although herbicides have been used in an attempt to control macrophytes in the Delta, the extent to which such efforts have been effective is unclear.

Nonetheless, some transitional waters formerly dominated by macrophytes seem to have shifted to pelagic ecosystems through the long-term use of herbicides [100].

4.2. Community Model Predictions

Comparisons between predicted population responses of community variables considered in this modeling study and field data [24] were possible for phytoplankton, zooplankton, *P. amurensis*, and delta smelt. Consistent with the direction of change and sign determinacy for delta smelt predictions at the low X2 position (Figures 4–6), the relative abundance of subadult delta smelt in fall 2011 when X2 was at 74 km was a ten-fold higher compared to years with higher fall X2 positions since the decline of pelagic species in early 2000s [38]. As in the case of the stability for quantitatively simulated subsystems, the response of delta smelt to sustained outflow was influenced by the location of the LSZ and hence, the underlying community structure. When outflow caused X2 to reach 74 km in the Suisun region, a positive population response was suggested for delta smelt and such response had high sign determinacy despite the ambiguous response of *P. amurensis* under most outflow input scenarios.

Field data for phytoplankton in fall 2011 showed either similar or greater average phytoplankton biomass, as measured by chlorophyll-α, at the low-X2 position relative to the mid- and high-X2 positions [24]. Such field pattern could result from variability in the mass balance of phytoplankton production, total grazing, and transport [89,101]. A positive overall phytoplankton response to outflow input was only predicted for the models at the low-X2 position, and such prediction had low uncertainty or marginally low uncertainty for scenarios 2 and 4 (Figures 4–6). Hence, suggesting a conditional phytoplankton response to outflow.

Despite the high variability in zooplankton biomass contributed by calanoids—a key prey item of delta smelt—such biomass seemed slightly higher at the low X2 position in fall 2011 compared to years in which the LSZ was found at higher X2 positions [24]. A positive response of zooplankton was predicted in the LSZ at the three X2 positions but such predictions only had low uncertainty at the low-X2 position under most outflow scenarios (Figures 4–6). Importantly, considering the larger area of the LSZ when X2 is located at 74 km relative to higher X2 positions (Figure 1), the potential carrying capacity for species and trophic levels could be significantly higher when the LSZ overlaps Suisun Bay in the fall.

The field abundance of *P. amurensis* in the Suisun Bay region was lower during October 2011 when X2 was at 75 km compared to the same location in the lower-outflow year 2010, when fall X2 was at 85 km, which is consistent with a negative local response of this species to outflow ([56], their Table 1), and with the low uncertainty in the direction of change under scenario 1 for the low-X2 position (Figures 4–6). In contrast, the uncertain direction of change of *P. amurensis* under other outflow input scenarios at the low X2 position implies the negative direct effect of outflow on *P. amurensis* would be offset by potential outflow-induced increases in its food supply. Because the approximate center of distribution and abundance of the clam *P. amurensis* is Suisun Bay and the downstream embayment, San Pablo Bay ([56,67], their Figure 3) its abundance is likely higher when the LSZ overlaps Suisun Bay relative to LSZ locations further upstream. Moreover, changes in its benthic abundance would not be as responsive to changes in the salinity field compared to pelagic species.

4.3. Delta Smelt Abundance under Different Ranges of X2

The predicted response for delta smelt in community models was generally consistent with the patterns of relative abundance for subadult delta smelt across different ranges of X2 (Figure 8). With the exception of 2011 and 2017, no years since 2000 have shown the average of X2 in September–October located in the Suisun region. However, as in the case of year 2017, low values of X2 in fall since 1967 have not always been associated with high relative abundance for subadult delta smelt (Figure 8). This suggest that additional abiotic factors can be important in controlling the abundance of delta smelt in some years. Unlike 2011, the summer 2017 was warmer than usual, which could have reduced the survival for juvenile delta smelt (e.g., [11,38]). Importantly, since year 2000 there has been a substantially lower occurrence of years where the average position of X2 in September–October was ≤75 km in the upper SF Estuary (11.1%) compared to the period 1967–1999 (39.4%). A contributing factor to such difference is the long-term increase in freshwater diversions, both upstream of the SF Estuary and in the south Delta. For example, total water diversions from 1986 to 2005 held X2 upstream of 71 km nearly 80% of the time. In contrast, assuming no upstream water diversions, X2 would have been maintained upstream of 71 km only 50% of the time [13].

The predicted response for delta smelt in community models was generally consistent with the reported decline of abiotic habitat quality indices of at higher positions of X2 (e.g., [10,54]). The present study suggests a key role of community structure in the LSZ given its influence on community stability and response to outflow. Hence, supporting the hypothesis that fall outflow exerts a positive influence on the abundance of subadult delta smelt when the X2 is positioned at 74 km [24]. When considering the four outflow scenarios for each of the three X2 positions, a generally similar response of community variables was apparent across outflow input scenarios, and this was suggested despite the overall ambiguity of subsystems at the mid- and high-X2 positions (Figures 5 and 6). This implies the propagation of outflow inputs through multiple feedbacks in the community could make it difficult to distinguish specific outflow input mechanisms among years.

4.4. Community Models and Prediction Metrics

Because qualitative models sacrifice precision for generality and realism [71], the qualitative models used in this study were intended to provide insights on the community stability and the complex mechanisms mediating the response of delta smelt and its community to fall outflows, rather than quantifying changes in the abundance of community variables in the upper SF Estuary. Given the large differences in qualitatively derived stability among X2 positions, the use of $\Delta\hat{p}$ in quantitatively specified subsystems seemed a useful community metric to evaluate both the overall direction of change of community variables and the determinacy in the overall direction of change across subsystems with the same topology but varying interaction strengths. Although estimating the direction of change and the degree of uncertainty in the response of community variables to multiple inputs entails several steps (e.g., determining qualitative direction of change of community variables, the absolute feedback matrix and the weighted predictions from each element of the $Adj - A$, [47]), the present study showed that the response of community variables to multiple inputs can be alternatively obtained by defining the press inputs for each variable in a modified community matrix (A^P). This enabled readily estimation of the qualitative response of each community variable, and their corresponding weighted predictions. In the case of communities without zero absolute feedback matrix elements, it is also feasible to define signed weighted predictions to facilitate comparisons between qualitative models and quantitative simulations. Moreover, the use of matrix A^P can provide independent verification of detailed computations involving inputs to several community variables. For quantitatively specified communities, the use of matrix A^P in combination with $\Delta\hat{p}$ seemed particularly useful to estimate the determinacy in the direction of change for each community variable under simulated press inputs and interaction strengths among community variables. Although the signed weighted feedback seemed a useful measure of uncertainty for qualitatively stable models, and was significantly related to the estimated uncertainty of quantitatively specified models (Figure 7B), quantitative simulations showed that interaction strength can greatly influence the stability of community models with the same sign structure of stable qualitative models. This in turn could influence the determinacy of the predicted direction of change following a perturbation. However, because the determinacy level indicated by $\Delta\hat{p}$ accounted for the percent of non-stable models in simulation analyses, the use of $\Delta\hat{p}$ can complement qualitative analyses of the community matrix and help to evaluate hypotheses about community stability and their response to perturbations.

Despite the differences in objectives and methods in past modeling studies related to pelagic fishes and their communities in Upper SF Estuary, independent analyses of flow-related metrics (e.g., X2, salinity, specific conductance, water diversions) have revealed that hydrological conditions play a major ecological role through their interactions with multiple natural and anthropogenic forces (e.g., [6,9,102]). For example, generalized additive models (GAM) have shown association between spatial and temporal patterns in specific conductance and turbidity and long-term declines in habitat quality for delta smelt and other pelagic fishes (e.g., [54,99]). Additional GAM-based habitat suitability scenarios for delta smelt suggested changes in outflow due to future development and climate change could lead to increased habitat declines for this species across a broad range of hydrological conditions [10]. A Bayesian-based multivariate autoregressive model (MAR) revealed increased spring and fall X2 could be linked to the decline of several pelagic fishes and their zooplankton prey since the early 2000s [55]. That study further suggested negative influence of spring and winter south Delta diversions and high summer water temperature on delta smelt. Another MAR model evaluating mechanisms of species invasions and their impacts in the Upper SF Estuary suggested salinity intrusion was a primary pressure facilitating species invasions and emphasized a central role introduced species in altering multi-trophic interactions [57]. Individual-based models for delta smelt have shown a combination of abiotic and biotic factors could control its population, including hydrodynamics, water temperature, zooplankton density, long-term changes in the prey-base, entrainment losses due to water diversions in the south Delta, and salinity (e.g., [103–105]). Moreover, hydrodynamic and statistical modeling showed the decline in phytoplankton in the upper SF Estuary could be largely explained by the combined impacts of increased freshwater diversions from the south Delta since the late 1960s, and the filter-feeding activity of the clam *P. amurensis* [106]. Hydrodynamic and statistical modeling also revealed that winter–spring diversions altered the salinity habitat for several species of pelagic fishes, and these changes were significantly associated to their corresponding interannual population trends [9]. Thus, these modeling studies lend further support to the connection between hydrological alteration and multiple processes contributing to the decline of pelagic fishes in the upper SF Estuary [13,38,102].

Given the physical variability of estuarine ecosystems (e.g., [6,13,15]), consideration of the variability in the salinity field seems particularly relevant when evaluating hydrodynamic forcing on estuarine communities. This study is the first qualitative analysis of the community matrix evaluating the role of freshwater flows on aquatic communities, and the first qualitative modeling study in transitional waters based on a hydrological model [75]. Qualitative community models in other estuaries have examined other types of disturbances, for example an intertidal community composed of native and introduced species (Yaquina Bay, Oregon, USA) suggested weak or absent overall feedback, which, to some extent could have buffered such community from impacts of introduced species [74]. In another estuarine community (Willapa Bay, Washington, USA), qualitative networks suggested ocean acidification could stimulate production of phytoplankton and eelgrasses [37]. Yet, the opposite predicted responses for two shellfish species were sensitive to the assumed perturbation on primary producers. This highlights the value of qualitative modeling to identify likely community outcomes while recognizing the uncertainties of complex ecological interactions. As in the case of the SF Estuary, other estuarine ecosystems have been primarily modeled quantitatively. For example, an oxygen-balance model coupled to fish growth models in a range of estuary types in South-Western Australia indicated that climate change could induce declines in river discharge, exacerbating the impacts of reduced rainfall and water diversions on hypoxia in these estuaries [107]. Such increased hypoxia could then magnify the significant reported differences in habitat quality for the growth and body condition of the teleost *Acanthopagrus butcheri* in those estuaries. A coastal ecosystem model (Integrated Compartment Model) in southern Louisiana, USA [108] suggested that even doubling current riverine flows would be insufficient to fully counteract the impacts of future sea level rise on forested wetlands, implying that habitat expansion for marine fishes would result in habitat contraction for freshwater fishes. Application of the ecosystem model AQUATOX 3.1 in the Minho Estuary, NW coast of Iberian Peninsula [109] revealed that estuarine production was strongly dependent on hydrodynamics, with temperature and river flow showing antagonistic interactions on macroinvertebrate communities. Given the increasing threats to transitional water ecosystems, the integration of qualitative and quantitative models to assess ecosystems responses to multiple perturbations could further promote ecologically sound management in these ecosystems.

4.5. Management Implications

Based on the modeled outflow-mediated community responses and the relations between subadult delta smelt abundance and fall X2 inferred from this study, the sustainability of this species could be compromised in part by the high prevalence of low- and mid- fall-outflows. Given the tendency of the delta smelt population to increase in high outflow years, the population could still be resilient provided the limited outflow in most years is not further reduced by the high percent of upstream water diversions [6,13], particularly in non-wet years [9]. While full understanding of the mechanisms by which outflow influences the abundance of a species like delta smelt does not seem necessary for conservation purposes, there is great value in continued experimental and modeling evaluations on the role of hydrological alteration and other perturbations to mitigate cumulative stressors in the upper SF Estuary, including those anticipated from climate change [10,11], provided such evaluations are implemented using sound ecological and adaptive management guidelines (e.g., [110–112]). Considering the long-term population declines for delta smelt and other native species like longfin smelt, *Spirinchus thaleichthys*, and their record low abundances since the drought beginning in 2012 [7,46], the positive influence of outflow in other seasons for delta smelt and other species (e.g., [7,9,38,102]) suggests the need to integrate flow adaptive management actions across species, seasons and hydrological conditions not restricted to wet years.

5. Conclusions

This study supports the conclusion that fall outflow in the upper SF Estuary tends to increase the abundance of delta smelt when the position of X2 overlaps Chipps Island (X2 = 74 km), but not when X2 is at 81 km or further upstream. The stability patterns of three modeled subsystems of delta smelt in the upper SF estuary support the hypothesis that fall outflows that maintain the X2 position at 74 km are more likely to result in a stable community structure supporting the delta smelt population compared subsystems located at higher X2 positions. Community models suggested a dual role of outflow in controlling the population dynamics delta smelt in the upper SF Estuary, first by determining the geographical location of the LSZ and the underlying community structure, and second by exerting press perturbations on community variables influencing biological rates, as mediated by trophic and non-trophic interactions. Outflow seemed to control community interactions and stability patterns at each X2 position by forcing the overlap among pelagic and benthic species and trophic levels based on their salinity-dependent and geographically-dependent distributions. Integration of flow adaptive management actions across seasons, species, and hydrological conditions could help to address the declining population trends for delta smelt and other native fish populations in the upper SF Estuary. Qualitative community models can complement ecological syntheses, hydrodynamic models and quantitative ecological models to inform ecologically sound management of aquatic ecosystems.

Author Contributions: The findings and conclusions of this study are those of the author and do not necessarily represent the views of the U.S. Fish and Wildlife Service.

Funding: This study was funded by the U.S. Bureau of Reclamation under agreement R15PG00046.

Acknowledgments: This study was based on the collective knowledge of the many scientists involved in the fall outflow studies in the San Francisco Estuary. Michael MacWilliams (Delta Modeling Associates) provided hydrodynamic model results for the salinity field and Janet Thompson (USGS) contributed valuable information on introduced clams. Jeff Dambacher (CSIRO) and Hans Luh (Oregon State University) provided valuable assistance with software codes. Julie Day, Kim Webb and Leo Polansky (U.S. Fish and Wildlife Service), and Anke Mueller-Solger (US Geological Survey), provided useful suggestions. Two anonymous reviewers provided constructive critical suggestions that improved this study.

Conflicts of Interest: The author declare no conflict of interest. Reference to trade names does not imply endorsement by the U.S. Government.

Appendix A

Table A1. Qualitative interactions for modeled community variables defined in Table 1 at the high-, mid-, and low-X2 positions in the upper San Francisco Estuary. (References shown in parentheses).

High-X2 Variables	DS	ZO	PH	ED	PR	CF	MA
DS	−1 [71]	1 [41]	0	−1 [64]	−1 [92]	0	−1 [70]
ZO	−1 [41]	−1 [71]	1 [113]	−1 [114]	0	0	−1 [69]
PH	0	−1 [113]	−1 [71]	−1 [85]	0	−1 [115]	−1 [80]
ED	0	0	0	−1 [71]	0	1 [116]	−1 [117]
PR	1 [92]	0	0	0	−1 [71]	0	−1 [80]
CF	0	0	1 [115]	−1 [100]	0	−1 [71]	−1 [118]
MA	0	0	0	−1 [119]	0	−1 [120]	−1 [71]

Mid-X2 Variables	DS	ZO	PH	PA	PR	CF	MA
DS	−1 [71]	1 [41]	0	0	−1 [92]	0	−1 [70]
ZO	−1 [41]	−1 [71]	1 [113]	−1 [66,88]	0	0	−1 [69]
PH	0	−1 [113]	−1 [71]	−1 [60]	0	−1 [115]	−1 [80]
PA	0	1 [66,88]	1 [60]	−1 [71]	0	0	−1 [118]
PR	1 [92]	0	0	0	−1 [71]	0	−1 [80]
CF	0	0	1 [115]	0	0	−1 [71]	−1 [118]
MA	0	0	0	0	0	−1 [120]	−1 [71]

Low-X2 Variables	DS	ZO	PH	PA	PR
DS	−1 [71]	1 [41]	0	0	−1 [92]
ZO	−1 [41]	−1 [71]	1 [113]	−1 [66,88]	0
PH	0	−1 [113]	−1 [71]	−1 [60]	0
PA	0	1 [66,88]	1 [60]	−1 [71]	0
PR	1 [92]	0	0	0	−1 [71]

References

1. Townend, I.H. Identifying change in estuaries. *J. Coast. Conserv.* **2004**, *10*, 5–12. [CrossRef]
2. Lotze, H.K.; Lenihan, H.S.; Bourque, B.J.; Bradbury, R.H.; Cooke, R.G.; Kay, M.C.; Kidwell, S.M.; Kirby, M.X.; Peterson, C.H.; Jackson, J.B.C. Depletion, degradation, and recovery potential of estuaries and coastal seas. *Science* **2006**, *312*, 1806–1809. [CrossRef]
3. Borja, A.; Dauer, D.M.; Elliott, M.; Simenstad, C.A. Medium- and long-term recovery of estuarine and coastal ecosystems: Patterns, rates and restoration effectiveness. *Estuaries Coasts* **2010**, *33*, 1249–1260. [CrossRef]
4. Gillson, J. Freshwater flow and fisheries production in estuarine and coastal systems: Where a drop of rain is not lost. *Rev. Fish. Sci.* **2011**, *19*, 168–186. [CrossRef]
5. Grimaldo, L.F.; Sommer, T.; Van Ark, N.; Jones, G.; Holland, E.; Moyle, P.B.; Herbold, B.; Smith, P. Factors affecting fish entrainment into massive water diversions in a tidal freshwater estuary: Can fish losses be managed? *N. Am. J. Fish. Manag.* **2009**, *29*, 1253–1270. [CrossRef]
6. Cloern, J.E.; Jassby, A.D. Drivers of change in estuarine-coastal ecosystems: Discoveries from four decades of study in San Francisco Bay. *Rev. Geophys.* **2012**, *50*. [CrossRef]
7. Hobbs, J.A.; Moyle, P.B.; Fangue, N.; Connon, R.E. Is extinction inevitable for delta smelt and longfin smelt? An opinion and recommendations for recovery. *SFEWS* **2017**, *15*. [CrossRef]
8. Cohen, A.N.; Carlton, J.T. Accelerating invasion rate in a highly invaded estuary. *Science* **1998**, *279*, 555–558. [CrossRef] [PubMed]
9. Castillo, G.C.; Damon, L.J.; Hobbs, J.A. Community patterns and environmental associations for pelagic fishes in a highly modified estuary. *Mar. Coast. Fish.* **2018**, *10*, 508–524. [CrossRef]
10. Feyrer, F.; Newman, K.; Nobriga, M.; Sommer, T. Modeling the effects of future outflow on the abiotic habitat of an imperiled estuarine fish. *Estuaries Coasts* **2011**, *34*, 120–128. [CrossRef]
11. Komoroske, L.M.; Connon, R.E.; Lindberg, J.; Cheng, B.S.; Castillo, G.; Hasenbein, M.; Fangue, N.A. Ontogeny influences sensitivity to climate change stressors in an endangered fish. *Conserv. Physiol.* **2014**, *2*, cou008. [CrossRef] [PubMed]

12. Schlacher, T.A.; Wooldridge, T.H. Ecological responses to reductions in freshwater supply and quality in South Africa's estuaries: Lessons for management and conservation. *J. Coast. Conserv.* **1996**, *2*, 115–130. [CrossRef]

13. Moyle, P.B.; Bennett, W.A.; Fleenor, W.E.; Lund, J.R. Habitat variability and complexity in the upper San Francisco Estuary. *SFEWS* **2010**, *8*. [CrossRef]

14. Reinert, T.R.; Peterson, J.T. Modeling the effects of potential salinity shifts on the recovery of striped bass in the Savannah River Estuary, Georgia-South Carolina, United States. *Environ. Manag.* **2008**, *41*, 753–765. [CrossRef] [PubMed]

15. Guenther, C.B.; MacDonald, T.C. Comparison of estuarine salinity gradients and associated nekton community change in the Lower St. Johns River Estuary. *Estuaries Coasts* **2012**, *35*, 1443–1452. [CrossRef]

16. Wissel, B.; Fry, B. Tracing Mississippi River influences in estuarine food webs of coastal Louisiana. *Oecologia* **2005**, *144*, 659–672. [CrossRef] [PubMed]

17. Vinagre, C.; Salgado, J.; Cabral, H.N.; Costa, M.J. Food web structure and habitat connectivity in fish estuarine nurseries—Impact of river flow. *Estuaries Coasts* **2011**, *34*, 663–674. [CrossRef]

18. Estevez, E.D. Review and assessment of biotic variables and analytical methods used in estuarine inflow studies. *Estuaries* **2002**, *25*, 1291–1303. [CrossRef]

19. Sakaris, P.C. A review of the effects of hydrologic alteration on fisheries and biodiversity and the management and conservation of natural resources in regulated river systems. In *Current Perspectives in Contaminant Hydrology and Water Resources Sustainability*; Bradley, P.M., Ed.; Tech: Rijeka, Croatia, 2013; pp. 273–297.

20. Rowell, K.; Flessa, K.W.; Dettmana, D.L.; Román, M.J.; Gerberc, L.R.; Findley, L.T. Diverting the Colorado River leads to a dramatic life history shift in an endangered marine fish. *Biol. Conserv.* **2008**, *141*, 1138–1148. [CrossRef]

21. Purtlebaugh, C.H.; Allen, M.C. Relative abundance, growth, and mortality of five Age-0 estuarine fishes in relation to discharge of the Suwannee River, Florida. *Trans. Am. Fish. Soc.* **2010**, *139*, 1233–1246. [CrossRef]

22. Livingston, R.J. Trophic response of estuarine fishes to long-term changes of river runoff. *Bull. Mar. Sci.* **1997**, *60*, 984–1004.

23. Kimmerer, W.J. Effects of freshwater flow on abundance of estuarine organisms: Physical effects or trophic linkages? *Mar. Ecol. Prog. Ser.* **2002**, *243*, 39–55. [CrossRef]

24. Brown, L.R.; Baxter, R.; Castillo, G.; Conrad, L.; Culberson, S.; Erickson, G.; Feyrer, F.; Fong, S.; Gehrts, K.; Grimaldo, L.; et al. *Synthesis of Studies in the Fall Low-Salinity Zone of the San Francisco Estuary, September–December 2011*; US Geological Survey Scientific Investigations Report 2014–5041; US Geological Survey: Reston, VA, USA, 2014.

25. Wootton, J.T.; Emmerson, M. Measurement of interaction strength in nature. *Annu. Rev. Ecol. Evol. Syst.* **2005**, *36*, 419–444. [CrossRef]

26. Lane, P.; Levins, R. The dynamics of aquatic systems. The effects of nutrient enrichment on model plankton communities. *Limnol. Oceanogr.* **1977**, *22*, 454–471.

27. Dambacher, J.M.; Li, H.W.; Rossignol, P.A. Qualitative predictions in model ecosystems. *Ecol. Model.* **2003**, *161*, 79–93. [CrossRef]

28. Fox, J.W. Current food web models cannot explain the overall topological structure of observed food webs. *Oikos* **2006**, *115*, 97–109. [CrossRef]

29. Levins, R. The qualitative analysis of partially specified systems. *Ann. N. Y. Acad. Sci.* **1974**, *231*, 123–138. [CrossRef] [PubMed]

30. Li, H.W.; Moyle, P.B. Ecological analysis of species introductions into aquatic ecosystems. *Trans. Am. Fish. Soc.* **1981**, *110*, 772–782. [CrossRef]

31. Dambacher, J.M.; Luh, H.K.; Li, H.W.; Rossignol, P.A. Qualitative stability and ambiguity in model ecosystems. *Am. Nat.* **2003**, *161*, 876–888. [CrossRef]

32. Hosack, G.R.; Hayes, K.R.; Dambacher, J.M. Assessing model structure uncertainty through an analysis of system feedback and Bayesian networks. *Ecol. Appl.* **2008**, *18*, 1070–1082. [CrossRef]

33. Li, H.W.; Rossignol, P.A.; Castillo, G. Risk analysis of species introduction: Insights from qualitative modeling. In *Non-Indigenous Fresh Water Organisms in North America; Vectors of Introduction, Biology and Impact*; Claudi, R., Leach, J., Eds.; Lewis Press: Boca Raton, FL, USA, 1999; pp. 431–447.

34. Montano-Moctezuma, G.; Li, H.W.; Rossignol, P.A. Alternative community structures in a kelp-urchin community: A qualitative modeling approach. *Ecol. Model.* **2007**, *205*, 343–354. [CrossRef]

35. Dambacher, J.M.; Brewer, D.T.; Dennis, D.M.; Macintyre, M.; Foale, S. Qualitative modelling of gold mine impacts on Lihir Island's socioeconomic system and reef-edge fish community. *Environ. Sci. Technol.* **2007**, *41*, 555–562. [CrossRef]

36. Ramos-Jiliberto, R.; Garay-Narváez, L.; Medina, M.H. Retrospective qualitative analysis of ecological networks under environmental perturbation: A copper-polluted intertidal community as a case study. *Ecotoxicology* **2012**, *21*, 234–243. [CrossRef] [PubMed]

37. Reum, J.C.P.; Ferriss, B.E.; McDonald, P.S.; Farrell, D.M.; Harvey, C.J.; Klinger, T.; Levin, P.S. Evaluating community impacts of ocean acidification using qualitative network models. *Mar. Ecol. Prog. Ser.* **2015**, *536*, 11–24. [CrossRef]

38. IEP (Interagency Ecological Program). *An Updated Conceptual Model of Delta Smelt Biology: Our Evolving Understanding of an Estuarine Fish*; Interagency Ecological Program for the San Francisco Bay/Delta Estuary. Technical Report 90; IEP: Sacramento, CA, USA, 2015.

39. Jassby, A.D.; Kimmerer, W.J.; Monismith, S.G.; Armor, C.; Cloern, J.E.; Powell, T.M.; Schubel, J.R.; Vendlinski, T.J. Isohaline position as a habitat indicator for estuarine populations. *Ecol. Appl.* **1995**, *5*, 272–289. [CrossRef]

40. Bennett, W.A. Critical assessment of the delta smelt population in the San Francisco Estuary, California. *SFEWS* **2005**, *3*. [CrossRef]

41. Slater, S.B.; Baxter, R.D. Diet, prey selection, and body condition of age-0 delta smelt, *Hypomesus transpacificus*, in the Upper San Francisco Estuary. *SFEWS* **2014**, *12*. [CrossRef]

42. Department of Fish and Wildlife. *State and Federally Listed endangered and Threatened Animals of California*; State of California, Natural Resources Agency, Department of Fish and Wildlife, Biogeographic Data Branch, California Natural Diversity Database: Sacramento, CA, USA, 2018.

43. U.S. Fish and Wildlife Service. *Formal Endangered Species Act Consultation on the Proposed Coordinated Operations of the Central Valley Project (CVP) and State Water Project (SWP)*; U.S. Fish and Wildlife Service, California and Nevada Region: Sacramento, CA, USA, 2008.

44. Kimmerer, W.J.; MacWilliams, M.L.; Gross, E.S. Variation of fish habitat and extent of the low-salinity zone with freshwater flow in the San Francisco Estuary. *SFEWS* **2013**, *11*. [CrossRef]

45. Sommer, T.; Mejia, F.H.; Nobriga, M.L.; Feyrer, F.; Grimaldo, L. The spawning migration of delta smelt in the upper San Francisco Estuary. *SFEWS* **2011**, *9*. [CrossRef]

46. Moyle, P.B.; Brown, L.R.; Durand, J.R.; Hobbs, J.A. Delta smelt: Life history and decline of a once-abundant species in the San Francisco Estuary. *SFEWS* **2016**, *14*. [CrossRef]

47. Dambacher, J.M.; Li, H.W.; Rossignol, P.A. Relevance of community structure in assessing indeterminacy of ecological predictions. *Ecology* **2002**, *83*, 1372–1385. [CrossRef]

48. Cloern, J.E. Habitat connectivity and ecosystem productivity: Implications from a simple model. *Am. Nat.* **2007**, *169*, E21–E33. [CrossRef] [PubMed]

49. Sommer, T.; Armor, C.; Baxter, R.; Breuer, R.; Brown, L.; Chotkowski, M.; Culberson, S.; Feyrer, F.; Gingras, M.; Herbold, B.; et al. The collapse of pelagic fishes in the upper San Francisco Estuary. *Fisheries* **2007**, *32*, 270–277. [CrossRef]

50. Arthur, J.F.; Ball, M.D.; Baughman, S.Y. Summary of federal and state water project environmental impacts in the San Francisco Bay-Delta estuary, California. In *The San Francisco Bay: The Ecosystem. Further Investigations into the Natural History of San Francisco Bay and Delta with Reference to the Influence of Man*; Hollibaugh, J.T., Ed.; Friesen Printers: Altona, MB, Canada, 1996; pp. 445–495.

51. California Department of Water Resources. Dayflow Data. Available online: https://water.ca.gov/Programs/Environmental-Services/Compliance-Monitoring-And-Assessment/Dayflow-Data (accessed on 14 May 2019).

52. CDEC (California Data Exchange Center). The California Data Exchange Center. California Department of Water Resources. Available online: http://cdec.water.ca.gov (accessed on 29 April 2019).

53. IEP (Interagency Ecological Program). Portal to IEP Data and Metadata. Available online: https://water.ca.gov/Programs/Environmental-Services/Interagency-Ecological-Program/Data-Portal (accessed on 29 April 2019).

54. Feyrer, F.; Nobriga, M.L.; Sommer, T.R. Multidecadal trends for three declining fish species: Habitat patterns and mechanisms in the San Francisco Estuary, California, USA. *Can. J. Fish. Aquat. Sci.* **2007**, *64*, 723–734. [CrossRef]

55. MacNally, R.; Thomson, J.R.; Kimmerer, W.J.; Feyrer, F.; Newman, K.B.; Sih, A.; Bennett, W.A.; Brown, L.; Fleishman, E.; Culberson, S.D.; et al. Analysis of pelagic species decline in the upper San Francisco Estuary using multivariate autoregressive modeling (MAR). *Ecol. Appl.* **2010**, *20*, 1417–1430. [CrossRef]

56. Peterson, H.A.; Vayssieres, M. Benthic assemblage variability in the upper San Francisco Estuary: A 27-year retrospective. *SFEWS* **2010**, *8*. [CrossRef]

57. Kratina, P.; MacNally, R.; Kimmerer, W.J.; Thomson, J.R.; Winder, M. Human-induced biotic invasions and changes in plankton interaction networks. *J. Appl. Ecol.* **2014**, *51*, 1066–1074. [CrossRef]

58. Santos, M.J.; Anderson, L.W.; Ustin, S.L. Effects of invasive species on plant communities: An example using submersed aquatic plants at the regional scale. *Biol. Invasions* **2011**, *13*, 443–457. [CrossRef]

59. Durand, J.; Fleenor, W.; McElreath, R.; Santos, M.J.; Moyle, P. Physical controls on the distribution of the submersed aquatic weed *Egeria densa* in the Sacramento–San Joaquin Delta and implications for habitat restoration. *SFEWS* **2016**, *14*. [CrossRef]

60. Nichols, F.H.; Thompson, J.K.; Schemmel, L.E. Remarkable invasion of San Francisco Bay (California, USA) by the Asian clam *Potamocorbula amurensis*. II. Displacement of a former community. *Mar. Ecol. Prog. Ser.* **1990**, *66*, 95–101. [CrossRef]

61. Winder, M.; Jassby, A.D. Shifts in zooplankton community structure: Implications for food web processes in the Upper San Francisco Estuary. *Estuaries Coasts* **2011**, *34*, 675–690. [CrossRef]

62. Nobriga, M.L.; Feyrer, F.; Baxter, R.D.; Chotkowski, M. Fish community ecology in an altered river delta: Species composition, life history strategies, and biomass. *Estuaries* **2005**, *28*, 776–785. [CrossRef]

63. Light, T.; Grosholz, T.; Moyle, P. *Delta Ecological Survey (Phase I): Nonindigenous Aquatic Species in the Sacramento-San Joaquin Delta, a Literature Review*; U.S. Fish and Wildlife Service: Stockton CA, USA, 2005.

64. Ferrari, M.C.; Ranåker, L.; Weinersmith, K.L.; Young, M.J.; Sih, A.; Conrad, J.L. Effects of turbidity and an invasive waterweed on predation by introduced largemouth bass. *Environ. Biol. Fishes* **2014**, *97*, 79–90. [CrossRef]

65. Hestir, E.L.; Schoellhamer, D.H.; Greenberg, J.; Morgan-King, T.; Ustin, S.L. The Effect of submerged aquatic vegetation expansion on a declining turbidity trend in the Sacramento-San Joaquin River Delta. *Estuaries Coasts* **2016**, *39*, 1110–1112. [CrossRef]

66. Kimmerer, W.J. Response of anchovies dampens effects of the invasive bivalve *Corbula amurensis* on the San Francisco Estuary foodweb. *Mar. Ecol. Prog. Ser.* **2006**, *324*, 207–218. [CrossRef]

67. Carlton, J.T.; Thompson, J.K.; Schemel, L.E.; Nichols, F.H. Remarkable invasion of San Francisco Bay (California, USA) by the Asian clam *Potamocorbula amurensis*. I. Introduction and dispersal. *Mar. Ecol. Prog. Ser.* **1990**, *66*, 81–94. [CrossRef]

68. Lehman, P.W.; Boyer, G.; Hall, C.; Waller, S.; Gehrts, K. Distribution and toxicity of a new colonial *Microcystis aeruginosa* bloom in the San Francisco Bay Estuary, California. *Hydrobiologia* **2005**, *541*, 87–90. [CrossRef]

69. Ger, K.A.; Teh, S.J.; Goldman, C.R. Microcystin-LR toxicity on dominant copepods *Eurytemora affinis* and *Pseudodiaptomus forbesi* of the upper San Francisco Estuary. *Sci. Total Environ.* **2009**, *407*, 4852–4857. [CrossRef] [PubMed]

70. Acuña, S.; Baxa, D.; Teh, S. Sublethal dietary effects of microcystin producing *Microcystis* on threadfin shad, *Dorosoma petenense*. *Toxicon* **2014**, *60*, 1191–1202. [CrossRef]

71. Puccia, C.J.; Levins, R. *Qualitative Modeling of Complex Systems. An Introduction to Loop Analysis and Time Averaging*; Harvard University Press: Cambridge, MA, USA, 1985.

72. Levins, R. *Evolution in Changing Environments. Some Theoretical Explorations. Monographs in Population Biology*; McArthur, R.H., Ed.; Princeton University Press: Princeton, NJ, USA, 1968.

73. Puccia, C.J.; Levins, R. Qualitative Modeling in Ecology: Loop Analysis, Signed Digraphs and Time Averaging. In *Qualitative Simulation Modeling and Analysis*; Fishwick, P.A., Luker, P.A., Eds.; Advances in Simulation; Springer: New York, NY, USA, 1991; pp. 119–143.

74. Castillo, G.C.; Li, H.W.; Rossignol, P.A. Absence of overall feedback in a benthic estuarine community: A system potentially buffered from impacts of biological invasions. *Estuaries* **2000**, *23*, 275–291. [CrossRef]

75. MacWilliams, M.L.; Bever, A.J.; Gross, E.S.; Ketefian, G.S.; Kimmerer, W.J. Three-dimensional modeling of hydrodynamics and salinity in the San Francisco Estuary: An evaluation of model accuracy, X2, and the low–salinity zone. *SFEWS* **2015**, *13*. [CrossRef]

76. Castillo, G. Annotated Bibliography of the Delta Smelt (*Hypomesus transpacificus*). Programmatic Review of Delta Smelt Program Elements (2005–06). Interagency Ecological Program. Available online: https://www.researchgate.net/publication/332725919 (accessed on 28 April 2019).

77. Sommer, T.; Mejia, F.H. A place to call home: A synthesis of delta smelt habitat in the upper San Francisco Estuary. *SFEWS* **2013**, *11*. [CrossRef]

78. Peterson, M.S. A conceptual view of environment-habitat-production linkages in tidal river estuaries. *Rev. Fish. Sci.* **2003**, *11*, 291–313. [CrossRef]

79. Lane, P.A. Preparing Marine Plankton Data Sets for Loop Analysis. ESA Supplement. Available online: http://esapubs.org/archive/ecol/E067/001/suppl-1B.pdf (accessed on 29 April 2019).

80. Lehman, P.W.; Teh, S.J.; Boyer, G.L.; Nobriga, M.L.; Bass, E.; Hogle, C. Initial impacts of *Microcystis aeruginosa* blooms on the aquatic food web in the San Francisco Estuary. *Hydrobiologia* **2010**, *637*, 229–248. [CrossRef]

81. Simberloff, D.; Dayan, T. The guild concept and the structure of ecological communities. *Annu. Rev. Ecol. Syst.* **1991**, *22*, 115–143. [CrossRef]

82. Lucas, L.V.; Cloern, J.E.; Thompson, J.K.; Monsen, N.E. Functional variability of habitats within the Sacramento-San Joaquin Delta: Restoration implications. *Ecol. Appl.* **2002**, *12*, 1528–1547.

83. Lopez, C.B.; Cloern, J.E.; Schraga, T.S.; Little, A.J.; Lucas, L.V.; Thompson, J.K.; Burau, J.R. Ecological values of shallow-water habitats: Implications for restoration of disturbed ecosystems. *Ecosystems* **2006**, *9*, 422–440. [CrossRef]

84. Brown, L.R.; Michniuk, D. Littoral fish assemblages of the alien-dominated Sacramento–San Joaquin Delta, California, 1980–1983 and 2001–2003. *Estuaries Coasts* **2007**, *30*, 186–200. [CrossRef]

85. Yarrow, M.; Marin, V.H.; Finlayson, M.; Tironi, A.; Delgado, L.E.; Fisher, F. The ecology of *Egeria densa* Planchón (Liliopsida: Alismatales): A wetland ecosystem engineer? *Rev. Chil. Hist. Nat.* **2009**, *82*, 299–313. [CrossRef]

86. Lehman, P.W.; Boyer, G.; Satchwell, M.; Waller, S. The influence of environmental conditions on the seasonal variation of Microcystis cell density and microcystins concentration in San Francisco Estuary. *Hydrobiologia* **2008**, *600*, 187–204. [CrossRef]

87. Alpine, A.E.; Cloern, J.E. Trophic interactions and direct physical effects control phytoplankton biomass and production in an estuary. *Limnol. Oceanogr.* **1992**, *37*, 946–955. [CrossRef]

88. Kimmerer, W.J.; Gartside, E.; Orsi, J.J. Predation by an introduced clam as the probable cause of substantial declines in zooplankton in San Francisco Bay. *Mar. Ecol. Prog. Ser.* **1994**, *113*, 81–93. [CrossRef]

89. Kimmerer, W.J.; Parker, A.E.; Lidström, U.; Carpenter, E.J. Short-term and interannual variability in primary production in the low-salinity zone of the San Francisco Estuary. *Estuaries Coasts* **2012**, *35*, 913–929. [CrossRef]

90. Sobczak, W.V.; Cloern, J.E.; Jassby, A.D.; Muller-Solger, A.B. Bioavailability of organic matter in a highly disturbed estuary: The role of detrital and algal resources. *Proc. Natl. Acad. Sci. USA* **2002**, *99*, 8101–8105. [CrossRef] [PubMed]

91. Jassby, A.D. Phytoplankton in the upper San Francisco Estuary: Recent biomass trends, their causes and their trophic significance. *SFEWS* **2008**, *6*. [CrossRef]

92. Nobriga, M.L.; Loboschefsky, E.; Feyrer, F. Common predator, rare prey: Exploring juvenile striped bass predation on delta smelt in California's San Francisco Estuary. *Trans. Am. Fish. Soc.* **2014**, *142*, 1563–1575. [CrossRef]

93. Novak, M.; Wootton, J.T.; Doak, D.F.; Emmerson, M.; Estes, J.A.; Tinker, M.T. Predicting community responses to perturbations in the face of imperfect knowledge and network complexity. *Ecology* **2011**, *92*, 836–846. [CrossRef] [PubMed]

94. Gantmacher, F.R. *Applications of the Theory of Matrices*; Dover Publications. Inc.: Mineola, NY, USA, 2005.

95. Nakajima, H. Sensitivity and stability of flow networks. *Ecol. Model.* **1992**, *62*, 123–133. [CrossRef]

96. Luh, H.K. Oregon State University. Loop Analysis Program in Microsoft.NET 3.5 Framework. Available online: http://ipmnet.org/loop/loopanalysis.aspx (accessed on 12 March 2019).

97. California Department of Fish and Wildlife. Fall Midwater Trawl Index. Available online: http://www.dfg.ca.gov/delta/data/fmwt/indices.asp (accessed on 12 March 2019).

98. Maceda-Veiga, A. Towards the conservation of freshwater fish: Iberian Rivers as an example of threats and management practices. *Rev. Fish. Biol. Fisher.* **2013**, *23*, 1–22. [CrossRef]

99. Nobriga, M.L.; Sommer, T.R.; Feyrer, F.; Fleming, K. Long-term trends in summertime habitat suitability for delta smelt, *Hypomesus transpacificus. SFEWS* **2008**, *6*, 6. [CrossRef]

100. Yamamuro, M. Herbicide-induced macrophyte-to-phytoplankton shifts in Japanese lagoons during the last 50 years: Consequences for ecosystem services and fisheries. *Hydrobiologia* **2012**, *699*, 5–19. [CrossRef]

101. Kimmerer, W.J.; Thompson, J.K. Phytoplankton growth balanced by clam and zooplankton grazing and net transport into the low-salinity zone of the San Francisco Estuary. *Estuaries Coasts* **2014**, *37*, 1202–1218. [CrossRef]

102. Bennett, W.A.; Moyle, P.B. Where have all the fishes gone? Interactive factors producing fish declines in the Sacramento-San Joaquin Estuary. In *The San Francisco Bay: The Ecosystem. Further Investigations into the Natural History of San Francisco Bay and Delta with Reference to the Influence of Man*; Hollibaugh, J.T., Ed.; Friesen Printers: Altona, MB, Canada, 1996; pp. 519–542.

103. Rose, K.A.; Kimmerer, W.J.; Edwards, K.P.; Bennett, W.A. Individual-based modeling of delta smelt population dynamics in the upper San Francisco Estuary I. Model description and baseline results. *Trans. Am. Fish. Soc.* **2013**, *142*, 1238–1259. [CrossRef]

104. Rose, K.A.; Kimmerer, W.J.; Edwards, K.P.; Bennett, W.A. Individual-based modeling of delta smelt population dynamics in the upper San Francisco Estuary II. Alternative baselines and good versus bad years. *Trans. Am. Fish. Soc.* **2013**, *142*, 1260–1272. [CrossRef]

105. Kimmerer, W.J.; Rose, K.A. Individual-based modeling of delta smelt population dynamics in the upper San Francisco Estuary III. Effects of entrainment mortality and changes in prey. *Trans. Am. Fish. Soc.* **2018**, *147*, 223–243. [CrossRef]

106. Hammock, B.G.; Moose, S.P.; Sandoval Solis, S.; Goharian, E.; Swee, J.T. Hydrodynamic modeling coupled with long-term field data provide evidence for suppression of phytoplankton by invasive clams and freshwater exports in the San Francisco Estuary. *J. Environ. Manag.* **2019**, *63*, 703–717. [CrossRef] [PubMed]

107. Cottingham, A.; Huang, P.; Hipsey, M.R.; Hall, N.G.; Ashworth, E.; Williams, J.; Potter, I.C. Growth, condition, and maturity schedules of an estuarine fish species change in estuaries following increased hypoxia due to climate change. *Ecol. Evol.* **2018**, *8*, 7111–7130. [CrossRef]

108. Baustian, M.M.; Clark, F.R.; Jerabek, A.S.; Wang, Y.; Bienn, H.C.; White, E.D. Modeling current and future freshwater inflow needs of a subtropical estuary to manage and maintain forested wetland ecological conditions. *Ecol. Indic.* **2018**, *85*, 791–807. [CrossRef]

109. Martins, I.; Antunes, C.; Dias, E.; Campuzano, F.J.; Pinto, L.; Santos, M.M.; Antunes, C. Antagonistic effects of multiple stressors on macroinvertebrate biomass from a temperate estuary (Minho estuary, NW Iberian Peninsula). *Ecol. Indic.* **2019**, *101*, 792–803. [CrossRef]

110. Allen, C.R.; Gunderson, L.H. Pathology and failure in the design and implementation of adaptive management. *J. Environ. Manag.* **2011**, *92*, 1379–1384. [CrossRef]

111. Cartwright, J.; Caldwell, C.; Nebiker, S.; Knight, R. Putting flow-ecology relationships into practice: A decision-support system to assess fish community response to water-management scenarios. *Water* **2017**, *9*, 196. [CrossRef]

112. Arthington, A.H.; Bhaduri, A.; Bunn, S.E.; Jackson, S.E.; Tharme, R.E.; Tickner, D.; Young, B.; Acreman, B.; Baker, N.; Capon, S.; et al. The Brisbane declaration and global action agenda on environmental flows. *Front. Environ. Sci.* **2018**, *6*, 45. [CrossRef]

113. Gifford, S.M.; Rollwagen Bollens, G.C.; Bollens, S.M. Mesozooplankton omnivory in the upper San Francisco Estuary. *Mar. Ecol. Prog. Ser.* **2007**, *348*, 33–46. [CrossRef]

114. Burks, R.L.; Jeppesen, E.; Lodge, D.M. Macrophyte and fish chemicals suppress *Daphnia* growth and alter life-history traits. *Oikos* **2000**, *88*, 139–148. [CrossRef]

115. Lucy, F.E.; Karatayev, A.Y.; Burlakova, L.E. Predictions for the spread, population density, and impacts of *Corbicula fluminea* in Ireland. *Aquat. Invasions* **2012**, *7*, 465–474. [CrossRef]

116. Sousa, R.; Gutierrez, J.L.; Aldridge, D.C. Non-indigenous invasive bivalves as ecosystem engineers. *Biol. Invasions* **2009**, *11*, 2367–2385. [CrossRef]

117. Pflugmacher, S. Promotion of oxidative stress in the aquatic macrophyte *Ceratophyllum demersum* during biotransformation of the cyanobacterial toxin microcystin-LR. *Aquat. Toxicol.* **2004**, *70*, 169–178. [CrossRef]

118. Hwang, S.J.; Kim, H.S.; Park, J.H.; Kim, B.H. Effects of cyanobacterium *Microcystis aeruginosa* on the filtration rate and mortality of the freshwater bivalve *Corbicula leana*. *J. Environ. Biol.* **2010**, *31*, 483–488.

119. Gao, Y.N.; Dong, J.; Fu, Q.Q.; Wang, Y.P.; Chen, C.; Li, J.H.; Li, R.; Zhou, C.J. Allelopathic effects of submerged macrophytes on phytoplankton. *Allelopath. J.* **2017**, *40*, 1–22.
120. Liu, Y.; Xie, P.; Wu, X.P. Grazing on toxic and non-toxic *Microcystis aeruginosa* PCC7820 by *Unio douglasiae* and *Corbicula fluminea*. *Limnology* **2009**, *10*, 1–5. [CrossRef]

Article

Biological and Physical Effects of Brine Discharge from the Carlsbad Desalination Plant and Implications for Future Desalination Plant Constructions

Karen Lykkebo Petersen [1,2,*], **Nadine Heck** [3,4], **Borja G. Reguero** [3,4], **Donald Potts** [3,5], **Armen Hovagimian** [6] **and Adina Paytan** [3]

[1] Department of Earth and Planetary Sciences, University of California, Santa Cruz, 1156 High Street, Santa Cruz, CA 95064, USA

[2] Department of Ecology, Environment and Plant Science, Stockholm University, Svante Arrhenius väg 20A, 114 18 Stockholm, Sweden

[3] Institute of Marine Sciences, University of California, Santa Cruz, 1156 High St., Santa Cruz, CA 95064, USA; nheck@ucsc.edu (N.H.); borja_reguero@tnc.org (B.G.R.); potts@ucsc.edu (D.P.); apaytan@ucsc.edu (A.P.)

[4] The Nature Conservancy, 115 McAllister Way, Santa Cruz, CA 95060, USA

[5] Department of Ecology and Evolutionary Biology, University of California, Santa Cruz, 130 McAllister Way, Santa Cruz, CA 95060, USA

[6] Cowell College, University of California, Santa Cruz, 1156 High Street, Santa Cruz, CA 95064, USA; ahovagim@ucsc.edu

* Correspondence: lykkebo.petersen@su.se

Received: 9 December 2018; Accepted: 22 January 2019; Published: 25 January 2019

Abstract: Seawater reverse osmosis (SWRO) desalination is increasingly used as a technology for addressing shortages of freshwater supply and desalination plants are in operation or being planned world-wide and specifically in California, USA. However, the effects of continuous discharge of high-salinity brine into coastal environments are ill-constrained and in California are an issue of public debate. We collected in situ measurements of water chemistry and biological indicators in coastal waters (up to ~2 km from shore) before and after the newly constructed Carlsbad Desalination Plant (Carlsbad, CA, USA) began operations. A bottom water salinity anomaly indicates that the spatial footprint of the brine discharge plume extended about 600 m offshore with salinity up to 2.7 units above ambient (33.2). This exceeds the maximum salinity permitted for this location based on the California Ocean Plan (2015 Amendment to Water Quality Control Plan). However, no significant changes in the assessed biological indicators (benthic macrofauna, BOPA-index, brittle-star survival and growth) were observed at the discharge site. A model of mean ocean wave potential was used as an indicator of coastal mixing at Carlsbad Beach and at other locations in southern and central CA where desalination facilities are proposed. Our results indicated that to minimize environmental impacts discharge should target waters where a long history of anthropogenic activity has already compromised the natural setting. To ensure adequate mixing of the discharge brine desalination plants should be constructed at high-energy sites with sandy substrates, and discharge through diffusor systems.

Keywords: SWRO desalination; brine discharge; osmotic stress; coastal monitoring; impacted coastal systems

1. Introduction

Freshwater demand is increasing worldwide due to a variety of factors, including population growth, agricultural expansion and environmental changes [1]. At the same time, natural freshwater resources are declining in quantity and quality [2]. In California, increasing agricultural activity and population growth have diminished natural groundwater reservoirs, resulting in substantial land subsidence and seawater intrusion [3]. Recurrent droughts that limit recharge of water reservoirs are common and are expected to increase in frequency, duration and intensity [4,5] exacerbating the problem. Seawater desalination by reverse osmosis (SWRO) is increasingly being seen as a way to counter freshwater shortages, and several desalination plants are being built or proposed along the Southern and Central California coastline [6,7]. Ten small seawater reverse osmosis (SWRO) desalination facilities with individual capacities from 30×10^3 to 10×10^6 L day^{-1} (combined capacity ~20×10^6 L day^{-1}) [8] are currently operating along the California coast, and one large-scale plant with a capacity of 180×10^6 L day^{-1}, started operation in Carlsbad (Southern California) in December 2015. Seven additional large-scale facilities have been proposed (one is under construction) with capacities of 40–500×10^6 L day^{-1} [8] in response to reoccurring droughts and increasing demand for water resources.

SWRO facilities draw coastal water as feed and continuously produce high salinity brine effluent that is typically diluted and discharged back into coastal environments. The discharge occurs either directly at the shoreline through outfall channels, or further from shore through pipes or diffusor systems [9–11]. Californian regulations require that the brine discharged not exceed two salinity units above ambient levels within 100 m offshore from the discharge point [12]. The efficiency of water mixing and the footprint of the discharged brine depend on (1) dilution prior to discharge and hence the final density of the discharging brine, (2) local coastal conditions (waves, currents and bathymetry), and (3) the design of the discharge method (channel, pipe or diffusor). To comply with these regulations, SWRO facilities need to control the salinity of the brine (by dilution prior to discharge) and select a discharge design that ensures easy mixing of brine with the surrounding seawater under the specific oceanographic conditions at the discharge location.

Water mixing potential at the discharge zone is usually evaluated prior to operation by using hydrodynamic computer models to ensure compliance [12]. While the California Ocean Plan (2015 Amendment to Water Quality Control Plan) specifies a salinity impact zone extending no more than 100 m offshore, the Carlsbad Desalination Plant received an exception to policy to extend its salinity impact zone to 200 m offshore, due to its high capacity [12]. For the Carlsbad Plant, the hydrodynamic model assumed a starting salinity of 42 at the outfall, and predicted that salinity would decline to 35.5 (ambient 33.2) at a distance of 196 m offshore (i.e., ~2 salinity units above ambient within the 200 m permitted limit) [13]. The brine produced at the Carlsbad SWRO facility is diluted by mixing with seawater used for cooling at a co-located power plant, and hence decreasing the brine salinity and increasing water temperature, hence reducing the brine density to increase the mixing potential and prevent bottom ponding of a high density brine.

Despite increasing use of SWRO desalination worldwide as well as in California, impacts of brine effluent discharge on the living organisms and ecosystems in the coastal environments are ill-constrained [14,15]. Past research on pelagic phytoplankton and benthic microbes, seagrasses, polychaetes and corals demonstrate that salinity tolerances are highly variable among species and also dependent on the magnitude of the salinity increase and exposure time [14,16–21]. For instance, seagrasses have low thresholds with a detectable mortality at salinity of 5% above ambient levels, whereas coral growth is not impacted at salinity as high as 10% above ambient [22–24]. Relative abundances and growth rates of phytoplankton, zooplankton, and benthic bacteria also do not seem to be significantly impacted at salinities of 10% above ambient, but community structure often changes [16,17,25,26]. The Benthic Opportunistic Polychaetes and Amphipods index (BOPA-index) is commonly used as an indicator of the level of "disturbance" to benthic communities in areas impacted by pollution [27–30], or other stressors such as changes in salinity [31]. This index is based on an inverse relationship between the abundances of sensitive amphipods and opportunistic polychaetes [27,29,30,32]. The index value specifies an ecological status ranging from "good" to "bad", where good is defined as an area dominated by sensitive species, and bad an area dominated mainly by opportunistic species.

Coastal California is a highly productive zone supported by upwelling of nutrient-rich sub-surface water. This productivity supports large kelp beds, productive rocky reefs with high biodiversity, and rich plankton communities that serve as food for numerous fish, seabirds, whales and dolphins. However, the population density along the coast in California is high (26 million people living in coastal counties) and many costal settings have been impacted by a wide range of anthropogenic activities including dredging, shipping, sewage discharge, eutrophication, commercial and recreational fishing and more transitional waters [33]. Along the state coastline, 124 marine protected areas (MPA's) are providing refuge for the ecosystem (total coverage of 16% of state water) [34,35]. Toxicity testing with high-salinity seawater has been conducted on a few key Californian rocky-reef species (i.e., *Haliotis rufescens* (Red Abalone), *Strongylocentrotus purpuratus* (Purple Urchin) and *Dendraster excentricus* (Sand Dollar)) and all proposed SWRO desalination facilities are required to use hydrological modeling to estimate the impact area of discharging brine [12,36]. However, uncertainties and concerns persist regarding potential impacts of SWRO desalination brine on coastal environments, especially among coastal users and the general public [37,38].

This study characterizes the spatial footprint of the discharge plume from Carlsbad Desalination Plant, both chemically and biologically, with the ultimate goal of informing legislators, regulators, plant managers and the public about appropriate locations and discharge methods for future SWRO plants and illustrates the need for monitoring. The extend and impact of the discharge zone is characterized by using (1) water samples collected at and around the outfall channel of the Carlsbad Desalination Plant, before and after the plant became operational (in December 2015), (2) biological surveys of benthic epifauna, (3) a BOPA analysis around the discharge zone, and (4) a laboratory bioassay with brittle stars (*Ophiothrix spiculata*). The study finds the brine plume to extend beyond the 200 m impact zone allowed in the California Ocean Plan (2015 Amendment to Water Quality Control Plan) but finds no significant impact on the benthic ecology. Using a model of coastal wave energy at Carlsbad Beach and in Southern and Central California, possible impact of future desalination plants in these California coastal zones is assessed.

2. Methods

2.1. Study Area

The Carlsbad Desalination Plant, built and operated by Poseidon Water, is the first and currently only large-scale SWRO desalination facility in California. Operations began in December 2015 with a capacity of 180×10^6 L day^{-1}. The plant is located in an industrial area at the southern end of the Agua Hedionda Lagoon, adjacent to the Encina Power Station (Figure 1A). Seawater enters the lagoon through a dredged channel at the north end of the lagoon, about 1 km from the seawater intake used by the power station as cooling water (since 1954). The desalination plant diverts ~10% of this water released by the power plant post-cooling for SWRO (Figure 1B). The brine effluent is then returned to the cooling water further downstream, resulting in a 1:10 dilution of the brine before the mixture is discharged to the ocean through a 10–15 m wide channel (outfall) between two rocky walls extending ~50 m offshore at Carlsbad Beach (Figure 1A) [39]. Carlsbad Beach is a relatively high-energy beach with wave heights averaging ~2.5 m in winter and ~1.5 m in summer [40–43]. The nearshore habitat is dominated by a sandy bottom with scattered small rocky reefs and seagrass patches at the northern end of the beach. The nearshore region is shallow with depths of 5–10 m up to 800 m offshore, and with deeper water (~20 m) starting ~1 km offshore [41].

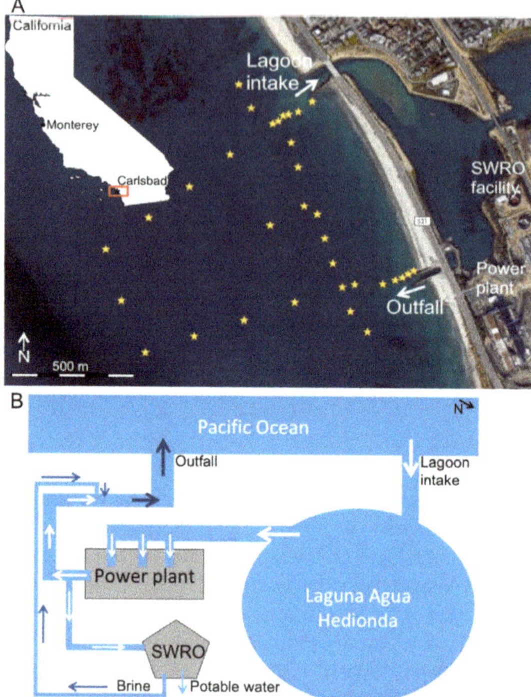

Figure 1. (**A**) Aerial view of the study site around Carlsbad Desalination Plant. The SWRO plant and power plant, intake and outfall channels, and sampling sites (stars) are marked; (**B**) schematic of water flow from the Pacific Ocean to Laguna Aqua Hedionda and the intake and discharge of water by the power plant and SWRO facility (not to scale). White arrows indicate seawater flow and dark-blue arrows represent discharging brine and its mixture with the power plant's seawater return flow.

2.2. Sample Collection and Biological Surveys

Water and sediment samples were collected, and benthic surveys were conducted twice in the year prior to the start of SWRO operation (pre-operation), and twice in the first year after brine discharge began (post-operation). Pre-operation samples were collected in December 2014 and September 2015, and post-operation samples were collected in May (five months post-operation) and November 2016 (11 months post-operation). Samples were collected by scuba divers, starting ~100 m offshore (closer if conditions allowed). Samples were taken along two transects perpendicular to the beach at 25 m intervals out to 200 m, and then at 250, 300, 400, 600, 800 and 1000 m from shore. Sampling was conducted close to shore as the chemical footprint and possible biological impacts were expected to be greatest close to the discharge site and quickly dissipate with distance due to mixing. A transect parallel to shore was conducted as close to the beach as possible (~200 m offshore), with samples collected every 100 m from 200 m south of the outfall to 200 m north of the lagoon entrance (Figure 1A). Bottom and surface water samples were collected at each sampling site in 1 L acid-cleaned, sample-rinsed, Nalgene HDPE bottles. Water temperatures were measured on the boat, immediately after collection, with a YSI 30SCT probe (YSI, Yellow Springs, OH, USA). Water samples were kept in a cooler until they were filtered (using various filters, see Section 2.3 below) and processed, within 4 h after collection. Sediment samples (from the upper 10 cm) were collected in 250 mL plastic jars by pushing the jars into the sediment and samples were kept frozen until analysis (see Section 2.3).

Benthic surveys of epifauna were conducted along two continuous perpendicular transects each 1 m wide, from ~100 to 200 m offshore, and then extended with 1×1 m quadrats (1 m^2) at ~100 m intervals from 200 to 1000 m offshore. Along the continuous transects, a 100 m measuring tape was laid on the sea floor and all visible benthic organisms within 0.5 m of each side of the tape were recorded for each 10 m long section (10 m^2). At stations running parallel to shore, organisms were recorded in ten randomly distributed 1 m^2 quadrats per station. Visible benthic epifauna were enumerated separately by two scuba divers, and counts were calculated to species abundance per m^2 and grouped in counts close to the discharge channel (0–200 m offshore) and in the surrounding area (250–1000 m offshore). Organisms were assigned to 10 categories: Porifera (sponges), Anthozoa, Gastropoda, Bivalvia, Cephalopoda, Polychaeta, Echinodermata, Arthropoda, fish, macroalgae and seagrass.

2.3. Water and Sediment Analyses

Water samples were analyzed for salinity, chlorophyll *a*, nutrients (PO_4, NO_3 and SiO_2), particulate organic matter (POM), dissolved organic carbon (DOC) and trace metals. More detail is provided below:

Salinity: 50 mL of seawater were pumped through a Guildline Portasal salinometer (Osil, Hampshire, UK) and salinity recorded with an accuracy of ± 0.001.

Chlorophyll *a*: 250 mL of seawater were filtered onto GF/F filters (0.7 µm) (Whatman, Chicago, IL, USA) and kept frozen in the dark until further analysis. The filters were immersed in vials with 7 mL of 90% acetone and extracted at 2 °C for 12 h before measurements. Chlorophyll *a* (Chl *a*) concentrations were measured using a TD-700 fluorometer (Turner Designs, San Jose, CA, USA) calibrated with Chl *a* according to the EPA 445 method, using CS-5-60 and CS-2-64 glass filters (Turner Designs, San Jose, CA, USA).

Nutrients: 50 mL of seawater were filtered through 0.2 µM filters and kept frozen until analysis. Samples were analyzed for soluble reactive PO_4 (SRP) [44] and for NO_3 and SiO_2. Samples were processed on a QuikChem FIA+ 8000 Series autoanalyzer (Lachat Instruments, Milwaukee, WI, USA) using methods 31-115-01-3-A (PO_4), 31-114-27-1-B (NO_3) and 31-107-04-1-A (SiO_2). Sample reproducibility was 0.5%, with a detection limit of 0.15 µM for all analytes.

Dissolved organic carbon: 30 mL of seawater were collected in pre-ashed glass vials (450 °C for 4 h). The seawater was acidified to pH 2 using concentrated trace metal grade HCl, and analyzed on a Shimadzu TOC-V autosampler (Shimadzu, Columbia, MD, USA) with prepared standards of 2.125 g KHP L^{-1}. Detection limit was 0.3 µM with 2 mL injections.

Particulate organic matter: 400 mL of seawater were filtered onto pre-ashed (450 °C for four hours) GF/F filters (0.7 µm) and samples kept frozen. The filters were dried for 48 h at 50 °C and pressed into tin capsules. The carbon to nitrogen molar ratio (C:N) and C and N isotope ratios were measured using isotope ratio mass spectrometry (CE instruments NC2500, Wigan, UK). Analyses were done with standards of pugel and acetanilide to an accuracy of ± 0.11‰.

Select major and trace elements: 50 mL of seawater was filtered (0.2 µM) and collected in acid-cleaned sample-rinsed LDPE bottles and acidified with trace metal grade HCl to pH 2 prior to analysis. Samples were diluted 500-fold and analyzed on an Element XR ICP-MS (Thermo Fisher Waltham, MA, USA) with standard curves prepared using NIST 1640a standard solution (National Institute of Standards and Technology, Gaithersburg, MD, USA). The average of two procedural blank values was used for blank correction. Concentrations of Li, Na, Mg, K, Ca, Mn, Fe, Sr and Ba were measured.

Infaunal organisms: Sediment samples were thawed and washed with 90% ethanol. The ethanol rinse was collected and all polychaetes and amphipods were counted immediately under a dissecting microscope. For polychaetes, individuals of the two most abundant families, Paraonidae and Capitellidae, were identified and counted; individuals from other families were counted as "other". For amphipods, the families Gammaridae, Hyperiidae, Caprellidae and Corophiidae were identified and counted. A BOPA-index was calculated for each sediment sample using methods described in reference [27] (Equation (1)), where fp_{op} is the proportion of polychaetes and f_a is the proportion of amphipods:

$$BOPA = log\left(\frac{fp_{op}}{f_a + 1} + 1\right). \tag{1}$$

Sediment grain size: The sediment was dried, and 75–90 g subsamples were sorted using twelve sieves (from 0.5 Φ to 4.75 Φ, Krumbein phi scale), and each fraction was weighed. The mean grain size and sorting factor were determined using methods described in [45].

2.4. Bioassay Experiment

A bioassay was conducted with brittle stars (*Ophiothrix spiculata*) collected in Monterey Bay (collected by Monterey Bay Abalone Co) and exposed to discharged desalination brine collected from the outfall channel of the Carlsbad Desalination Plant. The collected brine was filtered through a 0.2 µM filter and kept cool and in the dark with aquarium air pumps attached to maintain oxygen levels. Brittle stars were exposed to three different brine treatments: (1) Discharge water of salinity 37 collected from the discharge channel; (2) 1:1 dilution of the discharge water with ambient seawater to a salinity of 34; and (3) ambient seawater (control) with salinity of 33. There were 20 replicates for each treatment. Each replicate consisted of two, randomly selected, brittle stars maintained in a 500 mL glass culture dish containing 450 mL of treatment water and capped with a glass Petri dish. The culture dishes were kept in a water table with running seawater to maintain ambient temperature and were partly shielded to avoid direct light. Water in the culture dishes was changed every 2–3 days and the incubation lasted 5 weeks. Pre-incubation wet weight and arm length (estimated from photographs) were measured for all stars at the start of the experiment, and compared to wet weight, dry weight, body diameter and arm length at the end of the incubation. Each brittle star remaining alive at the end of the experiment was subjected to an agility test where the individual was flipped on its back and the time to return to normal orientation was measured. Wet weight was determined by placing a star on a tissue paper for 2 s before recording its weight (± 0.001 g). Body diameter and arm lengths were measured to the closest mm, and arm length was averaged over all 5 arms. The brittle stars were dried in a 50 °C oven for 48 h before obtaining dry weight.

2.5. Wave Energy Model

In coastal zones and estuaries, temporal and spatial variations in salinity are affected by changes in precipitation, evaporation and freshwater inflows, as well as by water mass mixing related to changes in circulation, waves and tides. Changes in salinity can have major effects on water density and water stratification, which, in turn, can modify circulation patterns. We focused on mixing along open coastlines with water depths similar to those at most discharge sites for proposed SWRO plants in California. We assumed mixing is controlled by wave dynamics and excluded tidal and other effects.

The mixing potential of wave action was quantified by calculating patterns of wave energy and orbital velocity for the mean conditions in each of four seasons, and then averaged them over a year. We used wave model data for the California continental shelf that provides information on wave heights, periods and orbital velocities at high resolution for the whole Californian coast, including mean and extreme (top 5%) wave parameters for each season from [42,43]. A set of 15 SWAN curvilinear grids was used to simulate wind-wave growth and propagation across the inner portion of the California continental shelf [42]. All grids had an average cross- and along-shore resolution of 30 to 50 m and 60 to 100 m, respectively.

An index of annual mixing potential, E was calculated by averaging the mean conditions for each season in Equation (2), where ρ is water density, g the gravitational pull, H is wave height and L wave length (see Supporting Information Equations (S1), (S2) and (S3)):

$$E_L = \frac{1}{8}\rho g H^2 L. \tag{2}$$

We considered seasonal variations in the potential for mixing because the North Pacific Ocean presents extremely large surface waves that can vary greatly between seasons and years [46,47]. Specifically, wave energy varies strongly between winter and summer storms and can also respond to large inter-annual climate variations, for example associated with El Niño events.

In shallow water, the orbital motions of water particles induced by surface waves extend down to the seabed. The resulting wave-induced orbital velocities near the seabed are considered to be representative measures of how waves influence the sea floor [42]. We used the orbital velocity as an indicator of mixing potential at the seabed, based on mean seasonal values from [43] for each season, and then averaged annually, similar to the wave energy (E_L).

Using monitoring data available from California Department of Fish and Wildlife [48] coastal ecosystem types were determined (seagrasses, kelp beds), habitats (rocky or sandy substrate), within a 2 km^2 radius of the outfall of each existing and proposed seawater desalination plant in Central and Southern California. We mapped the calculated seasonal wave energy together with the ecosystem and habitat type and compared water mixing potential at proposed SWRO sites and the ecological niches that may be affected by SWRO discharge.

2.6. Statistics and Geospatial Analyses

All statistical analyses were done using software R (R-studio, version 1.1.447). Salinity, temperature and other water chemical data were pooled in groups depicting the immediate area around the outfall channel (50–200 m offshore) and the surrounding area (250–1000 m offshore) for both surface and bottom water for all four sampling trips. The groups were compared between surface and bottom water with two-way ANOVA and post-hoc tested with Tukey-HSD with a significance level of $\alpha = 0.05$ (Table 1). Biological abundances of epifauna and infauna were similarly pooled and, along with brittle star measurements, analyzed with two-way ANOVAs and post-hoc Tukey-HSD tests ($\alpha = 0.05$). Salinity and temperature were mapped in ArcGIS (ESRI, version 10.2), using Inverse Distance Weighting (IDW) interpolation to visualize temperature and salinity variation. Average annual wave energy was mapped in ArcGIS 10.2 and combined with spatial data on coastal biology and marine habitats [48] including kelp beds, hard substrate (rocky reefs), two seagrasses (eel grass *Zostera*; surf grass, *Phyllospadix*) and marine protected areas (MPA's).

Table 1. Mean, maximum and minimum salinity and temperature (±S.D.), for surface and bottom water in the area immediately around the outfall mouth (50–200 m) and the surrounding area (250–1000 m) measured at the four sampling times. Post-hoc results (Tukey-HSD, $p < 0.05$) for mean salinity and temperature are indicated in small letters for surface water and capital letters for bottom water behind each mean value.

Sampling Time	Distance from Outfall Mouth (m)	Water Level	Salinity			Temperature		
			Mean	Max	Min	Mean	Max	Min
December 2014	50–200 (Immediate)	Surface (n = 5)	33.6 b (0.06)	33.6	33.5	18.4 b (0.7)	19.0	17.4
		Bottom (n = 4)	33.5 BC (0.01)	33.6	33.5	18.2 B (0.8)	19.3	17.1
	250–1000 (Surrounding)	Surface (n = 5)	33.4 b (0.1)	33.5	33.3	18.8 b (0.6)	19.5	18.4
		Bottom (n = 4)	33.5 BC (0.1)	33.7	33.4	19.1 B (0.3)	19.4	18.8
September 2015	50–200 (Immediate)	Surface (n = 8)	33.1 b (0.7)	33.5	31.5	24.9 a (0.7)	26.5	24.0
		Bottom (n = 6)	33.5 C (0.2)	33.9	33.0	24.5 A (0.3)	25.2	24.1
	250–1000 (Surrounding)	Surface (n = 8)	33.4 b (0.07)	33.5	30.4	24.3 a (0.5)	25.2	23.7
		Bottom (n = 6)	32.8 C (1.2)	33.5	33.3	24.4 A (0.5)	25.1	23.9
May 2016	Mouth of outfall	Surface	37.4 a	37.4	37.4	NA	NA	NA
	50–200 (Immediate)	Surface (n = 7)	33.9 b (0.7)	35.5	33.6	19.1 b (0.4)	19.4	18.2
		Bottom (n = 6)	34.9 A (0.7)	35.7	33.9	19.3 B (0.8)	19.8	17.8
	250–1000 (Surrounding)	Surface (n = 7)	33.4 b (0.5)	33.8	32.5	18.8 b (0.6)	19.4	17.9
		Bottom (n = 7)	34.4 AB (0.7)	35.2	33.5	17.9 B (1.4)	19.4	15.5
November 2016	50–200 (Immediate)	Surface (n = 7)	33.5 b (0.2)	33.9	33.4	19.1 b (0.3)	19.7	18.8
		Bottom (n = 6)	34.4 AB (0.2)	35.9	34.8	19.3 B (0.5)	20.0	18.6
	250–1000 (Surrounding)	Surface (n = 7)	33.5 b (0.2)	33.9	33.3	19.1 b (0.6)	19.9	18.0
		Bottom (n = 7)	34.4 A (0.8)	35.9	33.6	18.7 B (0.7)	19.5	17.3

3. Results and Discussion

3.1. Physical and Chemical Characterization of the Discharge Plume

Salinity and temperature (mean, minimum and maximum) in the immediate (50–200 m) and surrounding (250–1000 m) area around the discharge channel of all four sampling times are given in Table 1. Before the Carlsbad Desalination Plant began operations, ambient salinity at Carlsbad Beach was 33.2 ± 0.6 (pooling all data from December 2014 and September 2015 both surface and bottom water, n = 146). This value is consistent with previous reports of coastal salinity in this region [13]. Surface and bottom salinity for all four sampling times are mapped in Figure 2.

Post-operation measurements show surface water salinity of 37.4 right at the mouth of the outfall channel. In the immediate vicinity (50–200 m) outside the outflow channel, surface salinity averaged 33.9 and 33.5 in May and November 2016, respectively, with a maximum measured surface salinity of 35.5 at ~50 m from the mouth of the discharge channel (Table 1). Surface water salinity did not significantly differ between pre- and post-operation, except for the high salinity water right at the channel mouth ($p < 0.001$) (Table 1).

Bottom water salinity in the immediate vicinity (50–200 m) reached values of 35.9 at 200 m offshore, whereas the average salinity were 34.4 and 34.9 for May and November 2016, respectively, and was significantly higher than salinity in September 2015 ($p < 0.001$) (Table 1).

Further from the channel (250–1000 m offshore), surface salinity was near ambient levels (averaged 33.4 and 33.5 for May and November, respectively) and salinity did not differ significantly between pre- and post-operation (Table 1). The average bottom water salinity in the surrounding area (250–100 m) was 34.2 for both May and November (1 salinity unit higher than ambient), and the maximum bottom salinity of 35.9 was measured at 250 m offshore.

Both surface and bottom water temperatures were 1–2 °C warmer around the outfall than further offshore, or away from the discharge area during December 2014 and May and November 2016 but did not differ between these three sampling times (Table 1). During September 2015, temperatures throughout the coastal area were high due to El Niño conditions with ocean temperatures above normal [49] (Table 1). In May 2016, a thermocline was measured 800–1000 m offshore due to seasonal upwelling (Figure 3G), but the temperature difference between the discharge water and ambient coastal water did not otherwise create thermoclines.

The hydrological mixing model performed for Poseidon Water [13] to estimate the areal extent and dissipation of the plume (footprint) surrounding the outfall incorporated some physical properties of Carlsbad Beach (shallow water and high wave energy), and assumed an initial discharge salinity of 42 at the mouth of the channel. The model predicted sufficient mixing would be achieved to reduce salinity to levels approved for the plant within the 200 m offshore [12,13]. However, our field data indicate that, despite the lower salinity we measured at the outfall compare to that used in the hydrological model (37.4 vs. 42), a discrete salinity plume extends 600 m offshore from the mouth of the outfall, with salinity up to 35.9 observed beyond 200 m offshore (Figure 2C,D,G,H). This is 2.7 units above the mean ambient salinity (33.2) and exceeds the limit of two units over ambient within 200 m of the shore, specified in the California Ocean Plan (2015 Amendment to Water Quality Control Plan). These results raise concerns about the adequacy of the mixing model used, and the higher than expected salinity emphasizes the need for post-operation coastal monitoring. The higher temperature of the plume water (1–2 degrees above ambient) did not lower the density sufficiently to negate the higher salinity and a visible halocline developed close to the discharge channel.

Although the distinct salinity and temperature of the discharge plume were recognizable extending seaward from the discharge channel (Figure 2), the distribution and concentrations of DOC, POM, Chl *a* and nutrients (NO_3, PO_4 and SiO_2) showed no significant temporal or spatial patterns in the region and were not correlated to salinity (see Table S1 in Supporting Information). Concentrations of major elements (normalized to salinity) and trace elements in the plume also did not differ from those in other sampling sites in the vicinity. If our results are representative of SWRO operations at other sites, they suggest that the higher salinity within the plume (perhaps accompanied by higher temperature) would be the dominant chemical factor affecting coastal biota in terms of chemical impacts directly related to the discharge.

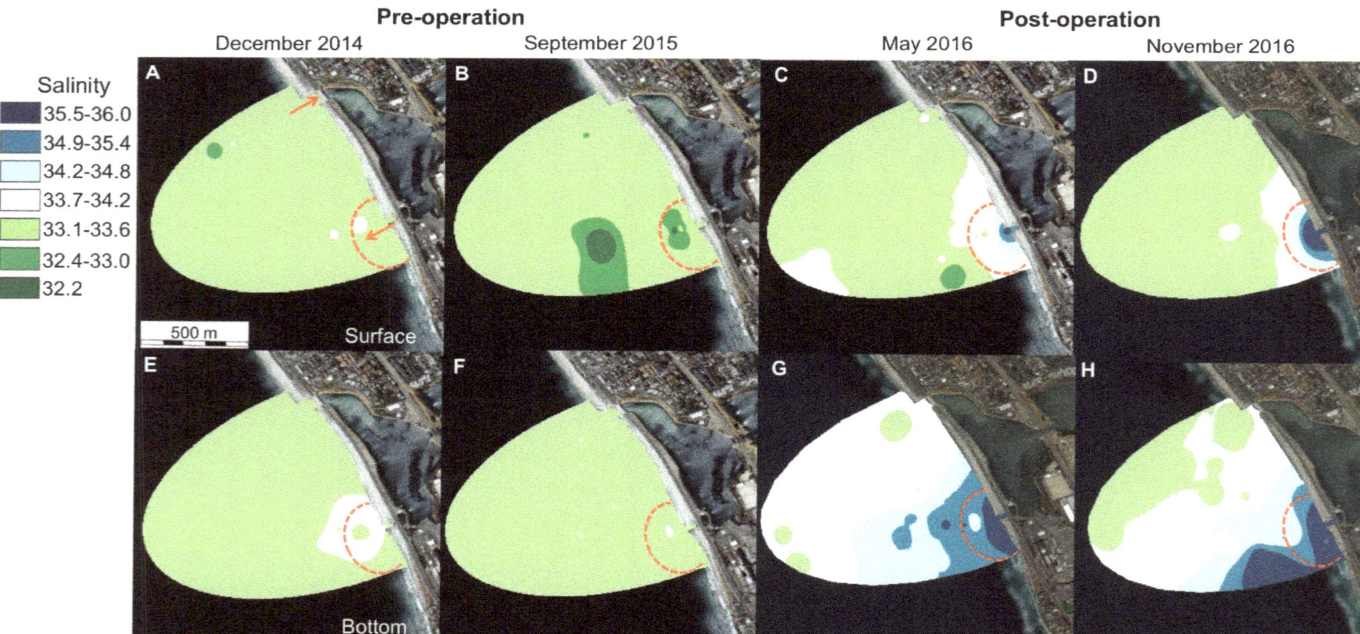

Figure 2. GIS-mapped salinity measurements for surface (**upper row**) and bottom water (**lower row**) for pre-operation (**A,B,E,F**) and post-operation (**C,D,G,H**). Arrows in (**A**) indicate the outfall location and lagoon intake area. The red semi-circle represents the 200 m distance set by California Ocean Plan (2015 Amendment to the Water Quality Control Plan).

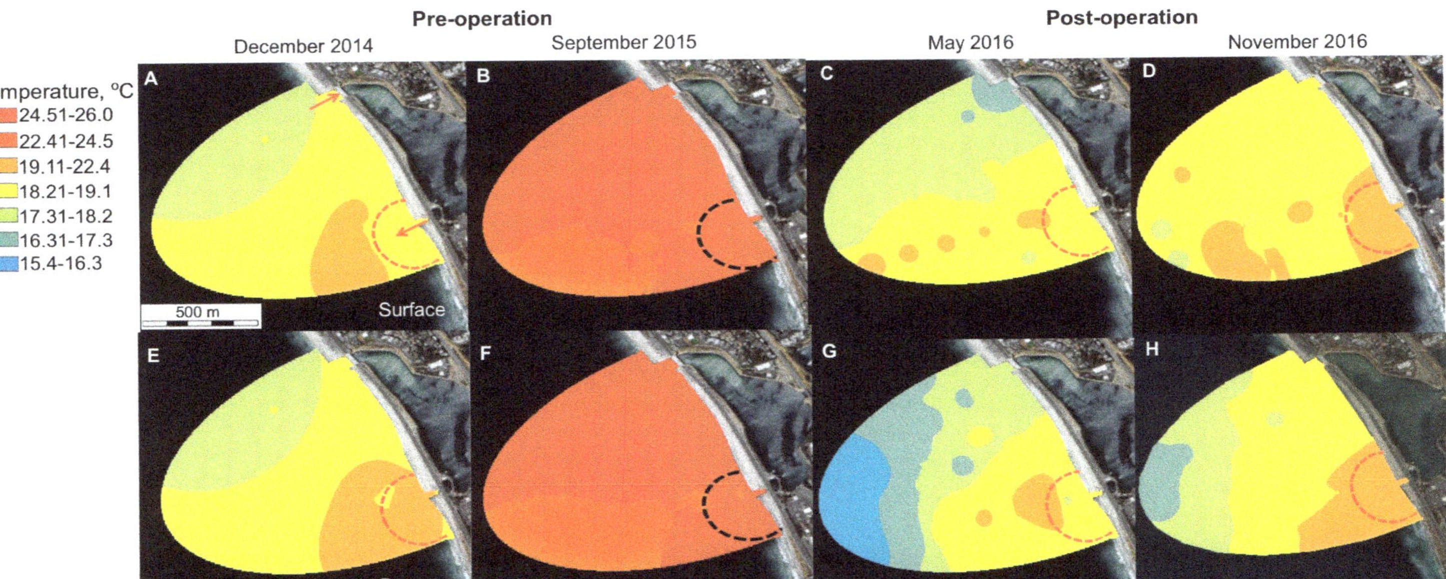

Figure 3. GIS-mapped temperature measurements for surface (**upper row**) and bottom water (**lower row**) for pre-operation (**A,B,E,F**) and post-operation (**C,D,G,H**). Arrows in (**A**) indicate locations of the outfall and lagoon intake. September 2015 was an El Niño year with higher average temperatures in the coastal zone (**B,F**). The red semi-circle (black for Sep 2015) represents the 200 m distance set by the California Ocean Plan (2015 Amendment to Water Quality Control Plan).

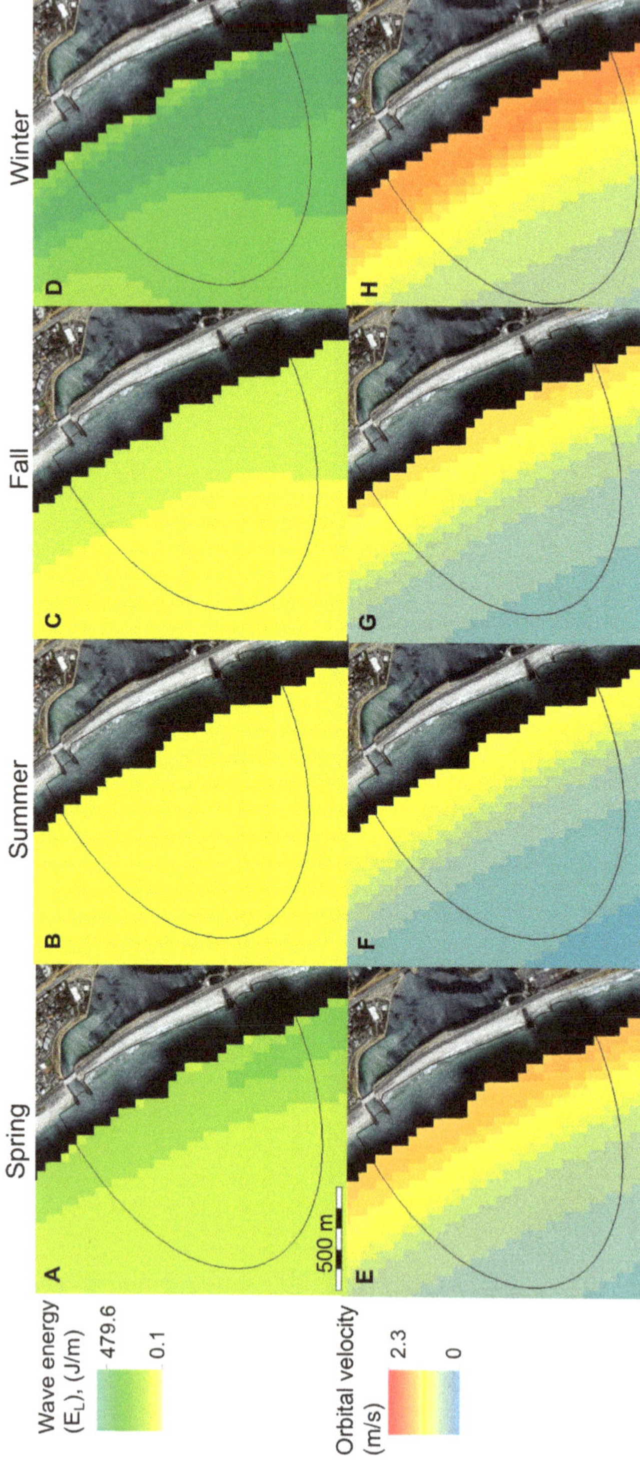

Figure 4. GIS-mapped averaged seasonal wave energy density (upper row **A–D**) and orbital velocity (lower row **E–H**) during spring, summer, fall and winter. The black parabola indicates the extent of the area of our field samples.

While the density of discharged brine and brine dispersion can be controlled by the design and operational practices of SWRO plants, coastal bathymetry and wave energy that also affect plume dispersal are not controlled through plant design. However, this study shows that these parameters should be considered when selecting discharge locations. Carlsbad Beach is a high energy site (Figure 4) where conditions appear optimal for brine dilution via extensive mixing in the nearshore area [13,40,41], yet the plume was clearly detected at least 600 m offshore. Hence, dilution via mixing at this site is inadequate. Wave energy and orbital velocities vary seasonally (Figure 4); therefore, wave-driven mixing potential should also vary throughout the year, with greater mixing in winter—however, we detected little difference in the brine plume's extent and properties between May and November 2016. This suggests that these seasonal mixing differences were not sufficient to alter the spatial extent of the brine plume. Further dilution of the brine prior to discharge (currently 1:10 with cooling seawater) would be required to lower the plume density and increase mixing potential. For future plants, this could be achieved if greater volumes of water are used for cooling in a co-located power plant or if combined with other freshwater discharges like sewage discharge. As physical and oceanographic properties are not adjustable, use of diffusor systems instead of open channels offers a solution for increasing dilution and mixing of SWRO discharges. At several SWRO facilities worldwide, implementation of diffusors has successfully increased brine mixing [50,51]. Continuous monitoring over several months or years would be necessary to determine whether the Carlsbad salinity plume always extends far from shore, or if it was due to special conditions at the time of our sampling.

3.2. Benthic Macrofauna and Ecotoxicological Study of Brine Exposure

The benthic epifauna had higher proportions of species at the northern end of the beach than at the southern end and around the discharge channel (see Supporting Information Figure S1), mainly due to different habitat types (small rocky reef in the north versus sandy bottom elsewhere). Accordingly, we only compared abundances at sampling sites with similar substrate pre- and post-operation.

The epifauna around the discharge area was dominated by tube-forming polychaetes (Onuphidae). Post-operation the abundance of Onuphidae was significantly higher in the immediate vicinity of the discharge channel (i.e., <200 m from the channel mouth) than in sandy areas away from the outfall ($p < 0.001$) (Figure 5A). The abundance of epifauna was significantly higher ($p < 0.001$) in the immediate vicinity of the discharge channel (0–200 m) compared to the abundance in the surrounding area (250–1000 m) in post-operation measurements (7.33 ± 3.8, 4.6 ± 0.9 and 1.4 ± 0.6, 1.6 ± 0.7 individuals m^{-2} for May and November 2016 in close vicinity and surrounding area, respectively (median $\pm$ SD)) (Figure 5A). This contrasts with pre-operation measurements, in which epifauna were significantly more abundant in the surrounding sandy areas (250–1000 m from discharge) than close to the channel (0.5 ± 1.3 individuals m^{-2} in the 0–200 m zone for both December and September and 7.3 ± 2.9 and 28 ± 15.7 for the 250–1000 m, December and September, respectively).

Infaunal organisms (in the upper ~10 cm of sediment) were generally more abundant in post-operation samples, but the difference was not significant (Figure 5B). The BOPA index close to the outfall channel was high (average of 0.23 ± 0.07), indicative of a poor ecological area with high disturbance. However, BOPA values did not differ between pre- and post-operation sampling times (Figure 5C).

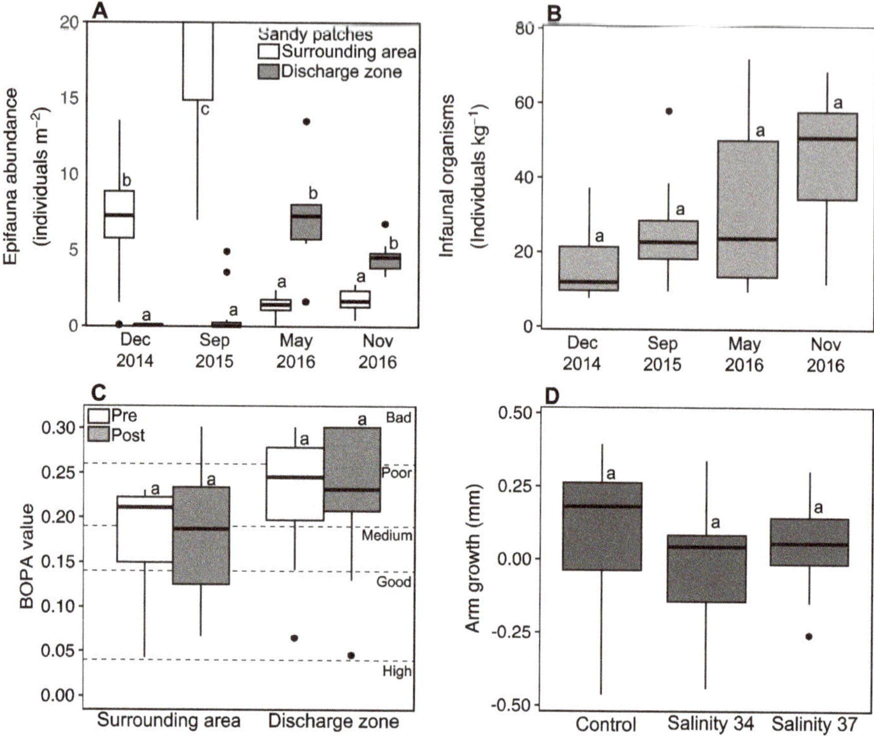

Figure 5. Box-whisker plots with significance letters above boxes refering to ANOVA followed by a Tukey-HSD test with significance of $p < 0.05$. (**A**) Median values for epifauna abundance in the immediate discharge zone (50–200 m from shore) and in the surrounding sandy areas (250–1000 m offshore), and (**B**) infaunal abundance in the area around the outfall channel (100–1000 m offshore) for pre-operation (December 2014 and September 2015) and post-operation (May and November 2016). Box-whiskers plot of (**C**) BOPA-values of pre- and post-operation at the sandy areas in the discharge zone (50–200 m from shore around the outfall) and in the surrounding sandy areas. The horizontal dashed lines indicating the ecological rank from good (high biodiversity) to bad (low biodiversity for mainly opportunistic species). Box-whiskers plot of the brittle stars' arm growth (**D**) after five weeks of incubation in brine from the outfall of Carlsbad Desalination Plant.

Grain size analysis of the sediment surrounding the outfall gave a sorting value of $\sigma = 0.8 \pm 0.2$ both pre- and post-operation, indicating moderately sorted sand implying high flow rate from the outfall channel which was indeed easy to observe from shore [45] (see Supporting Information Figure S2). Grain size away from the outfall has a larger fraction of fine grains being less impacted by channel flow. It is likely that the focused discharge of cooling water from the power plant for the past ~65 years has resulted in sand winnowing and extensive local disturbance of the sediment and benthos, and the addition of brine from the desalination plant does not induce any measurable acute toxicity for either epifaunal or infaunal organisms within the plume. In fact, the increase in polychaetes abundance at the outfall channel may be due to differential survival of organisms in the salinity plume, perhaps due to reduction of predation by larger organisms that may avoid the salinity plume [31,52]. While the lower abundance of epifauna in sandy areas away from the outfall zone supports this idea, we cannot rule out possible effects of natural seasonal variability or differential species recruitment. Particularly, the occurrence of El Niño during 2015 could have been a main driver for changes in abundances and recruitment.

Regardless of the specific causes, the already disturbed area from the cooling water discharge from the power plant, and the generally low biodiversity of organisms in the sandy environment (see Supporting Information Figure S1), result in minimal direct impacts of Carlsbad Desalination Plant on the ecology of benthic organisms still inhabiting this area, despite the salinity plume extending >200 m offshore.

In the incubation experiment, the brittle star growth (arm length) was not significantly different between the treatments (Figure 5D, $p > 0.06$). There also was no impact on mortality. Brittle stars are robust organisms present in virtually every marine habitat [53], and our data suggest that adult brittle stars can live at higher salinity than ambient, including salinity expected very close to discharge plumes [53]. In sandy environments like Carlsbad Beach, brittle stars are important for causing bioturbation in the upper sediment that enables rapid nutrient and oxygen flows to the sediment [54,55]. The growth and development of other echinoderms endemic to coastal California (*Strongylocentrotus purpuratus* (Purple Urchins) and *Dendraster excentricus* (Sand Dollar)) also are not significantly impacted by salinity increases up to 37 (~10% above ambient salinity of ~33) [36].

Observations around other desalination facilities worldwide have linked decreased abundances of epi- and infaunal organisms to increased salinity, typically when it reaches 5% above ambient [17,23,56]. Our field observations have not detected measurable negative changes (e.g., decrease) when comparing the pre- and post-operation conditions. As noted above, we speculate that this is likely because the site was already impacted by discharge from the power plant. We have not investigated organisms that are present in other settings in coastal California (for example, kelp or sea grasses) and it is possible that other organisms may be sensitive to the 3–5% above ambient salinity seen within the plume at Carlsbad Beach.

3.3. Future Desalination Plants in California

Rocky reefs, kelp beds and seagrass beds are widespread along the coast of California, and these highly productive ecosystems can be sensitive to small changes in their environments [57,58]. Previous experiments have shown higher mortality and lower sporulation in the macroalga *Ulva pertusa* when salinity was increased by 5–10% over ambient [25]. Similarly, seagrasses are particularly sensitive to salinity increases [20,23,59]. Our data indicate that SWRO discharge has little impact on organisms inhabiting sandy seafloors, particularly in non-pristine anthropogenic impacted areas, and such sites should be preferable locations for discharge of brine from new SWRO facilities. In addition, selecting sites with high wave action will increase mixing and dilution of brine discharges; discharge should be made offshore and in areas subjected to wave action that favors mixing.

At Carlsbad Beach, despite the relatively high average wave energy conditions (Figure 4 and Supporting Information Table S2), wave-driven mixing was not sufficient to reduce salinity to less than two units above ambient within 200 m from the discharge point. This suggests that meeting California's requirement for maintaining salinity of no more than two units above ambient within 100 m from the discharge site [12] at new SWRO facilities, particularly large ones (e.g., Camp Pendleton), will need more efficient discharge methods such as offshore diffusor systems, rather than a point discharge near the beach. Our modeling of proposed sites in Southern California (Camp Pendleton and Dana Point) shows they have lower wave energy than Carlsbad, and hence wave mixing is likely to be less efficient close to shore (Figure 6). At Camp Pendleton, the coastal substrate is dominated by sand, which will likely minimize possible ecological effects of the brine discharge; however, the coastal area surrounding Dana Point is dominated by rocky reefs that may hold great biodiversity, and caution should be taken when discharging brine at this site.

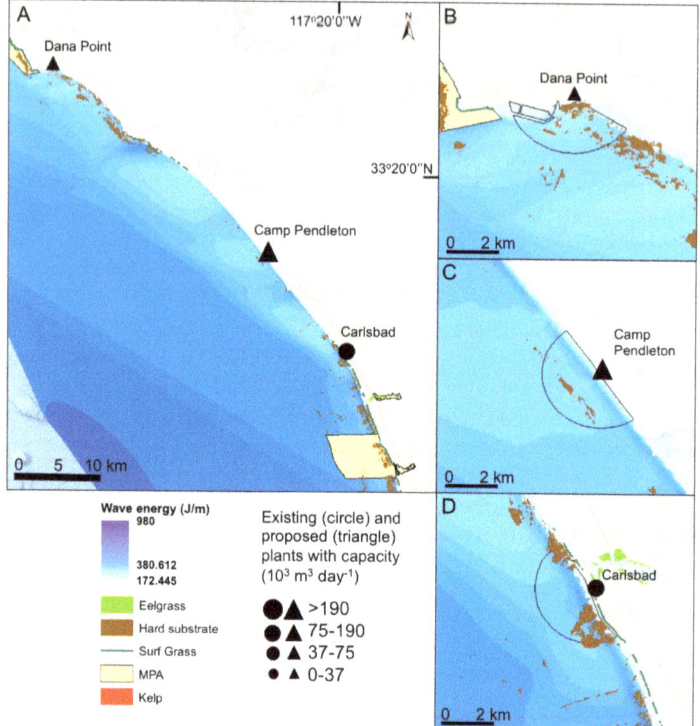

Figure 6. Average wave energy and benthic habitat for southern California with existing (circles) and proposed (triangles) desalination facilities (**A**). Zoomed perspective is shown in **B–D**.

All proposed sites in Central California, except for the Monterey Bay Peninsula Plant, are lower wave energy environments than at Carlsbad Beach (Figure 7 and Table S2) and presumably, lower mixing potentials. All of these plants are designed to have lower capacities than the Carlsbad Plant and, for the proposed plant at Moss Landing, co-location with an existing power plant and combining brine discharge with cooling water to increase dilution prior to discharge is possible. The benthic habitat or the three proposed locations around Monterey Bay is a predominately sandy bottom, but the two sites close to Moss Landing are near Elkhorn Slough, a protected National Estuarine Research Reserve, where a more cautious approach with extended monitoring of the brine mixing may be needed (Figure 7).

It is paramount that all proposed future desalination plants have comprehensive modeling of mixing trends that include seasonal differences in tides, waves and wind conditions at those sites and study the mixing potential in detail to assess if other techniques for diffusion are needed. To reduce the footprint of brine plumes and their possible impacts, use of alternative discharge technologies, such as diffuser systems, should be considered and appropriate modeling of the discharge design used. Both models and practice indicate that diffuser systems increase the mixing of discharge brine with ambient seawater [50,56,60,61]. Our results also suggest that long-term monitoring of coastal areas in the vicinity of proposed plants will be needed both before and after construction and operation. Ecosystem state and the likelihood for negative impact has to be considered in decisions regarding the location of future plants with effort to ensure that discharge will occur in areas with already low biodiversity (e.g., sandy areas and not near rocky substrate, kelp or seagrass) and where previous disturbances have already resulted in selection for resilient organisms (e.g., co-locating with power plant discharge, dredged areas or other potential natural or man-made disturbances).

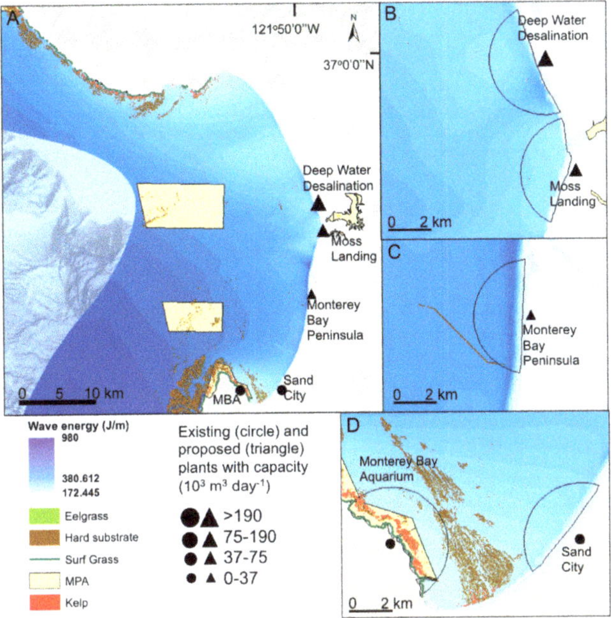

Figure 7. Average wave energy and benthic habitat for Monterey Bay with existing (circles) and proposed (triangles) desalination facilities (**A**). Zoomed perspective is shown in **B–D**.

4. Conclusions

A distinct salinity plume was observed around the outfall of Carlsbad Desalination Plant with a salinity increase of up to 2.7 units above ambient and extending 600 m offshore. This exceeds the maximum allowed salinity perturbation in the coastal zone as stated by the California Ocean Plan (2015 Amendment to Water Quality Control Plan). However, no significant impact was found on the benthic epifauna or infaunal composition and abundance in the same area, possibly due to an already disturbed conditions in this setting due to previous and ongoing discharge of post-cooling water from a local power plant. The coastal wave energy and mixing potential at Carlsbad Beach is high compared to other locations along the Southern and Central California coastline, and the lack of sufficient mixing of the brine at Carlsbad Beach raises concern regarding the efficiency of brine mixing at other proposed sites for desalination plants along the coast of California.

We recommend all future desalination facilities to be designed with optimal discharge options of either diffusor systems or a higher dilution of the brine prior to discharge. Furthermore, they should, where possible, be co-located with other water discharging facilities such as power plant or waste water treatment plants to restrict potential impacts to areas that are already disturbed from anthropogenic activities. Finally, we stress the need for long-term monitoring of coastal ecosystems in the area surrounding proposed and current desalination facilities to fully understand the ecological response to long-term exposure to salinity increase.

Supplementary Materials: The following are available online at http://www.mdpi.com/2073-4441/11/2/208/s1, Figure S1: Benthic macrofauna. Figure S2: Sediment grain size distribution, Table S1: Water chemical properties. Table S2: Brittle star growth. Table S3: Wave energy and benthic habitat estimation.

Author Contributions: Conceptualization, K.L.P., D.P. and A.P.; Data Curation, K.L.P. and A.P.; Formal Analysis, K.L.P., N.H., B.G.R. and A.H.; Funding Acquisition, D.P. and A.P. and K.L.P.; Investigation, K.L.P., D.P. and A.P.; Methodology, N.H. and B.G.R.; Project Administration, A.P. and K.L.P.; Resources, A.P.; Software, N.H. and B.G.R.; Supervision, D.P. and A.P.; Validation, D.P.; Visualization, K.L.P., N.H. and B.G.R.; Writing—Original Draft, K.L.P.; Writing—Review and Editing, K.L.P., N.H., B.G.R., D.P., A.H. and A.P.

Funding: This work was supported by National Science Foundation COASTAL SEES #1325649 awarded A.P and D.P. K.L.P. was partially supported by student fellowships from the Geological Society of America (#: 11215-16), the Weigel Scholarship for Coastal Studies, Myers Oceanographic and Marine Trust and the Explorers Club Youth Foundation (#: 83402).

Acknowledgments: We would further like to acknowledge the students and researchers of the Paytan Lab at UCSC, in particular Joseph Murray, Ana Martinez Fernandez, Kim Bitterwolf, Kyle Broach and Katie Roberts and the scientific divers and boat crew, especially Jacque Lord and Rich Walsh, for their extensive help and support in completing this work.

Conflicts of Interest: The authors declare no conflict of interest.

References

1. Grady, C.A.; Weng, S.-C.; Blatchley, E.R. Global potable water: Current status, critical problems, and future perspectives. In *Potable Water, The Handbook of Environmental Chemistry*; Younos, T., Grady, C.A., Eds.; Springer International Publishing: Cham, Switzerland, 2014; pp. 37–59.
2. UN-WWAP. *The United Nations World Water Development Report 2015: Water for a Sustainable World*; UN-WWAP: Paris, France, 2015; ISBN 978-92-3-100071-3.
3. Sneed, M.; Brandt, J.; Solt, M. Land subsidence along the Delta-Mendota Canal in the northern part of the San Joaquin Valley, California, 2003–2010. *Geol. Surv. Sci. Investig. Rep.* **2013**, *5142*, 87. [CrossRef]
4. Mann, M.E.; Gleick, P.H. Climate change and California drought in the 21st century. *Proc. Natl. Acad. Sci. USA* **2015**, *112*, 3858–3859. [CrossRef] [PubMed]
5. AghaKouchak, A.; Cheng, L.; Mazdiyasni, O.; Farahmand, A. Global warming and changes in risk of concurrent climate extremes: Insights from the 2014 California drought. *Geophys. Res. Lett.* **2014**, *41*, 8847–8852. [CrossRef]
6. Cooley, H.; Donnelly, K.; Ross, N.; Luu, P. Proposed Seawater Desalination Facilities in California, Pacific Inst. 2012. Available online: http://pacinst.org/wp-content/uploads/2013/02/full_report14.pdf (accessed on 9 November 2014).
7. California American Water. Monterey Peninsula Water Supply Project. 2015. Available online: http://www.watersupplyproject.org/home (accessed on 10 October 2017).
8. Pacific Institute. Existing and Proposed Seawater Desalination Plants in California. 2017. Available online: http://pacinst.org/publication/key-issues-in-seawater-desalination-proposed-facilities/ (accessed on 15 October 2017).
9. Voutchkov, N. Overview of seawater concentrate disposal alternatives. *Desalination* **2011**, *273*, 205–219. [CrossRef]
10. Lattemann, S.; Höpner, T. Environmental impact and impact assessment of seawater desalination. *Desalination* **2008**, *220*, 1–15. [CrossRef]
11. Shenvi, S.S.; Isloor, A.M.; Ismail, A.F. A review on RO membrane technology: Developments and challenges. *Desalination* **2015**, *368*, 10–26. [CrossRef]
12. Water Boards—CA EPA. *California Ocean Plan*; Water Boards—CA EPA: Sacramento, CA, USA, 2015.
13. Jenkins, S.A. *Revised Hydrodynamic Discharge Modeling Report*; Poseidon Water: Carlsbad, CA, USA, 2016.
14. Petersen, K.L.; Frank, H.; Paytan, A.; Bar-Zeev, E. Impacts of seawater desalination on coastal environments. In *Sustainable Desalination Handbook*, 1st ed.; Gude, V.G., Ed.; Butterworth-Heinemann: Oxford, UK, 2018; pp. 437–463.
15. Roberts, D.A.; Johnston, E.L.; Knott, N.A. Impacts of desalination plant discharges on the marine environment: A critical review of published studies. *Water Res.* **2010**, *44*, 5117–5128. [CrossRef]
16. Belkin, N.; Rahav, E.; Elifantz, H.; Kress, N.; Berman-Frank, I. The effect of coagulants and antiscalants discharged with seawater desalination brines on coastal microbial communities: A laboratory and in situ study from the southeastern Mediterranean. *Water Res.* **2017**, *110*, 321–331. [CrossRef]
17. Frank, H.; Rahav, E.; Bar-Zeev, E. Short-term effects of SWRO desalination brine on benthic heterotrophic microbial communities. *Desalination* **2017**, *417*, 52–59. [CrossRef]

18. Röthig, T.; Ochsenkühn, M.A.; Roik, A.; van der Merwe, R.; Voolstra, C.R. Long-term salinity tolerance is accompanied by major restructuring of the coral bacterial microbiome. *Mol. Ecol.* **2016**, *25*, 1308–1323. [CrossRef]

19. Del-Pilar-Ruso, Y.; de-la-Ossa-Carretero, J.A.; Giménez-Casalduero, F.; Sánchez-Lizaso, J.L. Effects of a brine discharge over soft bottom polychaeta assemblage. *Environ. Pollut.* **2008**, *156*, 240–250. [CrossRef] [PubMed]

20. Fernández-Torquemada, Y.; Sánchez-Lizaso, J.L. Effects of salinity on leaf growth and survival of the Mediterranean seagrass *Posidonia oceanica* (L.) Delile. *J. Exp. Mar. Biol. Ecol.* **2005**, *320*, 57–63. [CrossRef]

21. Park, G.S.; Yoon, S.-M.; Park, K.-S. Impact of desalination byproducts on marine organisms: A case study at Chuja Island Desalination Plant in Korea. *Desalin. Water Treat.* **2011**, *33*, 267–272. [CrossRef]

22. Sandoval-Gil, J.M.; Marín-Guirao, L.; Ruiz, J.M. The effect of salinity increase on the photosynthesis, growth and survival of the Mediterranean seagrass *Cymodocea nodosa*. *Estuar. Coast. Shelf Sci.* **2012**, *115*, 260–271. [CrossRef]

23. Sánchez-Lizaso, J.L.; Romero, J.; Ruiz, J.; Gacia, E.; Buceta, J.L.; Invers, O.; Torquemada, Y.F.; Mas, J.; Ruiz-Mateo, A.; Manzanera, M. Salinity tolerance of the Mediterranean seagrass *Posidonia oceanica*: Recommendations to minimize the impact of brine discharges from desalination plants. *Desalination* **2008**, *221*, 602–607. [CrossRef]

24. Lykkebo, K.; Paytan, A.; Rahav, E.; Levy, O.; Silverman, J.; Barzel, O.; Potts, D.; Bar-zeev, E. Impact of brine and antiscalants on reef-building corals in the Gulf of Aqaba e Potential effects from desalination plants. *Water Res.* **2018**, *144*, 183–191. [CrossRef]

25. Yoon, S.J.; Park, G.S. Ecotoxicological effects of brine discharge on marine community by seawater desalination. *Desalin. Water Treat.* **2011**, *33*, 240–247. [CrossRef]

26. Belkin, N.; Rahav, E.; Elifantz, H.; Kress, N.; Berman-Frank, I. Enhanced salinities, as a proxy of seawater desalination discharges, impact coastal microbial communities of the eastern Mediterranean Sea. *Environ. Microbiol.* **2015**, *17*, 4105–4120. [CrossRef]

27. Dauvin, J.C.; Ruellet, T. Polychaete/amphipod ratio revisited. *Mar. Pollut. Bull.* **2007**, *55*, 215–224. [CrossRef]

28. Riera, R.; de-la-Ossa-Carretero, J.A. Response of benthic opportunistic polychaetes and amphipods index to different perturbations in coastal oligotrophic areas (Canary archipelago, North East Atlantic Ocean). *Mar. Ecol.* **2014**, *35*, 354–366. [CrossRef]

29. de-la-Ossa-Carretero, J.A.; Del-Pilar-Ruso, Y.; Giménez-Casalduero, F.; Sánchez-Lizaso, J.L. Testing BOPA index in sewage affected soft-bottom communities in the north-western Mediterranean. *Mar. Pollut. Bull.* **2009**, *58*, 332–340. [CrossRef] [PubMed]

30. Joydas, T.V.; Krishnakumar, P.K.; Qurban, M.A.; Ali, S.M.; Al-suwailem, A.; Al-Abdulkader, K. Status of macrobenthic community of Manifa—Tanajib Bay System of Saudi Arabia based on a once-off sampling event. *Mar. Pollut. Bull.* **2011**, *62*, 1249–1260. [CrossRef] [PubMed]

31. de-la-Ossa-Carretero, J.A.; Del-Pilar-Ruso, Y.; Loya-Fernández, A.; Ferrero-Vicente, L.M.; Marco-Méndez, C.; Martinez-Garcia, E.; Giménez-Casalduero, F.; Sánchez-Lizaso, J.L. Bioindicators as metrics for environmental monitoring of desalination plant discharges. *Mar. Pollut. Bull.* **2016**, *103*, 313–318. [CrossRef] [PubMed]

32. Giangrande, A.; Licciano, M.; Musco, L. Polychaetes as environmental indicators revisited. *Mar. Pollut. Bull.* **2005**, *50*, 1153–1162. [CrossRef] [PubMed]

33. Halpern, B.S.; Kappel, C.V.; Selkoe, K.A.; Micheli, F.; Ebert, C.M.; Kontgis, C.; Crain, C.M.; Martone, R.G.; Shearer, C.; Teck, S.J. Mapping cumulative human impacts to California Current marine ecosystems. *Conserv. Lett.* **2009**, *2*, 138–148. [CrossRef]

34. California Department of Fish and Wildlife. California Marine Protected Areas (MPA's). 2017. Available online: https://www.wildlife.ca.gov/Conservation/Marine/MPAs#40713287-mpas (accessed on 26 October 2017).

35. Graham, M.H.; Vásquez, J.A.; Buschmann, A.H. Global ecology of the giant kelp *Macrocystis*: From ecotypes to ecosystems. In *Oceanography and Marine Biology*, 45th ed.; Gibson, R.N., Atkinson, R.J.N., Gordon, J.D.M., Eds.; Taylor and Francis Group: Boca Ranton, FL, USA, 2007; pp. 39–88.

36. Phillips, B.M.; Anderson, B.S.; Siegler, K.; Voorhees, J.P.; Katz, S.; Jennings, L.; Tjeerdema, R.S. *Hyper-Salinity Toxicity Thresholds for Nine California Ocean Plan Toxicity Test Protocols Preliminary Draft Final Report*; California State Water Resources Control Board: Davis, CA, USA, 2012.

37. Heck, N.; Paytan, A.; Potts, D.C.; Haddad, B.; Petersen, K.L. Management preferences and attitudes regarding environmental impacts from seawater desalination: Insights from a small coastal community. *Ocean Coast. Manag.* **2018**, *163*, 22–29. [CrossRef]

38. Heck, N.; Petersen, K.L.; Potts, D.C.; Haddad, B.; Paytan, A. Predictors of coastal stakeholders' knowledge about seawater desalination impacts on marine ecosystems. *Sci. Total Environ.* **2018**, *639*, 785–792. [CrossRef]

39. Poseidon Water. The Carlsbad Desalination Project. 2015. Available online: http://carlsbaddesal.com (accessed on 26 October 2017).

40. NOAA Satellite and Information Service, National Data Buoy Center. Hist. Data Download 2014–2017. (n.d.). Available online: http://www.ndbc.noaa.gov/download_data.php?filename=46224h2014.txt.gz&dir=data/historical/stdmet/ (accessed on 8 June 2017).

41. Southern California Coastal Ocean Observation System. Bathymetry Map of Carlsbad. 2015. Available online: http://www.sccoos.org/data/bathy/?r=7 (accessed on 8 June 2015).

42. Erikson, H.L.; Storlazzi, C.D.; Golden, N.E. Wave Height, Peak Period, and Orbital Velocity for the California Continental Shelf. U.S. Geology Survery Data Set. 2014. Available online: http://dx.doi.org/10.5066/F7125QNQ (accessed on 26 October 2017).

43. Erikson, H.L.; Storlazzi, C.D.; Golden, N.E. Modeling Wave and Seabed Energetics on the California Continental Shelf. U.S. Geological Survey Summary of Methods to Accompany Data Release. 2014. Available online: http://dx.doi.org/10.5066/F7125QNQ (accessed on 26 October 2017).

44. Murphy, J.; Riley, J.P. A modified single solution method for the determination of phosphate in natural waters. *Anal. Chim. Acta* **1962**, *27*, 31–36. [CrossRef]

45. Folk, R.L. A Review of grain-size parameters. *Sedimentology* **1966**, *6*, 73–93. [CrossRef]

46. Reguero, B.G.; Losada, I.J.; Méndez, F.J. A global wave power resource and its seasonal, interannual and long-term variability. *Appl. Energy* **2015**, *148*, 366–380. [CrossRef]

47. Barnard, P.L.; Short, A.D.; Harley, M.D.; Splinter, K.D.; Vitousek, S.; Turner, I.L.; Allan, J.; Banno, M.; Bryan, K.R.; Doria, A.; et al. Coastal vulnerability across the Pacific dominated by El Niño/Southern Oscillation. *Nat. Geosci.* **2015**, *8*, 801. [CrossRef]

48. California Department of Fish and Wildlife. Marine Region GIS Downloads. 2015. Available online: https://www.wildlife.ca.gov/Conservation/Marine/GIS/Downloads (accessed on 26 October 2017).

49. Jacox, M.G.; Hazen, E.L.; Zaba, K.D.; Rudnick, D.L.; Edwards, C.A.; Moore, A.M.; Bograd, S.J. Impacts of the 2015–2016 El Niño on the California Current System: Early assessment and comparison to past events. *Geophys. Res. Lett.* **2016**, *43*, 7072–7080. [CrossRef]

50. Ahmad, N.; Baddour, R.E. A review of sources, effects, disposal methods, and regulations of brine into marine environments. *Ocean Coast. Manag.* **2014**, *87*, 1–7. [CrossRef]

51. Del-Pilar-Ruso, Y.; Martinez-Garcia, E.; Giménez-Casalduero, F.; Loya-Fernández, A.; Ferrero-Vicente, L.M.; Marco-Méndez, C.; de-la-Ossa-Carretero, J.A.; Sánchez-Lizaso, J.L. Benthic community recovery from brine impact after the implementation of mitigation measures. *Water Res.* **2015**, *70*, 325–336. [CrossRef] [PubMed]

52. Iso, S.; Suizu, S.; Maejima, A. The lethal effect of hypertonic solutions and avoidance of marine organisms in relation to discharged brine from a destination plant. *Desalination* **1994**, *97*, 389–399. [CrossRef]

53. Stöhr, S.; O'Hara, T.D.; Thuy, B. Global diversity of brittle stars (Echinodermata: Ophiuroidea). *PLoS ONE* **2012**, *7*, e00319040. [CrossRef]

54. Blicher, M.; Sejr, M. Abundance, oxygen consumption and carbon demand of brittle stars in Young Sound and the NE Greenland shelf. *Mar. Ecol. Prog. Ser.* **2011**, *422*, 139–144. [CrossRef]

55. Vopel, K.; Thistle, D.; Rosenberg, R. Effect of the brittle star *Amphiura filiformis* (Amphiuridae, Echinodermata) on oxygen flux into the sediment. *Limnol. Oceanogr.* **2003**, *48*, 2034–2045. [CrossRef]

56. de-la-Ossa-Carretero, J.A.; Del-Pilar-Ruso, Y.; Loya-Fernández, A.; Ferrero-Vicente, L.M.; Marco-Méndez, C.; Martinez-Garcia, E.; Sánchez-Lizaso, J.L. Response of amphipod assemblages to desalination brine discharge: Impact and recovery. *Estuar. Coast. Shelf Sci.* **2016**, *172*, 13–23. [CrossRef]

57. Young, M.A.; Cavanaugh, K.C.; Bell, T.W.; Raimondi, P.T.; Edwards, C.A.; Drake, P.T.; Erikson, L.; Storlazzi, C. Environmental controls on spatial patterns in the long-term persistence of giant kelp in central California. *Ecology* **2015**, *86*, 45–60. [CrossRef]

58. Hamilton, S.L.; Caselle, J.E. Exploitation and recovery of a sea urchin predator has implications for the resilience of southern California kelp forests. *Proc. Biol. Sci.* **2015**, *282*, 20141817. [CrossRef] [PubMed]

59. Gacia, E.; Invers, O.; Manzanera, M.; Ballesteros, E.; Romero, J. Impact of the brine from a desalination plant on a shallow seagrass (*Posidonia oceanica*) meadow. *Estuar. Coast. Shelf Sci.* **2007**, *72*, 579–590. [CrossRef]
60. Missimer, T.M.; Maliva, R.G. Environmental issues in seawater reverse osmosis desalination: Intakes and outfalls. *Desalination* **2017**, *434*, 198–215. [CrossRef]
61. Fernández-Torquemada, Y.; Ferrero-vicente, L.; Gónzalez-correa, J.M.; Loya, A.; Ferrero, L.M.; Díaz-valdés, M.; Sánchez-lizaso, J.L. Dispersion of brine discharge from seawater reverse osmosis desalination plants. *Desalin. Water Treat.* **2009**, *5*, 137–145. [CrossRef]

Review

Lagoon Resident Fish Species of Conservation Interest According to the Habitat Directive (92/43/CEE): A Review on Their Potential Use as Ecological Indicator Species

Chiara Facca *, Francesco Cavraro, Piero Franzoi and Stefano Malavasi

Department of Environmental Sciences, Informatics and Statistics, Ca' Foscari University of Venice, via Torino 155, 30175 Venice, Italy; cavraro@unive.it (F.C.); pfranzoi@unive.it (P.F.); mala@unive.it (S.M.)
* Correspondence: facca@unive.it; Tel.: +39-(0)-41-234-7733

Received: 22 May 2020; Accepted: 17 July 2020; Published: 20 July 2020

Abstract: Transitional waters are fragile ecosystems with high ecological, social and economic values, that undergo numerous threats. According to the information provided by European Member States in the framework of the European Directive 92/43/EEC (Habitat Directive), the main threat to these ecosystems is represented by morphological and hydrological changes. The present work focuses on six lagoon fish species included in the Habitat Directive annex II (species requiring conservation measures: *Aphanius fasciatus, A. iberus, Knipowitschia panizzae, Ninnigobius canestrinii, Valencia hispanica* and *V. letourneuxi*) that spend their entire life cycle in the Mediterranean priority habitat 1150* "Coastal lagoons". The overview of the current scientific literature allowed us to highlight how the presence and abundance of these species may provide important indications on the conservation status of coastal lagoon habitats. In fact, their occurrence, distribution and biology depend on the presence of peculiar structures, such as salt marshes, small channels, isolated pools and oligohaline areas. Coastal lagoon fragmentation and habitat loss have led to a significant reduction in genetic diversity or local population extinction. Although *Aphanius* and gobies have been shown to survive in eutrophic environments, it is clear that they cannot complete their life cycle without salt marshes (mainly *Aphanius*) and wetland areas (mainly gobies).

Keywords: coastal lagoon; morphological alterations; habitat conservation; killifishes; gobies; Mediterranean Sea

---

1. Introduction

In 1992, the Directive 92/43/EEC, known as the Habitat Directive (HD), was issued to promote the "conservation of natural habitats and of wild fauna and flora" by ensuring bio-diversity (HD art. 2 par. 1) in the territory of European Member States. In Annex I, the Habitat Directive lists the sites of community interest, "whose conservation requires the designation of special areas of conservation". Among these sites, some are considered of "priority interest" because they are in danger of disappearance, such as the priority habitat 1150* "Coastal lagoons", defined as follows:

"Lagoons are expanses of shallow coastal saltwater, of varying salinity and water volume, wholly or partially separated from the sea by sand banks or shingle, or, less frequently, by rocks. Salinity may vary from brackish water to hypersalinity depending on rainfall, evaporation and through the addition of fresh seawater from storms, temporary flooding of the sea in winter or tidal exchange" [1].

These habitats are found along all European coasts and their extent in the Mediterranean Sea varies widely, from 2 ha of Chalikiopoulou Greek lagoon [2] to 57,000 ha of Italian lagoon of Venice.

In the Mediterranean biogeographic region (Table 1; Portugal excluded because it overlooks the Atlantic Ocean), there are 312 coastal lagoons included in the Natura 2000 network (the ecological network of sites to promote restoration and protection projects, in accordance with the HD's objectives, art. 3) [3] and, on average, their conservation status is unfavourable/bad (U2) [4]. The threats and pressures determining this evaluation are numerous and they can be linked both to the direct economic development of coastal areas (D03, E01, F02, J02) and to the effects of indirect activities that take place in the continental area (A02, A08; Table 2). These habitats, in fact, receive river inputs, enriched with nutrients and pollutants, leading to a substantial alteration of ecological balances and a general impairment of water quality [5]. However, the main threat is the direct destruction or reduction of habitats, due to the construction of infrastructure (e.g., ports), dredging of shipping channels and the creation of facilities for the regulation of hydrodynamism. Other threats include fishing, aquaculture, recreational development and the voluntary or accidental introduction of alien species, that has induced significant changes in environmental and ecological status [5]. From a management point of view, in almost all Member States, coastal lagoons are public property, except for a few special cases where some confined portions are privately managed [2].

Table 1. Number of "Coastal lagoons" (habitat code 1150*) in the Natura 2000 Network for EU Member States in the Mediterranean biogeographic region (Portugal excluded). Data were elaborated by consulting [3].

EU Member State in Mediterranean Region	Number of Natura 2000 Sites as Coastal Lagoon (HD Code 1150*)
Republic of Cyprus (CY)	1
Croatia (HR)	21
France (FR)	34
Greece (GR)	41
Italy (IT)	147
Malta (MT)	4
Slovenia (SI)	2
Spain (ES)	62
TOTAL	312

Table 2. Activities with highly important pressures and threats on "coastal lagoons" habitats throughout the EU. The "Pressures" and "Threats" columns show the number of Member States that have reported these activities as the ten most impacting (modified from the global report for the 2007–2012 reporting period for coastal lagoons [4]).

Code	Activity	Pressures	Threats
J02	Changes in water bodies conditions	21	22
H01	Pollution to surface waters	16	9
E03	Discharges (household/industrial)	7	7
F02	Fishing and harvesting aquatic resources	7	7
A08	Fertilisation in agriculture	5	5
H02	Pollution to groundwater	5	5
K02	Vegetation succession/Biocenotic evolution	5	7
A02	Modification of cultivation practices	4	
E01	Urbanisation and human habitation	4	7
H03	Pollution to marine waters	4	
L07	Storm, cyclone		5
D03	Shipping lanes and ports		4

From morphological and hydrodynamic points of view, each lagoon has peculiar characteristics closely related to the amount of freshwater inputs, tidal fluctuations and human interventions that, very often, have modified these areas either by land reclamation or by exploitation of fish resources or for navigation. The composite mosaic of the lagoon structure, with islands, sand barriers, salt marshes, wetlands, etc., determines a high heterogeneity from which we can derive the ability to support higher productivity, compared to adjacent seas. These characteristics, determined by the transition between continental and marine environments, may also favour the presence of endemic species of both freshwater and marine fauna [6].

However, not all organisms can tolerate these marked physio-chemical, spatial and temporal variations, so the biodiversity can be lower than that in the adjacent sea and freshwaters, because of the reduced euryvalence of most species. Among fish fauna, species can be categorized into functional guilds on the basis of the way they use transitional areas: to breed, to feed, as a nursery for juveniles, as refuge, as migratory routes between freshwaters and seas, for the entire life cycle [7–9].

The Annex II of HD includes "Animal and Plant Species of Community Interest whose conservation requires the designation of special conservation zones". Among these, more than 60 fish species are listed, including bony fishes and cyclostomes. Focusing the attention on the species recorded in the "Coastal Lagoon" habitats of the Mediterranean Sea, on the basis of data provided by Member States [3], it was possible to fill the list in Table 3 [3,10–14], further completed with the information on the Conservation status under the Habitat Directive and the red list assessments compiled by the International Union for the Conservation of Nature (IUCN) [13]. This list includes 26 species that could be grouped in functional guilds, according to the classification proposed by Potter et al. [9]: anadromous fishes (five species), freshwater fishes (17 species) and lagoon residents (four species). The present review aims at exploring the potential use of fish species of Community interest (sensu Habitat Directive art. 4, letter g), as indicators in the context of lagoon coastal habitats.

Ecological indicator species are defined as species whose presence and abundance provide information about ecosystems [15]. As the morphological alteration is one of the main threats for coastal lagoons (Table 2, [4]), the present paper describes the biology, habitat preference, distribution and conservation status of those fish species that are more strictly related to salt marshes and wetlands, in order to propose them as early warning signal of habitat loss. Bortone et al. [16] indicates that the guild of estuarine resident is the most suitable to be used as an indicator of lagoon environmental conditions. Therefore, four lagoon resident species (Table 3), that spend their entire life cycle in a transitional environment, were selected. Adapted to a naturally stressed ecosystem, these species can tolerate wide environmental variations and, hence, cannot be considered indicators of water quality. However, in the following sections, examples on how these species can be threatened by the loss or alteration of peculiar habitat structures will demonstrate their suitability as indicators. In this document, the term "ecological indicators" means the indicators that respond to environmental stressors related to changes in the structure of lagoon habitats, including morphological alterations, salinity changes, and presence of invasive alien species.

In the Habitat Directive, *Aphanius fasciatus* and *A. iberus* are referred to as belonging to the order Atheriniformes, family Cyprinodontidae, but the recent nomenclature places them in the order Cyprinodontiformes and the family is Aphaniidae [17]. Similarly, *Knipowitschia panizzae* and *Ninnigobius* (former *Pomatoschistus) canestrinii* are listed as Perciformes, family Gobiidae, while are now attributed to the order Gobiiformes, family Gonionellidae [12]. The most common species for the number of reports in the Mediterranean coastal lagoons of the Natura 2000 Network is *A. fasciatus* (109 sites), while the others are present in 20–30 sites (Table 3). None of these species are priority species, meaning a species "for the conservation of which the Community has particular responsibility in view of the proportion of their natural range" (HD art. 1 letter h) nor is present in other Annexes. Additionally, the two species belonging to the genus *Valencia* were discussed, since they are phylogenetically related to Aphanidae and have a high level of conservation priority. Moreover, *V. hispanica* and *V. letourneuxi* occur in small coastal lakes, tolerating low salinity conditions, and the former shares the endemic status with *A. iberus* along the Mediterranean Spanish coast.

Table 3. List of fish species in Habitat Directive's Annex II and their presence in the habitats 1150* Coastal lagoons in the Mediterranean Sea basin. The first column included the species as listed in HD, and the second column the currently accepted nomenclature, verified by consulting [10–12]. IUCN category abbreviations as follow: DD-Data deficient; LC-Least concern; NT-Near Threatened; VU-Vulnerable; EN-Endangered; CR-Critically Endangered; NA-Not Assessed. Conservation status abbreviations as follow: FV-Favourable; U1-Unfavourable/Inadequate; U2-Unfavourable/Bad; XX-unknown. The abbreviations for the Member States are shown in Table 1. The asterisk (*) indicates that the species is a priority species; (o) indicates species that are neither in Annex IV nor V; (V) indicates the species that are also in Annex V, but not in Annex IV. Species highlighted in grey are particularly related to Coastal lagoon conservation status. Data were elaborated by consulting [3,13,14].

Name of the Species in Annex II	Name of the Species Currently Accepted	Functional Group	IUCN	Conservation Status	CY	ES	FR	GR	HR	IT	MT	SI	Total
				PETROMYZONTIFORMES									
				Petromyzontidae									
Lampetra fluviatilis (V)	Lampetra fluviatilis (V)	Anadromous	LC	U2			2			2			4
Petromyzon marinus (o)	Petromyzon marinus (o)	Anadromous	LC	U2	1		4		1	9			15
				ACIPENSERIFORMES									
				Acipenseridae									
* Acipenser naccarii	*Acipenser naccarii	Anadromous	CR	XX						5			5
* Acipenser sturio	*Acipenser sturio	Anadromous	CR	U2				1					1
				CLUPEIFORMES									
				Clupeidae									
Alosa fallax (V)	Alosa fallax (V)	Anadromous	LC	U2		2	6	7	1	30			46
Alosa vistonica (V)	Alosa vistonica (V)	Freshwater	CR	U2				1					1
				SALMONIFORMES									
				Salmonidae									
Salmo marmoratus (o)	Salmo marmoratus (o)	Freshwater	LC	U2						1			1
				CYPRINIFORMES									
				Cyprinidae									
Alburnus albidus (o)	Alburnus albidus (o)	Freshwater	VU	U2					1	2			3
Aspius aspius (V)	Leuciscus aspius (V)	Freshwater	NA	XX				2					2
Barbus meridionalis (V)	Barbus meridionalis (V)	Freshwater	NT	U1		1	2						3
Barbus plebejus (V)	Barbus plebejus (V)	Freshwater	LC	XX						2			2
Chalcalburnus chalcoides (o)	Alburnus chalcoides (o)	Freshwater	LC	XX				1					1

Table 3. *Cont.*

Name of the Species in Annex II	Name of the Species Currently Accepted	Functional Group	IUCN	Conservation Status	CY	ES	FR	GR	HR	IT	MT	SI	Total
Chondrostoma soetta (o)	*Chondrostoma soetta* (o)	Freshwater	EN	XX						2			2
Rhodeus sericeus amarus (o)	*Rhodeus amarus* (o)	Freshwater	LC	XX				2					2
Rutilus pigus (V)	*Rutilus pigus* (V)	Freshwater	LC	XX						2			2
Rutilus rubilio (o)	*Sarmarutilus rubilio* (o)	Freshwater	NT	U1						2			2
Cobitidae													
Cobitis taenia (o)	*Cobitis taenia* (o)	Freshwater	LC	XX		9			1				10
Cobitis trichonica (o)	*Cobitis trichonica* (o)	Freshwater	EN	FV				1					1
SILURIFORMES													
Siluridae													
Silurus aristotelis (V)	*Silurus aristotelis* (V)	Freshwater	DD	FV				1					1
SCORPAENIFORMES													
Cottidae													
Cottus gobio (o)	*Cottus gobio* (o)	Freshwater	LC	U1			1						1
In directive annexes listed as ATHERINIFORMES, but currently accepted as CYPRINODONTIFORMES													
In directive annexes listed as Cyprinodontidae, but currently accepted as Aphaniidae													
Aphanius iberus (o)	*Aphanius iberus* (o)	Lagoon resident	EN	U1		29							29
Aphanius fasciatus (o)	*Aphanius fasciatus* (o)	Lagoon resident	LC	U1			10	20	1	72	4	2	109
In directive annexes listed as Cyprinodontidae, but currently accepted as Valenciidae													
* *Valencia hispanica*	**Valencia hispanica*	Freshwater	CR	U2		13							13
* *Valencia letourneuxi*	**Valencia letourneuxi*	Freshwater	CR	U1				2					2
In directive annexes listed as PERCIFORMES, but currently accepted as GOBIIFORMES													
In directive annexes listed as Gobiidae, but currently accepted as Gobionellidae													
Knipowitschia panizzae (o)	*Knipowitschia panizzae* (o)	Lagoon resident	LC	FV					3	22			25
Pomatoschistus canestrinii (o)	*Ninnigobius canestrinii* (o)	Lagoon resident	LC	FV					2	18			20

Despite the different conservation status of these six species, on the basis of the official European and International documents, related to the Habitat Directive and IUCN red list, these species may represent ecological indicator species of the complex mosaic of habitats characterizing the coastal lagoons in the Circum-Mediterranean area. The first goal of the present review was to collect and present three main components of the knowledge available on these species, on the basis of the current scientific literature: (1) Biology and Distribution, including the main information on population ecology and genetics, behaviour and life history (2) Conservation status, providing useful data to obtain some insights on the estimated conservation status of these species across their geographical range of distribution (3) Management, that is, interventions and actions conducted on these species, in terms of habitat conservation and restoration, including all projects carried out until present to enhance the favourable conservation status of their population.

The second goal was to overview and comparatively analyse this information, not only to provide a state of the art on the Conservation Biology of these species, but also to identify the potential future directions of research that may promote the use of these species as ecological indicators of morphological alterations in Mediterranean transitional water systems.

2. *Aphanius Fasciatus* (Valenciennes, 1821)—Order Cyprinodontiformes

2.1. Biology and Distribution

Individuals of *A. fasciatus* are small in size (maximum total length of female approx. 8–9 cm [18]) and have a marked sexual dimorphism, with males being smaller than females, characterized by yellow to yellow/orange caudal, dorsal and anal fin coloration (in some specimens, a vertical black band can also be present on the caudal fin) and 10–12 transversal brown bands alternating with straighter, silver-white bands on a greenish brown body [18]; females have 11–17 short dark brown bands along both sides of the body and the background is grey. The life cycle can last at maximum six to seven years [18,19] but, generally, sexual maturity is achieved in the first year of age [19], favouring a large number of generations and a rapid turnover of the population [20]. Females are, generally, more abundant than males, although the sex ratio displays significant seasonal variations: males, with their brighter colours and striking courtship behaviour, could suffer a higher mortality during the reproductive period [18,21,22]. It is omnivorous [18] and its diet varies with individuals' age and seasons: juveniles are planktonophagic while adults become benthivores [23]. Although *A. fasciatus* can be found in freshwaters, especially in the terminal sections of rivers, the most abundant populations live in coastal environments from brackish [24] to hyperhaline water bodies [20,25], both closed or subjected to tidal excursions. It is supposed that the reproduction is synchronized with the moon phases because adults exploit the rising spring tide to reach the most isolated canals within the salt marsh system [26]. During the courtship, the male, trying to exclude the competitors, pushes the female towards the optimal area for spawning, which is represented by submerged vegetation in small and isolated creeks inside salt marsh system. Males have no territorial behaviour and parental cares lack, so the cannibalism of eggs often occurs [27]. After hatching, which takes place 14 days after the spawning, when the tide height is again favourable to movements, the juveniles reach small bodies of water, isolated from open waters at low tide and, therefore, they are less at risk of predation [26]. Field observations on the life cycle of this species suggest that it is a sedentary fish throughout all of the year, preferring isolated areas [24], well protected from predation, avoiding the open lagoon [28]. The limited home range, the large demersal eggs, the absence of larval dispersal stages [19,21] and the fragmentation of brackish habitats lead to a genetic isolation with scarce gene flow [25] and a high degree of genetic divergence among populations [24], inducing some authors to hypothesize the presence (not verified) of different species within the complex *A. fasciatus* [20]. A marked genetic diversity among the populations of *A. fasciatus* has been observed in the east coast of the Adriatic Sea, where individuals of the saline in the Northern part were found to be genetically distinct from those in the south [29]. Likewise, populations from Sicilian, Sardinian and Adriatic coasts display significant osteological differentiation [30]. Bottleneck effects were documented in some Italian lagoons along the Tyrrhenian coast, where a high mortality rate due to particular events, such as dystrophic crises, significant tidal changes [31] or lack of maintenance in saltern environments [25], determined a dramatic genetic loss. However, except for these few documented cases, *A. fasciatus* populations are considered to be demographically stable, namely without recent founder events or recurrent bottlenecks [24]. In Italy, two main clusters were identified with a similar genetic structure: the first includes the Tyrrhenian populations above about latitude 41 °N and the second the southern Tyrrhenian and the Adriatic ones [32].

A. fasciatus is one of the most eurythermal and euryhaline species of the Mediterranean Sea, being able to withstand temperature changes between 4 and 40 °C and to reproduce at a salinity between 10 and 80 [20]. Despite this high capacity to adapt to extreme physical–chemical conditions, the genetic variation within-population is very small and is threatened by natural and anthropogenic pressures [25,32]; coupling this with the scarce gene flow among populations [24], the species survival can be locally compromised [33]. This situation was reported in the Maltese island, where only four populations were found in hydrologically isolated sites [34]. Furthermore, these populations were highly vulnerable as the abundance of adults and the percentage of juvenile survival were low [35].

It is widely distributed along the coasts of central Mediterranean basin; in the easternmost coasts it is scarcely abundant because *A. dispar* is present, even if, in some Egyptian lagoons, both species have been reported and hybrids have been observed [33]. In the westernmost areas, only *A. iberus* has always been recorded. According to the Natura 2000 network dataset, *A. fasciatus* is present along all Italian and Greek coasts, in Malta, Croatia and Slovenia and along the Corsican coasts (Table 3).

2.2. Conservation Status

At the level of the Mediterranean Sea, it is estimated that there are 153 populations [36], and the species is reported at 137 Natura 2000 sites, of which 109 are priority coastal lagoon habitats. In Italy, populations are recovering and in good condition [33], but the state of conservation at the international level is considered unfavourable/inadequate, although the species is not at risk of threat [37].

According to Natura 2000 dataset (Table 3), in Italy, *A. fasciatus* is present in 72 coastal lagoons, with a heterogeneous distribution. The populations of the northern Adriatic Sea are distributed in the lagoon systems from the Isonzo estuary to the Pialassa lagoon and appear constant over time but distinct from each other [33,38], with a highly variable density; in fact, depending on the season and local morphological structures, the density can range, as in the case of the Venice lagoon, from 2 ind/100 m^{-2} in salt marsh areas strongly influenced by tides to 205 ind/100 m^{-2} in almost-closed artificial ditches [28]. In the southern Adriatic Sea, two distinct populations, one in the northern and the other in the southern Apulia zone, may be the result of channelling works that have fragmented the habitats and created inhospitable intermediate areas [33]. Along the Tyrrhenian coast, it is present in Tuscany and partially in Latium, and it is believed that the absence of *A. fasciatus*, between Latium and Campania, may be the consequence of extensive land reclamation conducted in the first half of the 20th century [33]. *A. fasciatus* surely disappeared from several inland waters, where it was reported in the 1800s and until 1980, and recently it has been observed only in Sicilian ones [33,39]. Past and present data on its distribution and abundance in rivers are scarce, but it is supposed that *A. fasciatus*, although it can survive in freshwaters, has little affinity for this habitat [33]. In fact, freshwaters host a more diverse community than brackish zones, where more constraining conditions limit the taxonomic richness [40], and *A. fasciatus* was demonstrated to suffer from the presence of other species and to prefer marginal and isolated zones [28,33]. Moreover, in recent years, the freshwater species *Gambusia holbrooki* (order Cyprinodontiformes, family Poeciliidae) has determined the extinction of some *A. fasciatus* populations in Northern Italian rivers [33,39]. This invasive species was demonstrated to be highly aggressive in oligohaline waters, but to be less successful at higher salinity [41,42], inducing species that can tolerate wide salinity variations to move towards brackish waters.

Despite the reduction of the range in freshwaters, in the original brackish and coastal environments, there has been a recovery of *A. fasciatus*, with the reappearance of populations considered extinct, especially in the northern Tyrrhenian coast [33]. The factors favouring the recovery are unknown, but the most probable reason seems to be the reappearance of wetlands and the establishment of protected areas, including transitional water bodies [33].

In the Greek Porto Lagos lagoon, the relative abundance of *A. fasciatus* reached almost 10% of the ichthyofauna community, with catch per unit effort (CPUE) from 5 in May 1989 to 1572 in October 1989 [43]. More recent studies, also in other Greek lagoons, describe population dynamics and structure, but they do not report density or abundance data. However, considering the number of specimens caught during these studies, it can be supposed that the populations are rather abundant [19,21,23].

In Cyprus Akrotiri wetlands, *A. fasciatus* was demonstrated to have a "contraction-expansion" distribution depending on the annual hydrology. Based on observations between 2008 and 2015, the species' range was significantly contracted and, in 2016, the CPUE ranged from 12 to 931 [44].

In Croatia, until the early 2000s, the species was reported in several sites, while more recently, it has been found only in half of the historical basins. However, some new findings were recorded and some other habitats have to be investigated. The data available indicate that the distribution is discontinuous along the entire eastern Adriatic coast and the captures tend to be rather low [45].

Depending on the colonization range, the main, but not unique, threats for *A. fasciatus* are:

1. in freshwaters, especially, competition with the invasive species *G. holbrooki*, which is particularly aggressive at low salinity values [42,44]. Despite the fact that *G. holbrooki* can be found also at high salinity, as for example in Mar Menor (Spain; [46]) and in some French lagoons [47], its metabolism is particularly affected by the necessity to maintain the osmotic equilibrium. The effects on the community structure are evident due to an increase of sex ratio (M/F): females have a lower survival rate being larger than male and having higher reproductive investment [41]; moreover, Alcaraz et al. [48] demonstrated that, at salinity 25, its behaviour is less aggressive and its predation capacity slower than that of *A. fasciatus*;
2. in hyperhaline systems, the heavy hydromorphological modifications related to the salt production and extraction and/or the lack of maintenance leading to habitat degradation and significant genetic loss [25];
3. in lagoon systems, the loss of habitat complexity structure due to erosive phenomena and the consequent disappearance of salt marshes that represent the preferred habitat for refuge, reproduction and growth of both adults and juveniles [27,28,49].

2.3. Management

Some recent works were conducted to assess the ecological value of artificial habitats (ditches) for the species, with a view to manage and preserve these habitats as surrogates of saltmarsh areas [28,49]. In the Venice lagoon, abundant, well-structured, healthy and successful breeding populations were, in fact, observed in abandoned artificial ditches, once used for traditional fish farming, as small marinas or as a defence line during the two world wars [49]. These artificial ditches, located in marginal areas, completely isolated from the open lagoon, can offer protection from predators, high resources and determine different adaptive strategies. In the lagoon of Venice, it was observed that the populations of natural salt marshes had a higher mortality rate than those of artificial canals, in a probable relation with a higher predation pressure due to the presence of piscivore fishes. The different predatory pressure, combined also with a greater availability of food, has favoured the development of different life histories in the populations of artificial ditches, which live longer and invest more in growth and less in reproduction [28]. These artificial ditches can, therefore, be an important resource for the conservation of local populations, but it remains important to maintain natural ecosystems in their complexity.

The strong sedentary behaviour of the species can seriously compromise its ability to adapt to environmental changes in response to human pressure, because the gene flow is reduced and it cannot guarantee a sufficient genetic variability [24,25,32]. On the other hand, the strong relationship with the habitat is an important feature when choosing the species as an indicator of habitat conservation status. Its survival and reproductive rate, in fact, depends on the possibility to find sheltered areas with low hydrodynamism. On the whole, the salt marsh structures with small tidal creeks and pools have to be maintained and protected from erosion processes but, at the same time, the connectivity among similar habitats has to be guaranteed, avoiding the fragmentation, due to land reclamation or to the construction of infrastructures or to the dredging of large and deep canals. Although sedentary, the connectivity between salt marsh areas or even lagoon systems may allow some small migrations and, hence, a better gene flow.

Considering the water quality of some habitats where the species is generally abundant, it does not seem necessary to pay particular attention to the trophic status, whereas heavy metals and organic pollutants seem to affect the reproduction [50] and to cause spinal deformities [51].

The mosaic of habitats occupied by the species should be investigated and managed also in relation to potential competition and distribution of *G. holbrooki*. Specific management actions to reduce the impact of *G. holbrooki* on *A. fasciatus* were not found, but some authors described how to intervene with other native species, mainly in freshwater habitats. An efficient management plan of interventions could require the following actions:

(i)   actions aimed at decreasing the probability of habitat recolonization by *Gambusia* spp. from nearby locations, to avoid a rapid return of the species and, hence, the wastefulness of resources [52]. As an example, in the case of artificial, man-regulated habitats, the management of the water salinity could favour the autochthonous species survival and proliferation.

(ii) non-invasive eradication methods, although, to date, the only proven, cost-efficient eradication method is poisoning [52], but attention must be paid to the effect on native and non-target organisms. In any case, the eradication has to be scrupulously planned, because it can be more effective if it is carried out in area with limited probabilities of *Gambusia*'s recolonization [52];

(iii) actions aimed at increasing the intra-guild predation. It was found that adults of some native species can prey on *Gambusia*'s eggs and larvae [53]. Therefore, strengthening the native populations with the introduction of adult specimens may reduce *Gambusia*'s density.

3. *Aphanius iberus* (Valenciennes, 1846)—Order Cyprinodontiformes

3.1. Biology and Distribution

A. iberus is an endemic species of the Mediterranean coasts of Spain, often abundant in salt flats and brackish waters, but, sometimes, also present in freshwaters. It has a short life cycle (about 2 years), but a rapid growth rate and strong interannual variability in recruitment [54]. Several morphotypes are recognized, but the degree of differentiation is similar to that of other species of the genus *Aphanius*. The male has bluish grey, olive to bluish green body [18] with narrow silvery cross bands that also extend to the tail fin; the female has an olive green or bluish body with small black spots that tend to form stripes, one of which overlaps the lateral line. Males are also on average smaller (4.5 cm) than females (6.0 cm). The sex ratio is rather variable, but, in general, females are more abundant, due to male mortality after spawning [18]. As for the diet, it is omnivorous, combining both animal (mainly crustaceans) and plant of detritic origin [55]; as a generalist, it could be able to survive even the severe changes suffered by estuarine and coastal habitats [56]. In fact, the diet may vary depending on the availability of prey and, although it is generally considered to be a benthivorous micropredator, close correlations with organisms in the water column have been observed, particularly in juveniles. When available, harpacticoid copepods are the dominant prey above all for smaller fish; larger fish tend to add some larger prey without completely changing the trophic niche and feeding habitat with growth [55]. *A. iberus* is a gonochoric species with a reproductive strategy typical of unstable environments, as it reaches sexual maturity early, already a few month-olds, investing a lot of energy. The period of spawning varies depending on the latitudes, between May and August in the Ebro Delta (Catalonia, Spain) [57] and until October/November to the south, in the Mar Menor (Murcia, Spain) [58]. In one season, multiple spawning events take place [57]. Spawning and egg hatching can occur at salinities between 5 and 60 and temperatures 22–28 °C [18]. Populations of *A. iberus* from two areas in Catalonia (Spain) 300 km apart displayed significant genetic divergences [18]. Moreover, the species of *Aphanius* of the Atlantic coast of Spain, following genetic studies, has been recognized as *A. baeticus* and not as *A. iberus* [57].

3.2. Conservation Status

Along the Spanish coast, seven operational conservation units (OCUs) were distinguished with an overall low gene diversity [59]. The degree of isolation of populations is very high and it was observed that small-sized *A. iberus* can have a really restricted home range (approx. 250 m) [48]. The high sedentary behaviour has been accentuated by the fragmentation of habitats and the heavy changes suffered by the estuarine and coastal environments, such as the increase in intensive farming practices, the diversion of waterways, tourism pressure related to seasonal variability of domestic effluents and to nautical activities, the use of wetlands for rice cultivation, and aquaculture [5,55]. The tendency of *A. iberus* to prefer isolated areas is one of the main factors that threatens its survival in an environmental context in which human pressures have accentuated the fragmentation of the territory. The scarce gene flow, as in the case of *A. fasciatus*, can seriously compromise adaptive processes. In addition, habitat destruction (e.g., salt mines along the coast), water pollution and the introduction of exotic species, especially *Gambusia holbrooki*, are factors that need to be counteracted and to which conservation actions need to be directed. Laboratory and mesocosm experiments have identified the possible negative effects that *G. holbrooki* can have on native populations. The coupling of *G. holbrooki*'s aggressive behaviour toward *A. iberus*'s adults and the predation on its juveniles increases stress, reduces feeding rates and decreases reproductive activities [60]; although, in the wild, it tends to live more on the seabed and in saltier waters, and, hence, has fewer direct interactions with *G. holbrooki*.

In addition, continental populations, which live in small streams, are at risk of decline due to poor management of aquifers, which may experience periods of drought [57]. The combination of habitat degradation with the presence of invasive species puts this species at risk according to IUCN assessments and in an unfavourable/inadequate state of conservation [61]. The species is reported at 39 Natura 2000 sites but only 29 are coastal lagoons (Table 3) and they are all distributed along Mediterranean coasts of Spain.

3.3. Management

Restoration efforts aimed at increasing the connectivity of habitats in the Ebro Delta, after about 2–3 years did not show significant increases in the populations of *A. iberus*, which indeed decreased, probably due to climatic events (very low temperatures in winter and heavy rainfall with a consequent sudden change in salinity [54]). It is believed that the success of habitat restoration actions or introductions of individuals for repopulation would take about 10 years to determine whether the population will survive [54,57]. However, the reintroduction of *A. iberus*, also, may not be sufficient to reinforce the wild populations because the gene diversity is globally poor [59]. Alcaraz and García-Berthou [41] observed that the periodically flooded glasswort habitats presented a higher density of mature *A. iberus*, with the availability of both aquatic and terrestrial preys and a greater foraging efficiency. Beyond the preservation of the few areas where *A. iberus* is still present, the functional role of flooded habitats, such as the glasswort, may have significant implications to improve its conservation status [41]. *A. iberus* is demographically more fragile, and its distribution is more difficult to be managed than that of *A. fasciatus*. Despite this, more intervention projects (both national and international) have been dedicated to the management and the recovery of the Spanish endemism. Since 1996, three Spanish projects were co-funded by European LIFE programme aiming at the conservation of the species by means of habitat restoration (LIFE96 NAT/E/003118—introduced 20,000 specimen in Catalonia deltas [62]; LIFE99 NAT/E/006386—restored wetlands in Catalan Baix Ter coastal lagoon [63]; LIFE09 NAT/ES/000520—recovered Ebro Delta's Alfacada and Tancada lagoon hydrologic functioning [64]). Therefore, as for *A. fasciatus*, the habitat restoration is the first objective to be addressed, even though particular attention has to be paid to the introduction of alien species. At present, *G. holbrooki* does not appear to be a threat [60], but the introduction of further competitive species can be lethal. In fact, another potential competitor could be *Fundulus heteroclitus* [65,66], which can live also in brackish and estuarine waters, even though, at present, the effects of the interaction between the two species is not described.

4. *Knipowitschia panizzae* (Verga, 1841)—Order Gobiiformes

4.1. Biology and Distribution

K. pannizae is an euryhaline species, endemic in the Adriatic and the Ionian Sea, where it lives in shallow, well-vegetated environments such as streams, lakes, estuaries, lagoons and terminal stretches of rivers [67]. The body is grey-yellowish with darker and reticulated mottles; the females have a body and fins that are brighter than those of males and during spawning the females' bellies are yellow [68]. It is a small predator (a few centimetres in length), of no commercial interest [69], opportunist and generalist [70], that feeds, above all, on meio-fauna and juvenile macrofauna [71–74]. Reproduction takes place in brackish waters where the life cycle is completed in about 1 year, as adults disappear after spawning during their second summer of life [69,75]. Females are generally more abundant than males [75]; moreover, the female density increases during the reproductive period, but this observation can be biased by a lower susceptibility to catching of males that remain close under the nest [69].

The reproductive strategy is r-type (abbreviate iteroparity, sensu Miller [76]), suitable for unstable environments, which leads to a rapid increase of individuals, a long breeding season with multiple spawning, and females who release at each spawn about 100–150 eggs within the nests of different males. Males, that are territorial, carry out parental care and preferably use bivalve shells as shelter and a breeding site. Both the male and the female produce sounds, while the female approaches the nest [39]. The male emits sounds while remaining in the nest and the female produces them only when the ventral part is almost in contact with the male's head. The sounds cease when the female moves away, and are not produced during courtship outside the nest, nor during the spawning. Although it is not entirely clear the role of these "pre-deposition" sounds, they are supposed to influence the choice of the male by the female [77]. The species has sexual dimorphism that manifests, above all, in the nuptial coloration of the female with a conspicuous yellow/orange pigmentation in the ventral part. This coloration depends on the pigmentation of the dermis and is closely linked to the spawning: in fact, it disappears afterwards. The presence of nuptial coloration is a signal indicating the approaching spawning and its size, which varies from individual to individual, is a reliable indicator of the quantity of eggs that will be released [78]. Since the reproductive success of the male depends on the fertility of the female, it is believed that the conspicuousness of the nuptial coloration is the ornamental character on which the male bases its choice [79]. In fact, there is no evidence of competition between females and there is no correlation between body length and ventral coloration conspicuousness [78]. It is a carotenoid-based pigmentation, which can be considered indicative of the good health of the organism. Therefore, it is believed that the male chooses females with a bigger yellow ventral patch since they are in better condition and therefore are more fertile, preferring them also to larger females but with a less extensive coloration [79]. After breeding, adults die and only juveniles survive [70,75].

The high plasticity in the diet and the reproductive strategy make this species able to survive in very unstable environments. Very abundant populations of *K. panizzae* have been observed in hypertrophic basins, such as Comacchio lagoons, where dystrophic crises are frequent. This population is probably favoured by the great availability of preys before dystrophic crises, then, during the unfavourable conditions, individuals move away to areas with greater circulation and oxygenation [67]. It has also been observed, with laboratory experiments, that, as a survival strategy, in particular environmental conditions, intraspecific cannibalism can occur on juveniles or adults. Such events seem not to be determined by just food shortages but by a combination of factors: low availability of food in the reproductive period or poor quality of prey, as can happen during an anoxic crisis [74].

4.2. Conservation Status

K. panizzae is not a species considered to be at risk (for IUCN it is least concern and the conservation status is favourable) and can survive, as in the example of Comacchio, even in the case of high eutrophication, but the destruction of original habitats, mainly from a morphological point of view, can result in local population decline. In the Venice lagoon, *K. panizzae* was found to considerably prefer canals inside the salt marshes, compared to the open lagoon complex [80]. Its density, in fact, can vary significantly on both spatial and temporal scales: the annual mean in a saltmarsh creek was found to be 1374 ± 3167 ind/ha, while in a saltmarsh close to man-regulated canals it was 58 ± 80 ind/ha; during summer, in the saltmarsh creek, up to 14,000 ind/ha can be collected [69], while in autumn density can be 780 ± 1402 ind/ha [80]. The preference for more sheltered areas and possibly with macroalgal cover may depend, also, on the potential greater availability of bivalve molluscs of the genus *Cerastoderma*, whose shells are chosen as nest by males [39]. As a result, erosion and alteration of the natural habitat, with the decline or disappearance of the structures of salt marshes can compromise the population vitality [69].

K. panizzae was reported in 34 Natura 2000 sites, 25 of whose are coastal lagoons (Table 3) distributed along the Adriatic coast both on Italian and Balkan side [81,82]. However, recent phylogenetic studies, based on mitochondrial DNA analysis, have shown that *K. mrakovcici*, *K. radovici* and *K. panizzae* are not distinct species and are all *K. panizzae* [12], so the distribution range of the species should be reconsidered. The Adriatic and Ionian seas are the centre of diversity for sand gobies, with a total of 17 species belonging to five genera, including *Knipowitschia* and *Ninnigobius* [12]. The wide distribution of *K. panizzae* and the high ability to survive in extreme conditions allows us to suppose a good intraspecific genetic diversity, favoured also by the panmictic behaviour and by the presence of a free postlarval stage, favouring specimen dispersion with tide currents. However, there is a lack of specific studies comparing the genetic diversity among populations from different regions.

4.3. Management

Sand gobies are often dominant and they can be used as an indicator of habitat productivity. In Venice lagoon saltmarshes, in particular, their productivity was found to be higher than in other Mediterranean wetlands, reaching up to a total annual production >600 g ha^{-1} year^{-1} [70]. The main threat compromising this significant contribution to secondary production in coastal lagoons is represented by the destruction of saltmarshes. *K. panizzae* itself is not at risk of extinction, given its high adaptability. In Italy, in fact, accidental introductions are reported along the Tyrrhenian coast, in Trasimeno and Bolsena Lakes and in Sicily, with good survival [39,70,83]. The concern could be on the survival of local populations that have to be preserved, protecting or restoring saltmarsh systems. Moreover, the reproductive rate of *K. panizzae* could be reduced by the absence of the preferred bivalve shells for nest. Beyond the interventions to preserve coastal lagoon shallow habitat, river systems can also have a significant role in population survival, as demonstrated by the fact that the populations so far attributed to *K. mrakovcici* and *K. radovici* are *K. panizzae* and usually live in freshwater environments in the Croatian Krka and Neretve basins [12]. Since 2010, *K. panizzae* has been among the target species of three Italian projects co-funded by European LIFE programme (LIFE10 NAT/IT/000256—improved hydraulic circulation and as a consequence water quality in Italian coastal salt meadows [84]; LIFE12 NAT/IT/000331—contributed to the achievement of good ecological status by transplanting seagrasses in some areas of the Venice lagoon [85]; LIFE13 NAT/IT/000115—improved hydraulic conditions in Po delta's Sacca di Goro [86]).

On the other hand, particular attention has to be paid to the accidental introduction of *K. panizzae* in other basins, because it can become invasive and compromise the ecological equilibrium in fragile ecosystems [70], entering in competition with local species with a similar ecological niche.

5. *Ninnigobius canestrinii* (Ninni, 1883)—Order Gobiiformes

5.1. Biology and Distribution

Pomatoschistus canestrinii has recently been attributed to the genus *Ninnigobius*, which includes few species, but is phylogenetically well-distinguished by both *Pomatoschistus* and *Knipowitzschia* [12]. *N. canestrinii* is a small goby (about 6.5 cm), epibenthic, of no commercial interest and endemic of the Adriatic basin [39,67,69]. The lifespan is at most 16 months, and the spawners disappear after spawning [75]. The body is grey to bright brown, marked with very clear black points at least on the cheeks and the opercula; males are significantly larger than females, whose coloration is, on the whole, paler than that of the males, especially during the breeding period [68]. It lives preferentially in unvegetated sandy or muddy environments at the mouths of rivers or in the shallow lagoon areas [39,80], feeding on small meiobenthic invertebrates [75,87]. Its preferred habitat, characterized by low salinity, high turbidity, shallow waters and the presence of bivalve beds, is very similar to *K. panizzae*'s one [80,88]. However, *N. canestrinii* is known to prefer lower salinity and saltmarsh-dominated areas in the inner part of the Venice lagoon [89]. This should not depend on competition between the two species, because the choice of nest, reproduction times and predation tactics [39,80] are different, although there are similarities in the behaviour, including the production of sounds during the reproductive period [77,90]. *N. canestrinii* is less selective in the choice of the nest, which is created using shells, stones, bowls and any submerged object [39]. The male has a territorial behaviour for the defence of the nest and carries out parental care until the hatching of the eggs. Although the mechanism of sound production is not clear, it has been observed that, in the phase immediately before the spawning, the male "communicates" with the female, which in turn manifests a very active behaviour [90].

The species of the Gobionellidae family are often dominant in the communities of temperate and tropical transition environments [91] and, above all, the smaller ones play an important role in the trophic network as they favour the turnover of resources present in mobile sediments [69]. Well-structured populations are present in the lagoon of Venice [80,88] and in the basins of the Comacchio lagoons [67]. However, since this species does not tolerate salinity over 30, it is often also observed in oligohaline and freshwater environments [87], in some of which it was introduced accidentally, i.e., Trasimeno Lake in Italy [92] and reservoir of the Ričica River in Croatia [82,93]. Records were also made in Bosnia-Herzegovina, in the freshwaters of the coastal lake Svitava in the Neretve basin [94]. As the environmental characteristics of the Neretve Delta are potentially ideal as a habitat for *N. canestrinii*, Tutman et al. [94] hypothesize that it has always been present in that areas, but that in the past no special attention has been paid to investigate its presence.

5.2. Conservation Status

Compared to *K. pannizae*, *N. canestrinii* distribution is geographically more restricted (Table 3), abundant only in the Northern part of the Adriatic Sea, but it can reach a similar or higher density. In the Venice lagoon, in fact, its annual mean in the saltmarsh creek was found to be 1137 ± 2780 ind/ha (*K. pannizae* was 1374 ± 3167 ind/ha), while it resulted more abundant than *K. panizzae* in other saltmarsh zones, less sheltered but with lower salinity: *N. canestrinii* 239 ± 247 ind/ha compared to *K. pannizae* 58 ± 86 ind/ha [69]. Despite the fact that *N. canestrinii* shows a preference for saltmarsh creeks, it can be found also in more open areas, with biomass one order of magnitude higher than that of *K. pannizae* (*N. canestrinii* 115 ± 159 g/ha compared to *K. pannizae* 11 ± 16 g/ha) [69]. Considering the seasonal distribution, *N. canestrinii* was found to have similar density in both saltmarsh creeks (5.60 ± 15.61 ind/100 m^2 in summer and 3.26 ± 9.81 ind/100 m^2 in winter) and mudflats (3.32 ± 5.94 ind/100 m^2 in summer and 4.87 ± 11.58 ind/100 m^2 in winter) [80].

As an oligohaline species, *N. canestrinii* can be common also in freshwater habitats, where other factors could threat its local population survival. For Trasimeno Lake, for example, it has been hypothesized that the Eurasian copepod *Lernaea cyprinacea* could infest the host, generally belonging to the Gobionellidae family, causing its death [95]. At present, there is no evidence of parasitism on *N. canestrinii* and it is established that in brackish waters this risk is remote because *L. cyprinacea* does not tolerate even low salinity [95].

According to IUCN assessments, the threat level for *N. canestrinii* is least concern and the status is favourable [96]. Its presence is reported in 26 Natura 2000 sites, 20 of which are priority habitats 1150* (Table 3). The restricted range can determine limited gene flow, reducing the intraspecific genetic biodiversity, but no data are available in the literature describing the difference among populations. Although not considered to be an endangered species, its distribution and density should be carefully monitored, because it represents a good indicator of the conservation status of coastal lagoon habitats in its structural and morphological complexity (salts, small canals, shallow bottoms, etc.) and is important for nekton community biodiversity.

5.3. Management

The preservation of the habitats suitable for *N. canestrinii* survival is the main objective to be achieved by creating protection plans that regulate the human activities [87]. However, from 1992 to 2018, only four Italian projects co-funded by European LIFE programme, dedicated to restoration of coastal lagoon ecosystems, indicated *N. canestrinii* among target species that would benefit from the interventions ([84–86] and LIFE16 NAT/IT/000663—restoration of salinity gradient in an area of the Venice lagoon from salinity <5 towards <15 up to <25 [97]). One of them pays particular attention to the salinity values, introducing freshwaters and restoring the lagoon original gradient that would favour *N. canestrinii* increase [89,97]. Despite that it shares the same sub-habitat as *K. panizzae*, no competition seems to exist, whereas in the Po Delta, difficulties to access the structures to build the nest have been observed for the coexistence between *Pomatoschistus marmoratus* and *Salaria pavo* [18]. Therefore, particular attention has to be paid to keep not only the morphological structures of saltmarshes with their tidal creeks and pools, but also the bottom undisturbed to favour the presence of submerged objects, such as shells. On the fishing management plans, the activities impacting the bottoms have to be carefully assessed to avoid the sediment resuspension and the nest destruction. As an example, in the late 1990s, in the Venice lagoon, the clam harvesting activity with hydraulic and mechanical dredging systems that dug 10–30 cm deep furrows dramatically affected the biological equilibrium and the benthic communities [98,99].

Moreover, the introduction of allochthons Gobionellidae could be a serious threat due to the competition for resources.

6. *Valencia* spp.—Order Cyprinodontiformes

In Annex II of the Habitat Directive, other two Cyprinodontiformes species are listed belonging to the family Valenciidae: the Spanish endemism *Valencia hispanica* and the Greek *Valencia letourneuxi*. Both *V. hispanica* and *V. letourneuxi* are priority species (indicated with an asterisk in Annex II, under Article 1 letter h), whose preservation required special responsibility. They are freshwater species, but, living in coastal lakes or pounds and sometimes tolerating low salinity, can be a significant indicator of coastal lagoon status.

6.1. Valencia hispanica (Valenciennes, 1846)

6.1.1. Biology and Distribution

The distribution of *V. hispanica* is limited to the coast of eastern Spain, where it lives in small freshwater systems (pools and marshes) and occasionally in brackish areas, such as lagoons or estuaries. Adult males, on average smaller (<6.7 cm) than females (<7.1 cm), have a blue-grey coloration with narrow dark cross bars and yellow/orange edge on pectoral and caudal fins. They are slow-growing individuals who can live beyond 4 years for females and over 3 for males; sexual maturity is reached late, but multiple spawning can be observed in the period between April and June [100]. Although behavioural differences among individuals of the same population can occur, the general tendency of the species is to feed on 3 types of prey (Gammaridae, Sphaeromatidae and Chironomidae larvae) and preferably on living organisms, which live in the muddy/sandy vegetated seabed or on the surface of the water. The high specialization in the diet, even if prey of other groups can be available, seems to be one of the main constraints for the survival of the species [56]. In addition, other threats are added, such as habitat loss or fragmentation, the introduction of exotic species such as *Gambusia holbrooki*, and worsening water quality. *V. hispanica* and *G. holbrooki* can interact frequently because they tend to move vertically in the water column. Numerous *G. holbrooki* aggression phenomena have been observed, with negative consequences for *V. hispanica* courtship activities and feeding rates [60].

6.1.2. Conservation Status

According to Oliva-Paterna et al. [87], the number of natural populations of *V. hispanica* is 10, five of which are considered to be in a good state of conservation. Caiola et al. [55] reports four populations in the estuaries, but according to Natura 2000 reports, the species is present in 16 sites, 13 of which are coastal lagoons (Table 3). Despite the discrepancy in the information, the number of natural populations is still dramatically low, and, in fact, the threat level is critically endangered. The general conservation status is unfavourable/bad [101].

6.1.3. Management

In addition to international regulations, it is also protected at the national level, as it is one of the most endangered species of fish fauna at European level. Moreover, since 1993, in the framework of European action programmes, more than 100,000 individuals have been reintroduced in the Valencian region (2500 individuals in the framework of the LIFE04 NAT/ES/000048—aiming at recovering two permanently flooded freshwater pools [102]) and 35,000 specimens were re-introduced in the Ebro delta (Catalonian region; LIFE96 NAT/E/003118 [62]), but at present, the effects of such activities do not appear relevant to preserve the local populations [100].

6.2. *Valencia letourneuxi (Sauvage, 1880)*

6.2.1. Biology and Distribution

V. letourneuxi is distributed along the west Greek coast. It is a short-lived species that lives in temperate freshwater basins along the coast and can tolerate a salinity up to 4 [103]. Its original range was a basin in Corfu island (Greece), although it has not been recorded there in recent decades; however, it is widespread from the southern coast of Albania to the Peloponnese. Although it has also disappeared from other aquatic systems due to several causes (water pollution for dairy activities, drying of springs, etc.), the original wide distribution along the coast has prevented the complete extinction of the species [104]. The number of populations is in constant decline and the local density very low (i.e., in the Greek Kalamas river system, *V. letourneuxi* was found only in one site with a relative density <1%, [103]). Its typical habitats are spring streams with deep water and low current speed, with bottoms covered with an adequate bed of vegetation that provides food, protection and substrate for spawning [103].

6.2.2. Conservation Status

Beyond pollution and water extraction, the strong competition with *Gambusia holbrooki* is a significant threat [103,104]. Under IUCN assessments, the species is critically endangered and the conservation status is unfavourable/inadequate [105]. In coastal lagoon environments, it has been reported in two Greek sites (Table 3): the Delta Acheloou, Limnothalassa Mesolongiou—Aitolikou, Ekvoles Evinou, Nisoi Echinades, Nisos Petalas and the Limnes Voulkaria Kai Saltini.

6.2.3. Management

From 1993 to 2018, no European LIFE project was funded having as target species *V. letourneuxi*. Suggested interventions to improve the conservation status are nutrient load control, regulation of water extraction activities and, possibly, the reintroduction of adults for reproduction.

7. Discussion

The natural complexity of coastal lagoons is the result of a wide range of situations, mainly driven by the degree of confinement, the ionic composition and the salinity variations depending on evaporation vs. inputs from rivers and seas. This determines the absence of a clear or unidirectional salinity gradient and leads to a three-dimensional heterogeneity, influencing the species composition and favouring the dominance of a few ecological guilds, such as, for fish fauna, the marine migrants [106]. However, several studies demonstrated the significant role of lagoon geomorphological features to explain fish assemblages in terms of both productivity, taxonomic and functional diversity ([107–109] and references therein). According to the Ramsar convention, wetlands (including coastal lagoons) are "areas of marsh, fen, peatland or water, whether natural or artificial, permanent or temporary, with water that is static or flowing, fresh, brackish or salt, including areas of marine water the depth of which at low tide does not exceed six metres" [110]. Tidal wetlands have a significant role in selectively trapping fine sediments, influencing the water residence time and sequestering nutrients and pollutants [111]. In the present paper, the main focus is on the importance of fish fauna in two structures that characterize coastal lagoons of middle and high latitudes: salt marshes and mudflats. Marshes can be found in all aquatic environments where the erosive forces, produced by waves and currents, are weak and a sheltered and sedimentary environment prevails. Along coasts, the amplitude of the tidal regime determines the salt marsh structure, extension and the habitats within a salt marsh, that can be characterized by creeks and pools, which provide diverse habitats for aquatic vegetation and animals [112]. Nearby salt marshes, mud flats are the structures periodically inundated by the tide and their characteristics vary depending on the tide excursion. According to Wolanski and Elliott [111], "Mudflats located below neap high tide are wetted daily by the tides and the top layer of the mud is commonly soft and unconsolidated. Mudflats between the neap and spring tide high water marks are wetted on alternate weeks, whereas those in supratidal areas are tidally inundated only at the highest astronomical tides."

Multifactorial approaches are recommended to describe the ecological status of coastal lagoons [106], due to the high heterogeneity described above and also considering that the larger the lagoon, the greater the habitat diversity [108]. Traditional metrics or the indices developed in the framework of National and International laws mainly focus on the water quality or on rapid environmental changes. Often, the information on the rate and impact of erosive processes that affect lagoon morphology on a long-term scale is missed. Therefore, the use of indicator species can add additional support to decision-makers, targeting specific management actions to habitat restoration [15], focusing on morphological alterations, salinity changes, and the presence of invasive alien species. Ecological indicator species are sensitive to particular environment attributes [113], so their presence and abundance can provide information on it [15]. Despite that marine migrants, mainly juveniles, received much more attention in coastal lagoon research, due to their fishery importance, the few fish species of lagoon residents are more abundant [91,114–116] and can be more suitable indicators due to their low dispersal abilities and strong relationships with lagoon morphological structures during their entire life cycle. Some evidence on the relationship between morphological degradation and the status of nekton community highlighted that the effects of pressures acting on the lagoon morphology are stronger for resident species than for marine migrants [117]. As an example, the populations of *Atherina boyeri* (Risso, 1810), a small estuarine resident of commercial interest, resulted to be significantly affected by pressures deriving from the morphological (hydrological) alteration in Marano-Grado lagoon system (Northern Adriatic Sea) [118].

7.1. Cyprinodontiformes (Aphaniidae and Valenciidae)

Among the lagoon residents included in Annex II of the Habitat Directive here presented, *A. fasciatus* is recognized to be a typical component of the fish community of saltmarsh systems [114,119], which generally represent important transition areas in buffering flood phenomena [31] or in supporting the morphological complexity that counteracts the increasing inputs of seawaters. The in-depth study of *A. fasciatus* bio-ecology could provide the necessary tools to monitor and manage the erosion processes that affect its habitat.

The present overview highlights the occurrence of *A. fasciatus* in brackish to hyperhaline systems characterized by shelter habitats: such as small-sized intertidal creeks (200–250 m long, 2–4 m width), with a maximum depth of 0.7 m [49]. *A. fasciatus* seems to be able to tolerate eutrophic waters, but it needs small tidal creeks or sheltered canals to maintain successful reproductive and survival rates, indicating a significant relationship with the structure of saltmarshes. Moreover, sites with around 15% of organic matter were found to better support the local population than the zones with a percentage around 7% [28]. Locally, healthy populations were found to reach an annual mean higher than 100 ind/100 m^2 [28], while in the areas where a decreasing trend was observed, the catching accounts for few specimens [34,44]. The presence of aquatic vegetation coverage is another fundamental factor.

Beyond salt marsh degradation, and due to an increasing erosive process, habitat fragmentation also tends to accentuate population isolation and weaken their genetic heritage and, in the long term, adaptive capacity. The effects of habitat fragmentation and the consequent increase of isolation would be evident in the future, when the intraspecific genetic diversity will be drastically reduced.

Furthermore, *A. fasciatus* was demonstrated to be strongly related to artificial habitats, such as salt works and small man-made creeks [28,49], that are important components of the lagoon landscape, as the historical product of the interaction between traditional human activities and the lagoon coastal habitats. This species could, therefore, be considered indicator species for these peculiar components of the lagoon landscape, contributing to support its conservation and management.

The use of *A. fasciatus* as a sentinel of anthropogenic impacts in Mediterranean coastal lagoons is well-documented [51,120,121]. In fact, Kessabi et al. [51] believe that it may be a good indicator for detecting metal and organic pollutants by using mRNA biomarkers. Heavy metal pollution effects were, in fact, documented in both reproductive function [50] and spinal deformities [51]. Mossesso et al. [120] proposed *A. fasciatus* like a promising "sentinel organism" to detect the genotoxic impact of complex mixtures in coastal lagoon ecosystems.

As endemic species, *A. iberus*, that is, the most endangered species among lagoon residents, and *V. hispanica* and *V. letourneuxi*, which are priority species, can provide information on local habitats of high ecological value. In particular, *A. iberus* can be used as an indicator of the functional complexity of the lagoon ecosystem, with a habitat affinity similar to that of *A. fasciatus*, while the *Valencia* spp. are sensitive to hydrological variations and salinity increase. *A. iberus* was found to be more abundant far from the sea and to prefer vegetated areas [54]. *V. letourneuxi* was found to tolerate a salinity up to 4 in the wild [103]. Therefore, the increase of seawater intrusion due to infrastructural intervention or aquifer management could have further compromised its presence.

7.2. Gobiiformes, Gobionellidae

K. panizzae and *N. canestrinii* distribution is more limited than that of *A. fasciatus*; however, where present, they show important density, demonstrating an exclusive relationship with lagoon morphology. *K. panizzae* partially provides similar information as *A. fasciatus* does for saltmarsh systems, but it is also sensitive to alteration of mudflats (sensu Wolanski and Elliott [111]), where males usually prepare the nests. Therefore, the creation of infrastructures (such as navigable canals and water pipe digging) or their maintenance and, in general, the use of dredging systems represents a serious threat for the local population, ending in the destruction of breeding sites. Males of both species need hard structures and cavities to build their nest, such as shells or other artificial structures. This means that the substrate of their breeding habitats should be structurally complex.

Depending on the habitat structure, the literature demonstrates that the population density of *K. panizzae* can vary by even up to two orders of magnitude. Approximately, it can be supposed that a population with an annual density below 100 ind/100 m^2 should be monitored with particular care.

Despite being able to tolerate high eutrophicated systems [67] and suffering from the alteration and fragmentation of native habitats, due to the loss of typically lagoon morphologies as in the case of *A. fasciatus* and *K. panizzae*, *N. canestrinii* is particularly sensitive to salinity increase, its optimal salinity values ranging from 2 to 20 and never exceeding 30 [87]. Human-regulation of water exchanges is frequent in Mediterranean costal lagoons to meet the need for numerous anthropogenic activities, such as navigation and fishing and aquaculture, and to maintain the functioning of the ecosystem [122]. Modifying communication with the sea and dredging canals may lead to significant changes in retention time [123] and salinity values [122]. Moreover, the ongoing climate changes and the future sea-level rise can accentuate these processes, increase eutrophication and accelerate the erosion rate of wetlands [5,124]. Although the species seems to not be affected by high eutrophication levels, the salinity rise and the saltmarsh degradation remain a serious threat. In light of these considerations, the decrease of *N. canestrinii* populations could be a marker of environmental processes leading to higher salinity and a higher sea level, indicating a direction of change in coastal lagoons towards marine environments.

7.3. Conservation Management Issues

Consulting the European Union database and the literature:

(1) At the European level, few management actions have been undertaken since the issue of the HD. Despite the attention of European Union towards habitat conservation status, in the period from 1992 to 2018, 1691 projects were co-financed through the LIFE Financial Programme (Sub-Programme Environment—Nature and biodiversity). These projects were specifically dedicated to promoting conservation efforts of Natura 2000 Network sites and/or of Community interest species listed in Annex II of the Habitat Directive. Among the 1691 projects, 682 were carried out by the Member States which have coasts bordering the Mediterranean Sea and 60 to the priority habitat 1150*. Excluding the Atlantic lagoons of France and Spain, there were 52 projects left for Mediterranean lagoons and only 10 explicitly stated the objective of improving environmental conditions in relation to fish species of Community interest and especially those of the *Aphanius* species. From this information, it is not possible to understand if few proposals were submitted by Member States or if they were not targeting the LIFE Financial Programme aims, and, thus, were not financed, but the number sounds very low. Therefore, more attention towards the here-discussed species has to be paid in light of their role as potential indicators of ecosystem conservation status.

(2) At the national level, some interventions were dedicated to the Spanish endemisms *A. iberus* and *V. hispanica*, but with poor success.

(3)   In the European Union, the regulation 1143/2014 [125] entered into force on 1 January 2015, to set common rules that allow the Member States to coordinate actions with the aims of preventing, early detecting, rapidly eradicating and managing invasive species. The core of Regulation (EU) 1143/2014 is the list of Invasive Alien Species (IAS) of Union concern, which includes four fish species (three freshwaters and one marine species), subject to restrictions and measures for the early detection and rapid eradication. None of these species are of concern for coastal lagoon habitats, but, as an example, *Gambusia holbrookii*, that was demonstrated to be aggressive and competitive towards *Aphanius* spp. and *Valencia* spp. in oligohaline waters, is not in the list. *G. holbrooki* already drove the *A. fasciatus* population to extinction in some Italian inland waters [33] and *V. hispanica* is seriously at risk. Being among the 100 most invasive species at the international level [126] and demonstrating its aggressiveness, *Gambusia* should be included in the European list of IAS. At the moment, efficient solutions to the eradications of *G. holbrooki* are poorly documented, but efforts should be addressed to avoid further introduction.

(4)   In the European basin of the Mediterranean Sea, some areas of particular importance for the conservation of species are: western Greece, the Acheloos Delta, along the Balkan coast, the Neretve Delta, along the northwest coast of the Adriatic Sea, the system of coastal lagoons from Grado-Marano to the Po Delta, along the coast of Mediterranean Spain, the Ebro Delta [127].

(5)   The species discussed here are strongly related to local environmental features. Therefore, absolute density values indicating the limit between healthy condition and local extinction are difficult to be set at the Mediterranean basin level. The most important issue to be addressed is the local monitoring activity. The use of species included in the national and international regulations, as in the case here presented, could make their use as indicators more efficient, because:

    a.   data collection is compulsory, and in the case of Natura 2000, started in the mid-1990s and for all Member States, allowing the assessment of long-term trends on a large spatial scale;

    b.   data are comparable;

    c.   the access to funding can be facilitated.

8. Conclusions

The aim of the present paper was to summarise the information on the fish fauna of Community interest (sensu European Habitat Directive) in coastal lagoons and to highlight how they can provide indications on morphological conservation status of salt marshes and mudflats. The following was achieved:

(1)   The complex of species *Aphanius* + *Valencia* could be used as indicator of habitat alterations in coastal lagoons, focusing, in particular, on salt marsh complexity, habitat connectivity within the lagoon or among adjacent lagoon systems and the presence of a system of artificial habitats, such as salt-works and man-made creeks. The presence and abundance of these species indicate, in fact, a mosaic of isolated and closed habitats, with respect to the open lagoons. The complex system of small creeks and pools has a significant role, not only for lagoon residents, but also for many other fish fauna species, also of commercial interest, such as marine migrants, of which juveniles exploit the trophic resource in saltmarsh areas. Together with *A. fasciatus*, *K. panizzae*, *N. canestrinii*, it is possible to find *A. boyeri*, *Sparus aurata* (L.) and some species of the genus *Chelon* and of the family Sygnathidae [49]. *V. hispanica* was found in the same sampling site as *Anguilla anguilla* (L.), *Mugil cephalus* (L.), *Pomatoschistus microps* (Krøyer) and *A. boyeri* [55]. The lagoon residents are more sensible to local habitat alterations than marine migrants [101], therefore, the species belonging to *Aphanius* + *Valencia* genera can be considered umbrella species, through which protection and broader conservation objectives could be achieved also for other species [113], even of economic interest. Their decrease, as a consequence of habitat degradation due to sea level rise, excavations of canals and infrastructure development, can be recorded earlier

than that of other guilds. Mainly *A. fasciatus* responds to the definition of umbrella species, being highly detectable and correlating with species diversity [15].

(2) A decrease in the presence and abundance of the two small gobies, *K. panizzae* and *N. canestrinii*, could partially indicate the same alterations in terms of habitat complexity and connectivity as *Aphanius* + *Valencia*, but they also provide information on the structural changes of the lagoon substrate that they use for nesting. Moreover, *N. canestrinii* selects a salinity range between 2 and 20 and could be useful as an indicator of oligohaline habitats, that could be threatened by significant rise of sea influence on coastal lagoon hydrodynamism.

(3) Beyond the morphological alterations, *Aphanius* + *Valencia* species could be used, also, as ecological indicators in relation to the distribution, occurrence and abundance of the invasive *Gambusia* complex or of other alien species compromising the coastal lagoon biological community.

(4) Mainly *A. fasciatus* was demonstrated to be particularly sensible to heavy metal contamination.

Author Contributions: Conceptualization, C.F. and S.M.; writing—original draft preparation, C.F.; writing—review and editing, F.C., P.F., S.M. All authors have read and agree to the published version of the manuscript.

Funding: This research received no external funding.

Acknowledgments: The authors are thankful to the editors and the anonymous reviewers for their comments and suggestions.

References

1. European Commission. *Interpretation Manual of European Union Habitats*; EUR 28. E. C.; DG ENVIRONMENT: Bruxelles, Belgium, 2013; p. 144.

2. Cataudella, S.; Crosetti, D.; Massa, F. *Mediterranean Coastal Lagoons: Sustainable Management and Interactions among Aquaculture, Capture Fisheries and the Environment; Studies and Reviews. General Fisheries Commission for the Mediterranean. No 95*; FAO: Rome, Italy, 2015; p. 278.

3. Natura 2000 Network Viewer. Available online: https://natura2000.eea.europa.eu/ (accessed on 4 December 2019).

4. European Environment Agency. European Topic Centre on Biological Diversity—Report under the Article 17 of the Habitats Directive Period 2007–2012. Available online: https://nature-art17.eionet.europa.eu/article17/reports2012/ (accessed on 14 May 2020).

5. Newton, A.; Icely, J.; Cristina, S.; Brito, A.; Cardoso, A.C.; Colijn, F.; Riva, S.D.; Gertz, F.; Hansen, J.W.; Holmer, M.; et al. An overview of ecological status, vulnerability and future perspectives of European large shallow, semi-enclosed coastal systems, lagoons and transitional waters. *Estuar. Coast. Shelf Sci.* **2014**, *140*, 95–122. [CrossRef]

6. Kara, H.M.; Quignard, J.-P. *Fishes in Lagoons and Estuaries in the Mediterranean 1: Diversity, Bioecology and Exploitation*; John Wiley & Sons: Hoboken, NJ, USA, 2018; p. 265.

7. Elliott, M.; Whitfield, A.; Potter, I.C.; Blaber, S.J.M.; Cyrus, D.P.; Nordlie, F.G.; Harrison, T.D. The guild approach to categorizing estuarine fish assemblages: A global review. *Fish Fish.* **2007**, *8*, 241–268. [CrossRef]

8. Franco, A.; Elliott, M.; Franzoi, P.; Torricelli, P. Life strategies of fishes in European estuaries: The functional guild approach. *Mar. Ecol. Prog. Ser.* **2008**, *354*, 219–228. [CrossRef]

9. Potter, I.C.; Tweedley, J.R.; Elliott, M.; Whitfield, A. The ways in which fish use estuaries: A refinement and expansion of the guild approach. *Fish Fish.* **2013**, *16*, 230–239. [CrossRef]

10. World Register of Marine Species (WORMS). Available online: www.marinespecies.org (accessed on 4 December 2019).

11. Bianco, P.G.; Ketmaier, V. A revision of the *Rutilus* complex from Mediterranean Europe with description of a new genus, *Sarmarutilus*, and a new species, *Rutilus stoumboudae* (Teleostei: Cyprinidae). *Zootaxa* **2014**, *3841*, 379–402. [CrossRef]

12. Thacker, C.E.; Gkenas, C.; Triantafyllidis, A.; Malavasi, S.; Leonardos, I. Phylogeny, systematics and biogeography of the European sand gobies (Gobiiformes: Gobionellidae). *Zoöl. J. Linn. Soc.* **2018**, *185*, 212–225. [CrossRef]

13. International Union for Conservation of Nature. 2006 IUCN Red List of Threatened Species. 2006. Available online: http://www.iucnredlist.org/ (accessed on 1 March 2008).

14. European Environmental Agency. European Nature Information System. Available online: https://eunis.eea.europa.eu/ (accessed on 4 December 2019).

15. Gilby, B.L.; Olds, A.D.; Connolly, R.M.; Yabsley, N.A.; Maxwell, P.S.; Tibbetts, I.; Schoeman, D.S.; Schlacher, T.A. Umbrellas can work under water: Using threatened species as indicator and management surrogates can improve coastal conservation. *Estuar. Coast. Shelf Sci.* **2017**, *199*, 132–140. [CrossRef]

16. Bortone, S.A.; Dunson, W.A.; Greenawalt, J.M. Fishes as Estuarine Indicators. In *Estuar. Indicators*; CRC Press: Boca Raton, FL, USA, 2005; pp. 381–389.

17. Nelson, J.S.; Grande, T.C.; Wilson, M.V.H. *Fishes of the World*; Wiley: Hoboken, NJ, USA, 2016; p. 707.

18. Kara, H.M.; Quignard, J.-P. *Fishes in Lagoons and Estuaries in the Mediterranean 2*; Wiley: Hoboken, NJ, USA, 2019; p. 39.

19. Leonardos, I.; Sinis, A. Reproductive strategy of *Aphanius fasciatus* Nardo, 1827 (Pisces: Cyprinodontidae) in the Mesolongi and Etolikon lagoons (W. Greece). *Fish. Res.* **1998**, *35*, 171–181. [CrossRef]

20. Ferrito, V.; Pappalardo, A.M.; Canapa, A.; Barucca, M.; Doadrio, I.; Olmo, E.; Tigano, C. Mitochondrial phylogeography of the killifish *Aphanius fasciatus* (Teleostei, Cyprinodontidae) reveals highly divergent Mediterranean populations. *Mar. Boil.* **2013**, *160*, 3193–3208. [CrossRef]

21. Leonardos, I.; Sinis, A. Population age and sex structure of *Apahnius fasciatus* Nardo, 1827 (Pisces: Cyprinodontidae) in the Mesologi and Etolikon lagoons (W. Greece). *Fish. Res.* **1999**, *40*, 227–235. [CrossRef]

22. Cavraro, F.; Georgalas, V.; Malavasi, S.; Franzoi, P.; Torricelli, P. Population Structure and Reproductive Ecology of the Mediterranean Killifish Aphanius fasciatus (Nardo, 1827) in the Venice Lagoon. In Proceedings of the 4th European Conference on Coastal Lagoon Research, Montpellier, France, 14–18 December 2009; p. 279.

23. Leonardos, I. The feeding ecology of *Apahnius fasciatus* (Valenciennes, 1821) in the lagoonal system of Mesolongi (Western Greece). *Sci. Mar.* **2008**, *72*, 393–401. [CrossRef]

24. Maltagliati, F. Genetic divergence in natural populations of the Mediterranean brackish-water killifish Aphanius fasciatus. *Mar. Ecol. Prog. Ser.* **1999**, *179*, 155–162. [CrossRef]

25. Angeletti, D.; Cimmaruta, R.; Nascetti, G. Genetic diversity of the killifish *Aphanius fasciatus* paralleling the environmental changes of Tarquinia salterns habitat. *Genetica* **2010**, *138*, 1011–1021. [CrossRef]

26. Cavraro, F.; Varin, C.; Malavasi, S. Lunar-induced reproductive patterns in transitional habitats: Insights from a Mediterranean killifish inhabiting northern Adriatic saltmarshes. *Estuar. Coast. Shelf Sci.* **2014**, *139*, 60–66. [CrossRef]

27. Cavraro, F.; Torricelli, P.; Malavasi, S. Quantitative Ethogram of Male Reproductive Behavior in the South European Toothcarp *Aphanius fasciatus*. *Boil. Bull.* **2013**, *225*, 71–78. [CrossRef]

28. Cavraro, F.; Daouti, I.; Leonardos, I.; Torricelli, P.; Malavasi, S. Linking habitat structure to life history strategy: Insights from a Mediterranean killifish. *J. Sea Res.* **2014**, *85*, 205–213. [CrossRef]

29. Ćaleta, M.; Buj, I.; Miočić-Stošić, J.; Marčić, Z.; Mustafić, P.; Zanella, D.; Mrakovčić, M.; Mihinjač, T. Population genetic structure and demographic history of *Aphanius fasciatus* (Cyprinodontidae: Cyprinodontiformes) from hypersaline habitats in the eastern Adriatic. *Sci. Mar.* **2015**, *79*, 399–408. [CrossRef]

30. Ferrito, V.; Mannino, M.C.; Pappalardo, A.M.; Tigano, C. Morphological variation among populations of *Aphanius fasciatus* Nardo, 1827 (Teleostei, Cyprinodontidae) from the Mediterranean. *J. Fish Boil.* **2007**, *70*, 1–20. [CrossRef]

31. Maltagliati, F. Genetic monitoring of brackish-water populations: The Mediterranean toothcarp *Aphanius fasciatus* (Cyprinodontidae) as a model. *Mar. Ecol. Prog. Ser.* **2002**, *235*, 257–262. [CrossRef]

32. Cimmaruta, R.; Scialanca, F.; Luccioli, F.; Nascetti, G. Genetic diversity and environmental stress in Italian populations of the cyprinodont fish *Aphanius fasciatus*. *Oceanol. Acta* **2003**, *26*, 101–110. [CrossRef]

33. Valdesalici, S.; Langeneck, J.; Barbieri, M.; Castelli, A.; Maltagliati, F. Distribution of natural populations of the killifish *Aphanius fasciatus* (Valenciennes, 1821) (Teleostei: Cyprinodontidae) in Italy: Past and current status, and future trends. *Ital. J. Zoöl.* **2015**, *82*, 1–12. [CrossRef]

34. Deidun, A.; Diacono, I.; Tigano, C.; Schembri, P. Present distribution of the threatened killifish *Aphanius fasciatus* (Actinopterygii, Cyprinodonitdae) in the Maltese Islands. *Cent. Mediterr. Naturalist.* **2002**, *3*, 177–180.

35. Zammit Mangion, M.; Deidun, A.; Vassallo-Agius, R.; Magri, M. Management of threatened *Aphanius fasciatus* at Il-Maghluq, Malta. In Proceedings of the Tenth International Conference on the Mediterranean Coastal Environment, MEDCOAST 11, Rhodes, Greece, 25–29 October 2011.

36. Valdesalici, S.; Brahimi, A.; Freyhof, J. First record of *Aphanius almiriensis* from Italy and notes on the distribution of *Aphanius fasciatus* (Teleostei: Aphaniidae). *J. Appl. Ichthyol.* **2019**, *35*, 541–550. [CrossRef]

37. Mediterranean Toothcarp—*Aphanius fasciatus* (Valenciennes, 1821). Available online: https://eunis.eea.europa.eu/species/430 (accessed on 14 May 2020).

38. Bertoli, M.; Giulianini, P.G.; Chiti, J.; De Luca, M.; Pastorino, P.; Prearo, M.; Pizzul, E. Distribution and Biology of *Aphanius fasciatus* (Actinopterygii, Cyprinodontidae) in The Isonzo River Mouth (Friuli Venezia Giulia, Northeast Italy). *Turk. J. Fish. Aquat. Sci.* **2019**, *20*, 279–290. [CrossRef]

39. Gandolfi, G.; Zerunian, S.; Torricelli, P.; Marconato, A. *I Pesci Delle Acque Interne Italiane*; Istituto Poligrafico e Zecca dello Stato: Roma, Italy, 1991; p. 617.

40. Basset, A.; Barbone, E.; Elliott, M.; Li, B.-L.; Jorgensen, S.E.; Lucena-Moya, P.; Pardo, I.; Mouillot, D. A unifying approach to understanding transitional waters: Fundamental properties emerging from ecotone ecosystems. *Estuar. Coast. Shelf Sci.* **2013**, *132*, 5–16. [CrossRef]

41. Alcaraz, C.; García-Berthou, E. Life history variation of invasive mosquitofish (*Gambusia holbrooki*) along a salinity gradient. *Boil. Conserv.* **2007**, *139*, 83–92. [CrossRef]

42. Alcaraz, C.; Pou-Rovira, Q.; García-Berthou, E. Use of a flooded salt marsh habitat by an endangered cyprinodontid fish (*Aphanius iberus*). *Hydrobiologia* **2007**, *600*, 177–185. [CrossRef]

43. Koutrakis, E.T.; Kamidis, N.I.; Leonardos, I.D. Age, growth and mortality of a semi-isolated lagoon population of sand smelt, *Atherina boyeri* (Risso, 1810) (Pisces: Atherinidae) in an estuarine system of northern Greece. *J. Appl. Ichthyol.* **2004**, *20*, 382–388. [CrossRef]

44. Zogaris, S. Conservation Study of the Mediterranean Killifish *Aphanius fasciatus* in Akrotiri Marsh, (Akrotiri SBA, Cyprus)—Final Report. Darwin Project DPLUS034 "Akrotiri Marsh Restoration: A Flagship Wetland in the Cyprus SBAs BirdLife Cyprus"; Nicosia, Cyprus. Unpublished final report. 2017; 64.

45. Buj, I.; Zanella, D.; Marčić, Z.; Ćaleta, M.; Mustafić, P.; Mihinjač, T.; Mrakovčić, M. New Records of *Aphanius fasciatus* (Valenciennes, 1821) along the Eastern Coast of the Adriatic Sea in Croatia. *Croat. J. Fish.* **2015**, *73*, 124–127. [CrossRef]

46. Franco, A.; Pérez-Ruzafa, Á.; Drouineau, H.; Franzoi, P.; Koutrakis, E.; Lepage, M.; Verdiell-Cubedo, D.; Bouchoucha, M.; López-Capel, A.; Riccato, F.; et al. Assessment of fish assemblages in coastal lagoon habitats: Effect of sampling method. *Estuar. Coast. Shelf Sci.* **2012**, *112*, 115–125. [CrossRef]

47. Keith, P.; Persat, H.; Feunteun, É.; Allardi, J. *Les Poissons D'eau Douce de France*; Muséum National d'Histoire Naturelle: Paris, France, 2011; p. 552.

48. Alcaraz, C.; Bisazza, A.; García-Berthou, E. Salinity mediates the competitive interactions between invasive mosquitofish and an endangered fish. *Oecologia* **2007**, *155*, 205–213. [CrossRef]

49. Cavraro, F.; Zucchetta, M.; Malavasi, S.; Franzoi, P. Small creeks in a big lagoon: The importance of marginal habitats for fish populations. *Ecol. Eng.* **2017**, *99*, 228–237. [CrossRef]

50. Annabi, A.; Saïd, K.; Messaoudi, I. Heavy metal levels in gonad and liver tissues—Effects on the reproductive parameters of natural populations of *Aphanius facsiatus*. *Environ. Sci. Pollut. Res.* **2013**, *20*, 7309–7319. [CrossRef] [PubMed]

51. Kessabi, K.; Navarro, A.; Casado, M.; Said, K.; Messaoudi, I.; Piña, B. Evaluation of environmental impact on natural populations of the Mediterranean killifish *Aphanius fasciatus* by quantitative RNA biomarkers. *Mar. Environ. Res.* **2010**, *70*, 327–333. [CrossRef] [PubMed]

52. Nicol, S.; Sabbadin, R.; Peyrard, N.; Chadès, I. Finding the best management policy to eradicate invasive species from spatial ecological networks with simultaneous actions. *J. Appl. Ecol.* **2017**, *54*, 1989–1999. [CrossRef]

53. Henkanaththegedara, S.M.; Stockwell, C.A. Intraguild predation may facilitate coexistence of native and non-native fish. *J. Appl. Ecol.* **2014**, *51*, 1057–1065. [CrossRef]

54. Prado, P.; Alcaraz, C.; Jornet, L.; Caiola, N.; Ibanez, C. Effects of enhanced hydrological connectivity on Mediterranean salt marsh fish assemblages with emphasis on the endangered Spanish toothcarp (*Aphanius iberus*). *PeerJ* **2017**, *5*, e3009. [CrossRef]

55. Caiola, N.; Vargas, M.J.; De Sostoa, A. Feeding ecology of the endangered Valencia toothcarp, *Valencia hispanica* (Actinopterygii: Valenciidae). *Hydrobiologia* **2001**, *448*, 97–105. [CrossRef]

56. Alcaraz, C.; García-Berthou, E. Food of an endangered cyprinodont (*Aphanius iberus*): Ontogenetic diet shift and prey electivity. *Environ. Boil. Fishes* **2006**, *78*, 193–207. [CrossRef]

57. Oliva-Paterna, F.J.; Torralva, M.; Fernández-Delgado, C. Threatened Fishes of the World: *Aphanius iberus* (Cuvier & Valenciennes, 1846) (Cyprinodontidae). *Environ. Boil. Fishes* **2006**, *75*, 307–309. [CrossRef]

58. Oliva-Paterna, F.J.; Ruiz-Navarro, A.; Torralva, M.; Fernández-Delgado, C. Biology of the endangered cyprinodontid *Aphanius iberus* in a saline wetland (SE Iberian Peninsula). *Ital. J. Zoöl.* **2009**, *76*, 316–329. [CrossRef]

59. Araguas, R.M.; Roldán, M.I.; García-Marín, J.-L.; Pla, C. Management of gene diversity in the endemic killifish *Aphanius iberus*: Revising Operational Conservation Units. *Ecol. Freshw. Fish* **2007**, *16*, 257–266. [CrossRef]

60. Rincón, P. Interaction between the introduced eastern mosquitofish and two autochthonous Spanish toothcarps. *J. Fish Boil.* **2002**, *61*, 1560–1585. [CrossRef]

61. Spanish Cyprinodont—*Aphanius iberus* (Valenciennes, 1846). Available online: https://eunis.eea.europa.eu/species/431#protected (accessed on 14 May 2020).

62. LIFE96 NAT/E/003118. Available online: http://ec.europa.eu/environment/life/project/Projects/index.cfm?fuseaction=search.dspPage&n_proj_id=117&docType=pdf (accessed on 4 December 2019).

63. LIFE99 NAT/E/006386. Available online: http://ec.europa.eu/environment/life/project/Projects/index.cfm?fuseaction=search.dspPage&n_proj_id=562&docType=pdf (accessed on 4 December 2019).

64. LIFE09 NAT/ES/000520. Available online: http://ec.europa.eu/environment/life/project/Projects/index.cfm?fuseaction=search.dspPage&n_proj_id=3845 (accessed on 4 December 2019).

65. Bernardi, G.; Fernández-Delgado, C.; Gomez-Chiarri, M.; Powers, D.A. Origin of a Spanish population of *Fundulus heteroclitus* inferred by cytochrome b sequence analysis. *J. Fish Boil.* **1995**, *47*, 737–740. [CrossRef]

66. Oltra, R.; Todoli, R. Reproduction of the endangered killifish *Apanius iberus* at different salinities. *Environ. Biol. Fish* **2000**, *57*, 113–115. [CrossRef]

67. Lanzoni, M.; Gavioli, A.; Aschonitis, V.; Merighi, M.; Fano, E.A.; Castaldelli, G. Length-weight relationships of three estuarine species in the Comacchio Lagoon, Po River delta, Italy. *J. Appl. Ichthyol.* **2016**, *32*, 1284–1285. [CrossRef]

68. Haedrich, R.L. *Fishes of the North-Eastern Atlantic and the Mediterranean*; Whitehead, P.J.P., Bauchot, M.-L., Hureau, J.-C., Nielsen, J., Tortonese, E., Eds.; Unesco: London, UK, 1986; pp. 1019–1085.

69. Franco, A.; Franzoi, P.; Malavasi, S.; Zucchetta, M.; Torricelli, P. Population and habitat status of two endemic sand gobies in lagoon marshes – Implications for conservation. *Estuar. Coast. Shelf Sci.* **2012**, *114*, 31–40. [CrossRef]

70. Spinelli, A.; De Matteo, S.; Costagliola, A.; Giacobbe, S.; Kovačić, M. First record from Sicily of the Adriatic dwarf goby, *Knipowitschia panizzae* (Osteichthyes, Gobiidae), a threatened species or a threat for conservation? *Mar. Biodivers.* **2016**, *47*, 237–242. [CrossRef]

71. Franzoi, P.; Ceccherelli, V.U.; Rossi, R. Differenze di abitudini alimentari tra novellame e specie ittiche residenti in una laguna del Delta del Po. *Oebalia Suppl.* **1992**, *17*, 181–186.

72. Ceccherelli, V.U.; Mistri, M.; Franzoi, P. Predation Impact on the Meiobenthic Harpacticoid *Canuella perplexa* in a Lagoon of the Po River Delta, Italy. *Estuaries* **1994**, *17*, 283. [CrossRef]

73. Franzoi, P.; Marchetti, C. Regime alimentare di due specie di piccoli ghiozzi (Pisces, Gobiidae) in un ambiente salmastro del Delta del Po. In Proceedings of the Atti del VI Convegno Nazionale dell'Associazione Italiana Ittiologi Acque Dolci, Varese Ligure, Italia, 6–8 June 1996; pp. 128–142.

74. Marzano, F.N.; Gandolfi, G. Active cannibalism among adults of *Knipowitschia panizzae* (Pisces Gobiidae) induced by starvation and reproduction. *Ethol. Ecol. Evol.* **2001**, *13*, 385–391. [CrossRef]

75. Maccagnani, R.; Carrieri, A.; Franzoi, P.; Rossi, R. Osservazioni sulla struttura di popolazione ed il ruolo trofico di tre specie di gobidi (*Kinpowitschia panizzae*, *Pomatoschistus marmoratus*, *Pomatoschistus canestrinii*) in un ambiente del Delta del Po. *Nova Thalassia* **1985**, *7*, 373–378.

76. Miller, P.J. The tokology of gobioid fishes. In *Fish Reproduction: Strategies and Tactics*; Potts, G.W., Wooton, R.J., Eds.; Academic Press: London, UK, 1984; pp. 119–152.

77. Lugli, M.; Torricelli, P. Prespawning sound production in mediterranean sand-gobies. *J. Fish Boil.* **1999**, *54*, 691–694. [CrossRef]

78. Massironi, M.; Rasotto, M.B.; Mazzoldi, C. A reliable indicator of female fecundity: The case of the yellow belly in *Knipowitschia panizzae* (Teleostei: Gobiidae). *Mar. Boil.* **2005**, *147*, 71–76. [CrossRef]

79. Pizzolon, M.; Rasotto, M.B.; Mazzoldi, C. Male lagoon gobies, *Knipowitschia panizzae*, prefer more ornamented to larger females. *Behav. Ecol. Sociobiol.* **2007**, *62*, 521–528. [CrossRef]

80. Malavasi, S.; Franco, A.; Fiorin, R.; Franzoi, P.; Torricelli, P.; Mainardi, D. The shallow water gobiid assemblage of the Venice Lagoon: Abundance, seasonal variation and habitat partitioning. *J. Fish Boil.* **2005**, *67*, 146–165. [CrossRef]

81. Painzza's Goby—*Knipowitschia panizzae* (Verga, 1841). Available online: https://eunis.eea.europa.eu/species/518 (accessed on 14 May 2020).

82. Ćaleta, M.; Marčić, Z.; Buj, I.; Zanella, D.; Mustafić, P.; Duplić, A.; Horvatić, S. A Review of Extant Croatian Freshwater Fish and Lampreys. *Croat. J. Fish.* **2019**, *77*, 137–234. [CrossRef]

83. Zerunian, S. Pesci delle acque interne d'Italia. In *Quad. Cons. Natura, Min. Ambiente*; Ist. Naz. Fauna Selvatica: Roma, Italia, 2004; Volume 20, p. 257.

84. LIFE10 NAT/IT/000256. Available online: http://ec.europa.eu/environment/life/project/Projects/index.cfm?fuseaction=search.dspPage&n_proj_id=4065 (accessed on 4 December 2019).

85. LIFE12 NAT/IT/000331. Available online: http://ec.europa.eu/environment/life/project/Projects/index.cfm?fuseaction=search.dspPage&n_proj_id=4838 (accessed on 4 December 2019).

86. LIFE13 NAT/IT/000115. Available online: http://ec.europa.eu/environment/life/project/Projects/index.cfm?fuseaction=search.dspPage&n_proj_id=5057 (accessed on 4 December 2019).

87. Franco, A.; Fiorin, R.; Franzoi, P.; Torricelli, P. Threatened fishes of the world: *Pomatoschistus canestrinii* Ninni, 1883 (Gobiidae). *Environ. Boil. Fishes* **2005**, *72*, 32. [CrossRef]

88. Malavasi, S.; Gkenas, C.; Leonardos, I.; Torricelli, P.; McLennan, D.A. The phylogeny of a reduced 'sand goby' group based on behavioural and life history characters. *Zoöl. J. Linn. Soc.* **2012**, *165*, 916–924. [CrossRef]

89. Scapin, L.; Zucchetta, M.; Bonometto, A.; Feola, A.; Brusà, R.B.; Sfriso, A.; Franzoi, P. Expected Shifts in Nekton Community Following Salinity Reduction: Insights into Restoration and Management of Transitional Water Habitats. *Water* **2019**, *11*, 1354. [CrossRef]

90. Malavasi, S.; Valerio, C.; Torricelli, P. Courtship sounds and associated behaviours in the Canestrini's goby *Pomatoschistus canestrinii*. *J. Fish Boil.* **2009**, *75*, 1883–1887. [CrossRef] [PubMed]

91. Malavasi, S.; Fiorin, R.; Franco, A.; Franzoi, P.; Granzotto, A.; Riccato, F.; Mainardi, D. Fish assemblages of Venice Lagoon shallow waters: An analysis based on species, families and functional guilds. *J. Mar. Syst.* **2004**, *51*, 19–31. [CrossRef]

92. Freyhof, J. First record of *Pomatoschistus canestrinii* (Ninni, 1883) in lake Trasimeno. *Riverine Idrobiol.* **1998**, *37*, 107–108.

93. Jelić, D.; Špelić, I.; Žutinić, P. Introduced species community over-dominates endemic ichthyofauna of High Lika Plateau (Central Croatia) over a 100 year period. *Acta Zoöl. Acad. Sci. Hung.* **2016**, *62*, 191–216. [CrossRef]

94. Tutman, P.; Sanda, R.; Glamuzina, B.; Dulčić, J. First confirmed record of *Pomatoschistus canestrinii* (Ninni, 1883) (Gobiidae) from Bosnia and Herzegovina. *J. Appl. Ichthyol.* **2013**, *29*, 937–939. [CrossRef]

95. Ahnelt, H.; Konecny, R.; Gabriel, A.; Bauer, A.; Pompei, L.; Lorenzoni, M.; Sattmann, H. First report of the parasitic copepod *Lernaea cyprinacea* (Copepoda: Lernaeidae) on gobioid fishes (Teleostei: Gobonellidae) in southern Europe. *Knowl. Manag. Aquat. Ecosyst.* **2018**, *419*, 34. [CrossRef]

96. Canestrini's Goby—*Pomatoschistus canestrinii* (Ninni, 1883). Available online: https://eunis.eea.europa.eu/species/578 (accessed on 13 May 2020).

97. LIFE16 NAT/IT/000663. Available online: http://ec.europa.eu/environment/life/project/Projects/index.cfm?fuseaction=search.dspPage&n_proj_id=6223 (accessed on 4 December 2019).

98. Raicevich, S.; Granzotto, A.; Pastres, R.; Giovanardi, O.; Pranovi, F.; Libralato, S. Mechanical clam dredging in Venice lagoon: Ecosystem effects evaluated with a trophic mass-balance model. *Mar. Boil.* **2003**, *143*, 393–403. [CrossRef]

99. Sfriso, A.; Facca, C.; Marcomini, A. Sedimentation rates and erosion processes in the lagoon of Venice. *Environ. Int.* **2005**, *31*, 983–992. [CrossRef]

100. Oliva-Paterna, F.J.; Caiola, N.; Torralva, M. Threatened fishes of the world: *Valencia hispanica* (Valenciennes, 1846) (Valenciidae). *Environ. Boil. Fishes* **2009**, *85*, 275–276. [CrossRef]

101. Valencia Tooth-Carp—*Valencia hispanica* (Valenciennes, 1846). Available online: https://eunis.eea.europa.eu/species/624 (accessed on 13 May 2020).

102. LIFE04 NAT/ES/000048. Available online: http://ec.europa.eu/environment/life/project/Projects/index.cfm?fuseaction=search.dspPage&n_proj_id=2609 (accessed on 4 December 2019).

103. Perdikaris, C.; Paschos, I.; Gouva, E.; Giakoumi, S.; Pappas, E.; Kalogianni, E. Rapid population collapse of the critically endangered *Valencia letourneuxi* in Kalamas basin of Northwest Greece. *AACL Bioflux* **2010**, *3*, 69–76.

104. Economidis, P. Endangered freshwater fishes of Greece. *Boil. Conserv.* **1995**, *72*, 201–211. [CrossRef]

105. *Valencia letourneuxi* (Sauvage, 1880). Available online: https://eunis.eea.europa.eu/species/12498 (accessed on 13 May 2020).

106. Pérez-Ruzafa, Á.; Marcos, C.; Perez-Ruzafa, I.M.; Pérez-Marcos, M. Coastal lagoons: "transitional ecosystems" between transitional and coastal waters. *J. Coast. Conserv.* **2010**, *15*, 369–392. [CrossRef]

107. Pérez-Ruzafa, Á.; Mompeán, M.C.; Marcos, C. Hydrographic, geomorphologic and fish assemblage relationships in coastal lagoons. *Hydrobiologia* **2007**, *577*, 107–125. [CrossRef]

108. Franco, A.; Franzoi, P.; Torricelli, P. Structure and functioning of Mediterranean lagoon fish assemblages: A key for the identification of water body types. *Estuar. Coast. Shelf Sci.* **2008**, *79*, 549–558. [CrossRef]

109. Scapin, L.; Zucchetta, M.; Sfriso, A.; Franzoi, P. Local Habitat and Seascape Structure Influence Seagrass Fish Assemblages in the Venice Lagoon: The Importance of Conservation at Multiple Spatial Scales. *Chesap. Sci.* **2018**, *41*, 2410–2425. [CrossRef]

110. Ramsar Convention. Available online: https://www.ramsar.org/sites/default/files/documents/library/info2007-01-e.pdf (accessed on 19 June 2020).

111. Wolanski, E.; Elliott, M. *Estuar. Ecohydrology—An Introduction*, 2nd ed.; Elsevier: Amsterdam, The Netherlands, 2016; p. 321.

112. Zedler, J.B.; Bonin, C.L.; Larkin, D.J.; Varty, A. Salt marshes. In *Encyclopedia of Ecology*; Jørgensen, S.E., Fath, B.D., Eds.; Elsevier, B.V. Radarweg: Amsterdam, The Netherlands, 2008; Volume 29, pp. 3132–3141.

113. Siddig, A.A.; Ellison, A.M.; Ochs, A.; Villar-Leeman, C.; Lau, M.K. How do ecologists select and use indicator species to monitor ecological change? Insights from 14 years of publication in Ecological Indicators. *Ecol. Indic.* **2016**, *60*, 223–230. [CrossRef]

114. Franco, A.; Franzoi, P.; Malavasi, S.; Riccato, F.; Torricelli, P.; Mainardi, D. Use of shallow water habitats by fish assemblages in a Mediterranean coastal lagoon. *Estuar. Coast. Shelf Sci.* **2006**, *66*, 67–83. [CrossRef]

115. Scapin, L.; Zucchetta, M.; Facca, C.; Sfriso, A.; Franzoi, P. Using fish assemblage to identify success criteria for seagrass habitat restoration. *Web Ecol.* **2016**, *16*, 33–36. [CrossRef]

116. Scapin, L.; Redolfi Bristol, S.; Cavraro, F.; Zucchetta, M.; Franzoi, P. Fish fauna in the Venice lagoon: Updating the species list and reviewing the functional classification. *Ital. J. Freshw. Ichthyol.* **2019**, *5*, 271–277.

117. Zucchetta, M.; Scapin, L.; Cavraro, F.; Pranovi, F.; Franco, A.; Franzoi, P. Can the Effects of Anthropogenic Pressures and Environmental Variability on Nekton Fauna Be Detected in Fishery Data? Insights from the Monitoring of the Artisanal Fishery Within the Venice Lagoon. *Chesap. Sci.* **2016**, *39*, 1164–1182. [CrossRef]

118. Cavraro, F.; Bettoso, N.; Zucchetta, M.; D'Aietti, A.; Faresi, L.; Franzoi, P. Body condition in fish as a tool to detect the effects of anthropogenic pressures in transitional waters. *Aquat. Ecol.* **2019**, *53*, 21–35. [CrossRef]

119. Franzoi, P.; Franco, A.; Torricelli, P. Fish assemblage diversity and dynamics in the Venice lagoon. *Rend. Lincei* **2010**, *21*, 269–281. [CrossRef]

120. Mosesso, P.; Angeletti, D.; Pepe, G.; Pretti, C.; Nascetti, G.; Bellacima, R.; Cimmaruta, R.; Jha, A.N. The use of cyprinodont fish, *Aphanius fasciatus*, as a sentinel organism to detect complex genotoxic mixtures in the coastal lagoon ecosystem. *Mutat. Res. Toxicol. Environ. Mutagen.* **2012**, *742*, 31–36. [CrossRef] [PubMed]

121. Kessabi, K.; Annabi, A.; Navarro, A.; Casado, M.; Hwas, Z.; Said, K.; Messaoudi, I.; Piña, B. Structural and molecular analysis of pollution-linked deformities in a natural *Aphanius fasciatus* (Valenciennes, 1821) population from the Tunisian coast. *J. Environ. Monit.* **2012**, *14*, 2254. [CrossRef]

122. García-Oliva, M.; Marcos, C.; Umgiesser, G.; McKiver, W.; Ghezzo, M.; De Pascalis, F.; Pérez-Ruzafa, A. Modelling the impact of dredging inlets on the salinity and temperature regimes in coastal lagoons. *Ocean Coast. Manag.* **2019**, *180*, 104913. [CrossRef]

123. Ferrarin, C.; Ghezzo, M.; Umgiesser, G.; Tagliapietra, D.; Camatti, E.; Zaggia, L.; Sarretta, A. Assessing hydrological effects of human interventions on coastal systems: Numerical applications to the Venice Lagoon. *Hydrol. Earth Syst. Sci.* **2013**, *17*, 1733–1748. [CrossRef]

124. Lloret, J.; Marín, A.; Marín-Guirao, L. Is coastal lagoon eutrophication likely to be aggravated by global climate change? *Estuar. Coast. Shelf Sci.* **2008**, *78*, 403–412. [CrossRef]

125. Regulation (EU) 1143/2014. Available online: https://eur-lex.europa.eu/legal-content/EN/TXT/?qid=1483614313362&uri=CELEX:32014R1143 (accessed on 13 May 2020).

126. Global Invasive Species Database. Available online: http://www.iucngisd.org/gisd/100_worst.php (accessed on 4 December 2019).

127. Facca, C. Specie ittiche di interesse comunitario (allegato II della Direttiva 92/43/CEE) dell'habitat prioritario 1150* "Lagune costiere" del Mar Mediterraneo: Biologia, Gestione e Conservazione. In Proceedings of the 1st Level Professional Master in Wildlife Management, Venice, Italy, 4 December 2019. (In Italian)

MDPI

St. Alban-Anlage 66

4052 Basel

Switzerland

Tel. +41 61 683 77 34

Fax +41 61 302 89 18

www.mdpi.com

Water Editorial Office

E-mail: water@mdpi.com

www.mdpi.com/journal/water